AF581742

Preparation of Novel Nanomaterial and Its Application in Food Industry

Preparation of Novel Nanomaterial and Its Application in Food Industry

Editors

Hong Wu
Hui Zhang

MDPI • Basel • Beijing • Wuhan • Barcelona • Belgrade • Manchester • Tokyo • Cluj • Tianjin

Editors
Hong Wu
South China University of Technology
China

Hui Zhang
Zhejiang University
China

Editorial Office
MDPI
St. Alban-Anlage 66
4052 Basel, Switzerland

This is a reprint of articles from the Special Issue published online in the open access journal *Actuators* (ISSN 2076-0825) (available at: https://www.mdpi.com/journal/foods/special_issues/novel_nanomaterial_application).

For citation purposes, cite each article independently as indicated on the article page online and as indicated below:

LastName, A.A.; LastName, B.B.; LastName, C.C. Article Title. *Journal Name* **Year**, *Volume Number*, Page Range.

ISBN 978-3-0365-4409-0 (Hbk)
ISBN 978-3-0365-4410-6 (PDF)

Contents

About the Editors . . . . **vii**

Hong Wu and Hui Zhang
Preparation of Novel Nanomaterial and Its Application in Food Industry
Reprinted from: *Foods* **2022**, *11*, 1382, doi:10.3390/foods11101382 . . . . **1**

Liangliang Zhu, Hongshun Hao, Chao Ding, Hanwei Gan, Shuting Jiang, Gongliang Zhang, Jingran Bi, Shuang Yan and Hongman Hou
A Novel Photoelectrochemical Aptamer Sensor Based on CdTe Quantum Dots Enhancement and Exonuclease I-Assisted Signal Amplification for *Listeria monocytogenes* Detection
Reprinted from: *Foods* **2021**, *10*, 2896, doi:10.3390/foods10122896 . . . . **5**

Zeyu Xu, Yizhong Wang, Jiaran Zhang, Ce Shi and Xinting Yang
A Highly Sensitive and Selective Fluorescent Probe Using MPA-InP/ZnS QDs for Detection of Trace Amounts of Cu^{2+} in Water
Reprinted from: *Foods* **2021**, *10*, 2777, doi:10.3390/foods10112777 . . . . **25**

Jinyi Yang, Rui Si, Guangpei Wu, Yu Wang, Ruyu Fang, Fei Liu, Feng Wang, Hongtao Lei, Yudong Shen, Qi Zhang and Hong Wang
Preparation of Specific Nanobodies and Their Application in the Rapid Detection of Nodularin-R in Water Samples
Reprinted from: *Foods* **2021**, *10*, 2758, doi:10.3390/foods10112758 . . . . **37**

Yuanhang Yao, Jiaxing Jansen Lin, Xin Yi Jolene Chee, Mei Hui Liu, Saif A. Khan and Jung Eun Kim
Encapsulation of Lutein via Microfluidic Technology: Evaluation of Stability and In Vitro Bioaccessibility
Reprinted from: *Foods* **2021**, *10*, 2646, doi:10.3390/foods10112646 . . . . **51**

Yufeng Chen, Jingchong Peng, Yueqi Wang, Daniel Wadhawan, Lijun Wu, Xiaojing Gao, Yi Sun and Guobin Xia
Development, Characterization, Stability and Bioaccessibility Improvement of 7,8-Dihydroxyflavone Loaded Zein/ Sophorolipid/Polysaccharide Ternary Nanoparticles: Comparison of Sodium Alginate and Sodium Carboxymethyl Cellulose
Reprinted from: *Foods* **2021**, *10*, 2629, doi:10.3390/foods10112629 . . . . **65**

Zhichao Yang, Chaoyi Shen, Yucheng Zou, Di Wu, Hui Zhang and Kunsong Chen
Application of Solution Blow Spinning for Rapid Fabrication of Gelatin/Nylon 66 Nanofibrous Film
Reprinted from: *Foods* **2021**, *10*, 2339, doi:10.3390/foods10102339 . . . . **87**

Peng Wen, Teng-Gen Hu, Yan Wen, Ke-Er Li, Wei-Peng Qiu, Zhi-Lin He, Hong Wang and Hong Wu
Development of *Nervilia fordii* Extract-Loaded Electrospun PVA/PVP Nanocomposite for Antioxidant Packaging
Reprinted from: *Foods* **2021**, *10*, 1728, doi:10.3390/foods10081728 . . . . **99**

Xiangmei Li, Xiaomin Chen, Jinxiao Wu, Zhiwei Liu, Jin Wang, Cuiping Song, Sijun Zhao, Hongtao Lei and Yuanming Sun
Portable, Rapid, and Sensitive Time-Resolved Fluorescence Immunochromatography for On-Site Detection of Dexamethasone in Milk and Pork
Reprinted from: *Foods* **2021**, *10*, 1339, doi:10.3390/foods10061339 . . . . **113**

Rui Cui, Bifen Zhu, Jiatong Yan, Yuyue Qin, Mingwei Yuan, Guiguang Cheng and Minglong Yuan
Development of a Sodium Alginate-Based Active Package with Controlled Release of Cinnamaldehyde Loaded on Halloysite Nanotubes
Reprinted from: *Foods* **2021**, *10*, 1150, doi:10.3390/foods10061150 . . . . **127**

Wei Zhou, Yun Zhang, Ruyi Li, Shengfeng Peng, Roger Ruan, Jihua Li and Wei Liu
Fabrication of Caseinate Stabilized Thymol Nanosuspensions via the pH-Driven Method: Enhancement in Water Solubility of Thymol
Reprinted from: *Foods* **2021**, *10*, 1074, doi:10.3390/foods10051074 . . . . **141**

Mianhong Chen, Ruyi Li, Yuanyuan Gao, Yeyu Zheng, Liangkun Liao, Yupo Cao, Jihua Li and Wei Zhou
Encapsulation of Hydrophobic and Low-Soluble Polyphenols into Nanoliposomes by pH-Driven Method: Naringenin and Naringin as Model Compounds
Reprinted from: *Foods* **2021**, *10*, 963, doi:10.3390/foods10050963 . . . . **155**

Kun Feng, Meng-Yu Zhai, Yun-Shan Wei, Min-Hua Zong, Hong Wu and Shuang-Yan Han
Fabrication of Nano/Micro-Structured Electrospun Detection Card for the Detection of Pesticide Residues
Reprinted from: *Foods* **2021**, *10*, 889, doi:10.3390/foods10040889 . . . . **171**

Lingli Deng
Current Progress in the Utilization of Soy-Based Emulsifiers in Food Applications—A Review
Reprinted from: *Foods* **2021**, *10*, 1354, doi:10.3390/foods10061354 . . . . **187**

About the Editors

Hong Wu

Hong Wu, Ph.D., is Professor at School of Food Science and Engineering, South China University of Technology, China. She received a PhD in Biochemical Engineering from South China University of Technology in 2004. From May 2007 to July 2007, she carried out a collaborative research in Warwick University, UK; and she worked as a visiting scientist in Rensselaer Polytechnic Institute, USA from May 2010 to May 2011. Presently, her research focuses on the stabilization of functional compounds by various novel technologies and their application; screening, efficacy evaluation and application of new probiotics; and biosynthesis of structured lipids and their properties and bioactivities evaluation. She has published more than 150 papers and is currently Editorial Board member of Foods (MDPI).

Hui Zhang

Hui Zhang, Ph.D., is Professor at Department of Food Science and Nutrition, Zhejiang University, China. He received a PhD in Food Science from Zhejiang University in 2009, and carried out work as a visiting researcher at University of Hohenheim, Germany, from October 2010 to August 2012. He was a visiting scholar at the University of California Davis, USA, from August 2014 to August 2015. His research focuses on novel food materials (e.g., nanofibers, aerogels, emulsions) for bioactive encapsulation and controlled release. He is currently Associate Editor of the Journal of the Science of Food and Agriculture (Wiley), Editorial Board member of Food Packaging and Shelf Life (Elsevier), and serves as Advisory Board member of AOCS (American Oil Chemists' Society) China Section.

 foods

MDPI

Editorial

Preparation of Novel Nanomaterial and Its Application in Food Industry

Hong Wu [1,*] and Hui Zhang [2,*]

[1] School of Food Science and Engineering, South China University of Technology, Guangzhou 510000, China
[2] College of Biosystems Engineering and Food Science, Zhejiang University, Hangzhou 310058, China
* Correspondence: bbhwu@scut.edu.cn (H.W.); hubert0513@zju.edu.cn (H.Z.)

Nanotechnology has offered a wide range of opportunities for the development and application of structures, materials, or systems with new properties in the food industry in recent years. The developed nanomaterials could greatly improve not only food quality and safety but also the foods' health benefits. In this special issue, different nano-sized vehicles (e.g., nanoparticle, nanoliposome, nanofiber, nanobody) are reported as efficient bioactives delivery systems and sensitive detection materials.

To resolve the low chemical instability, poor water solubility, and intestinal efflux limitations and challenges of bioactive ingredients, constructing an effective delivery vehicle using food-grade polymers is supposed to be a novel and feasible strategy. For example, Chen et al. [1] encapsulated hydrophobic naringenin and naringin in nanoliposomes based on the gradual reduction in their water solubility after the pH changed to acidity. The naringenin-loaded nanoliposomes were predominantly nanometric (44.95–104.4 nm), negatively charged (−14.1 to −19.3 mV) and exhibited relatively high encapsulation efficiency (EE = 95.34% for 0.75 mg/mL naringenin within 1% w/v lecithin). Additionally, the naringenin-loaded nanoliposomes still maintained good stability during 31 days of storage at 4 °C. Zhou et al. [2] fabricated a caseinate-stabilized thymol nanosuspension by pH-driven methods. Thymol was extremely stable at pH 7.0–12.0 even after incubation for 24 h, which means the loss of thymol during the pH-driven process is negligible. The physicochemical properties of thymol nanosuspensions are highly dependent on the caseinate concentration. Caseinate could stabilize thymol nanosuspensions even at a relatively low caseinate concentration, and the loading capacity can be as high as 45.9%. Chen et al. [3] compared two polysaccharides [sodium alginate (ALG) and sodium carboxymethyl cellulose (CMC)] to establish zein/sophorolipid/ALG (ALG/S/Z) and zein/sophorolipid/ALG (CMC/S/Z) nanoparticles to encapsulate 7,8-dihydroxyflavone (7,8-DHF), respectively. They found that CMC/S/Z possessed lower polydispersity index, particle size and turbidity, but higher zeta potential, encapsulation efficiency and loading capacity compared to ALG/S/Z. Compared to zein/sophorolipid nanoparticles (S/Z), both ALG/S/Z and CMC/S/Z had better stability against low pH (pH 3~4) and high ionic strengths (150~200 mM NaCl). Apart from the above particle forms, nanofibers have emerged as a novel delivery system due to its simplicity and effectiveness. Electrospinning, solution blow spinning, and eletro-blow spinning are the most common techniques for continuously producing nanofibers with a fiber diameter range from sub-nanometers to micrometers. The electrospun films were also adopted to stabilize the sensitive bioactives. Cui et al. [4] fabricated an antimicrobial food packaging film with controlled release by loading cinnamaldehyde (CIN) on etched halloysite nanotubes (T-HNTs) and adding it to sodium alginate (SA) matrix. It was found that CIN could be successfully loaded into the T-HNTs and the addition of T-HNTs-CIN significantly improved the water vapor barrier properties and tensile strength of the film. Additionally, the SA/T-HNTs-CIN film could delay the release of CIN into fatty food simulation solution compared with that of SA/CIN film. Wen et al. [5] reported that the incorporation of a Nervilia fordii extract (NFE) in the

Citation: Wu, H.; Zhang, H. Preparation of Novel Nanomaterial and Its Application in Food Industry. *Foods* **2022**, *11*, 1382. https://doi.org/10.3390/foods11101382

Received: 29 April 2022
Accepted: 9 May 2022
Published: 11 May 2022

electrospun poly(vinyl alcohol) (PVA) and polyvinyl(pyrrolidone) (PVP) bio-composite film could retain its antioxidant capacity, avoiding the fish oil's oxidation (and thus extending its shelf life). Yang et al. [6] fabricated gelatin/nylon 66 (PA66) composite nanofibers by solution blow spinning (SBS). Morphology observations show that GA/PA66 composite films had a nano-diameter from 172.3 to 322.1 nm. Nylon 66 (PA66) was proved to improve the mechanical properties and the ability to resist dissolution of gelatin nanofibrous films. Another study was performed by microfluidics to encapsulate lutein to improve its bioaccessibility in the gastrointestinal tract [7]. Two types of oils (safflower oil (SO) and olive oil (OL)) were selected as a delivery vehicle for lutein, and two customized microfluidic devices (co-flow and combination-flow) were used. The results demonstrated that the types of oil and device do not affect the lutein bioaccessibility. Findings from this study may provide scientific insights into emulsion-based delivery systems that employ microfluidics for the encapsulation of bioactive compounds into foods. Finally, Deng [8] systematically summarized that soy-based emulsifiers are currently extensively studied and applied in the food industry for its applications in bioactive and nutrient delivery.

Another focus of the published articles in this special issue is the sensitive detection of various contaminants (pesticides, drug, copper (II) ions, pathogenic bacterium) associated with food safety through different nano-techniques. In the study by Feng et al. [9], a novel nano/micro-structured pesticide detection card was developed by combining electrospinning and hydrophilic modification, and its feasibility for detecting different pesticides was investigated [9]. This self-made detection card showed a 5-fold, 2-fold, and 1.5-fold reduction of the minimum detectable concentration for carbofuran, malathion, and trichlorfon, respectively, compared to the national standard values. In another study, Li et al. [10] created a portable, rapid, and sensitive time-resolved fluorescence immunochromatography for on-site detection of dexamethasone in milk and pork. A parallel experiment for 20 milk and 10 pork samples with LC-MS/MS was carried out to confirm the performance of the developed TRFM-ICA. The results of the two methods are basically the same. In addition, the nanobody, as an important tool in immunoassay for chemical contaminants, was developed and its efficiency was examined by detecting a secondary metabolite of cyanobacteria, namely nodularin (NOD-R) [11]. The ic-ELISA method based on the nanobody N56 was validated with spiked water sample and confirmed by UPLC–MS/MS, which indicated that the ic-ELISA established in this work is a reproducible detection assay for nodularin residues in water samples. Xu et al. [12] prepared a highly sensitive and selective fluorescence probe that used mercaptopropionic acid (MPA)-capped InP/ZnS quantum dots (MPA-InP/ZnS QDs) for the detection of trace amounts of Cu^{2+} in water. This probe exhibited an extremely low limit of detection of 0.22 nM. Meanwhile, a possible fluorescence-quenching mechanism was proposed in this study. In another study, to achieve the rapid detection of *Listeria monocytogenes*, Zhu et al. [13] used aptamers for the original identification and built a photoelectrochemical aptamer sensor using exonuclease-assisted amplification. In brief, tungsten trioxide (WO_3) was used as a photosensitive material, which was modified with gold nanoparticles to immobilize complementary DNA, and amplified the signal by means of the sensitization effect of CdTe quantum dots and the shearing effect of exonuclease I (Exo I) to achieve high-sensitivity detection. This strategy had a detection limit of 45 CFU/mL in the concentration range of 1.3×10^1–1.3×10^7 CFU/mL, providing a new way to detect *Listeria monocytogenes*.

Overall, these articles extend the knowledge on the application of nanomaterials in food nutrition and safety, promoting the development of nanotechnologies in food industry.

Finally, we would like to thank all of the authors for their submissions, and all of the referees for their valuable suggestions for improving the manuscripts.

Author Contributions: H.W. and H.Z. conceived and wrote this editorial. All authors have read and agreed to the published version of the manuscript.

Funding: This research received no external funding.

Conflicts of Interest: The authors declare no conflict of interest.

References

1. Chen, M.; Li, R.; Gao, Y.; Zheng, Y.; Liao, L.; Cao, Y.; Li, J.; Zhou, W. Encapsulation of Hydrophobic and Low-Soluble Polyphenols into Nanoliposomes by pH-Driven Method: Naringenin and Naringin as Model Compounds. *Foods* **2021**, *10*, 963. [CrossRef]
2. Zhou, W.; Zhang, Y.; Li, R.; Peng, S.; Ruan, R.; Li, J.; Liu, W. Fabrication of Caseinate Stabilized Thymol Nanosuspensions via the pH-Driven Method: Enhancement in Water Solubility of Thymol. *Foods* **2021**, *10*, 1074. [CrossRef] [PubMed]
3. Chen, Y.; Peng, J.; Wang, Y.; Wadhawan, D.; Wu, L.; Gao, X.; Sun, Y.; Xia, G. Development, Characterization, Stability and Bioaccessibility Improvement of 7,8-Dihydroxyflavone Loaded Zein/Sophorolipid/Polysaccharide Ternary Nanoparticles: Comparison of Sodium Alginate and Sodium Carboxymethyl Cellulose. *Foods* **2021**, *10*, 2629. [CrossRef]
4. Cui, R.; Zhu, B.; Yan, J.; Qin, Y.; Yuan, M.; Cheng, G.; Yuan, M. Development of a Sodium Alginate-Based Active Package with Controlled Release of Cinnamaldehyde Loaded on Halloysite Nanotubes. *Foods* **2021**, *10*, 1150. [CrossRef] [PubMed]
5. Wen, P.; Hu, T.G.; Wen, Y.; Li, K.E.; Qiu, W.P.; He, Z.L.; Wang, H.; Wu, H. Development of *Nervilia fordii* Extract-Loaded Electrospun PVA/PVP Nanocomposite for Antioxidant Packaging. *Foods* **2021**, *10*, 1728. [CrossRef] [PubMed]
6. Yang, Z.; Shen, C.; Zou, Y.; Wu, D.; Zhang, H.; Chen, K. Application of Solution Blow Spinning for Rapid Fabrication of Gelatin/Nylon 66 Nanofibrous Film. *Foods* **2021**, *10*, 2339. [CrossRef]
7. Yao, Y.; Lin, J.J.; Chee, X.Y.J.; Liu, M.H.; Khan, S.A.; Kim, J.E. Encapsulation of Lutein via Microfluidic Technology: Evaluation of Stability and In Vitro Bioaccessibility. *Foods* **2021**, *10*, 2646. [CrossRef] [PubMed]
8. Deng, L. Current Progress in the Utilization of Soy-Based Emulsifiers in Food Applications-A Review. *Foods* **2021**, *10*, 1354. [CrossRef] [PubMed]
9. Feng, K.; Zhai, M.Y.; Wei, Y.S.; Zong, M.H.; Wu, H.; Han, S.Y. Fabrication of Nano/Micro-Structured Electrospun Detection Card for the Detection of Pesticide Residues. *Foods* **2021**, *10*, 889. [CrossRef] [PubMed]
10. Li, X.; Chen, X.; Wu, J.; Liu, Z.; Wang, J.; Song, C.; Zhao, S.; Lei, H.; Sun, Y. Portable, Rapid, and Sensitive Time-Resolved Fluorescence Immunochromatography for On-Site Detection of Dexamethasone in Milk and Pork. *Foods* **2021**, *10*, 1339. [CrossRef] [PubMed]
11. Yang, J.; Si, R.; Wu, G.; Wang, Y.; Fang, R.; Liu, F.; Wang, F.; Lei, H.; Shen, Y.; Zhang, Q.; et al. Preparation of Specific Nanobodies and Their Application in the Rapid Detection of Nodularin-R in Water Samples. *Foods* **2021**, *10*, 2758. [CrossRef] [PubMed]
12. Xu, Z.; Wang, Y.; Zhang, J.; Shi, C.; Yang, X. A Highly Sensitive and Selective Fluorescent Probe Using MPA-InP/ZnS QDs for Detection of Trace Amounts of Cu^{2+} in Water. *Foods* **2021**, *10*, 2777. [CrossRef] [PubMed]
13. Zhu, L.; Hao, H.; Ding, C.; Gan, H.; Jiang, S.; Zhang, G.; Bi, J.; Yan, S.; Hou, H. A Novel Photoelectrochemical Aptamer Sensor Based on CdTe Quantum Dots Enhancement and Exonuclease I-Assisted Signal Amplification for *Listeria monocytogenes* Detection. *Foods* **2021**, *10*, 2896. [CrossRef]

 MDPI

Article

A Novel Photoelectrochemical Aptamer Sensor Based on CdTe Quantum Dots Enhancement and Exonuclease I-Assisted Signal Amplification for *Listeria monocytogenes* Detection

Liangliang Zhu [1], Hongshun Hao [1,2,*], Chao Ding [1], Hanwei Gan [1], Shuting Jiang [1], Gongliang Zhang [2], Jingran Bi [2], Shuang Yan [1] and Hongman Hou [2]

[1] Department of Inorganic Nonmetallic Materials Engineering, Dalian Polytechnic University, Dalian 116034, China; zhull72@163.com (L.Z.); dc1608658525@163.com (C.D.); ghw1717@163.com (H.G.); JJiangSHuTing@163.com (S.J.); yanye150@outlook.com (S.Y.)

[2] Liaoning Key Lab for Aquatic Processing Quality and Safety, School of Food Science and Technology, Dalian Polytechnic University, Dalian 116034, China; zhanggl1978@hotmail.com (G.Z.); bijingran1225@foxmail.com (J.B.); houhongman@dlpu.edu.cn (H.H.)

* Correspondence: beike1952@163.com

Abstract: To achieve the rapid detection of *Listeria monocytogenes*, this study used aptamers for the original identification and built a photoelectrochemical aptamer sensor using exonuclease-assisted amplification. Tungsten trioxide (WO_3) was used as a photosensitive material, was modified with gold nanoparticles to immobilize complementary DNA, and amplified the signal by means of the sensitization effect of CdTe quantum dots and the shearing effect of Exonuclease I (Exo I) to achieve high-sensitivity detection. This strategy had a detection limit of 45 CFU/mL in the concentration range of 1.3×10^1–1.3×10^7 CFU/mL. The construction strategy provides a new way to detect *Listeria monocytogenes*.

Keywords: photoelectrochemical detection; aptamer; *Listeria monocytogenes*; quantum dots; exonuclease I

Citation: Zhu, L.; Hao, H.; Ding, C.; Gan, H.; Jiang, S.; Zhang, G.; Bi, J.; Yan, S.; Hou, H. A Novel Photoelectrochemical Aptamer Sensor Based on CdTe Quantum Dots Enhancement and Exonuclease I-Assisted Signal Amplification for *Listeria monocytogenes* Detection. *Foods* **2021**, *10*, 2896. https://doi.org/10.3390/foods10122896

Academic Editors: Hong Wu and Hui Zhang

Received: 14 October 2021
Accepted: 16 November 2021
Published: 23 November 2021

1. Introduction

Foodborne illness induced by pollution is a serious safety issue [1]. Pathogens in food can cause the food to spoil and rot, and some can produce toxic substances that can lead to illness. In recent years, food-borne poisoning incidents due to *Listeria monocytogenes* have aroused widespread concern [2]. *Listeria monocytogenes* can survive and grow under both aerobic and anaerobic conditions. It grows in a wide temperature range and has strong resistance to alkali and salt [3]. It is widely found in milk; meat; aquatic products; and frozen, and cold storage, and ready-to-eat foods [4]. *Listeria monocytogenes* can cause meningitis and sepsis, which have high hospitalization rates and mortality [5–7]. Various countries have formulated relevant limit standards. In China, it is stipulated that *Listeria monocytogenes* cannot be detected in meat products (GB29921-2013), as is true in the United States [8].

Traditional microbial detection methods include separation, culture, detection, and other steps. Although the detection accuracy is high, the detection time is long and the subsequent detection steps are cumbersome, which is not conducive to rapid detection [9]. In recent years, many rapid detection methods have been developed for pathogenic bacteria. These are mainly immunological detection methods based on the enzyme-linked immunosorbent method and the enzyme-linked fluorescence analysis method [10], along with molecular biological detection based on PCR technology and loop-mediated isothermal amplification technology [9,11–13]. However, all these methods need special reagents, instruments, and equipment, which are operated by professionals. There is a very important, practical need to explore more convenient, sensitive, effective, and rapid detection methods. Because of their good sensitivity, rapidity, and easy operation, biosensors have been developed and widely used. They have been used in medicine, environmental testing, food safety, and other fields [14–19]. The photoelectrochemical (PEC) sensor is a new type of detection technology with development potential. Because of the separation of the excitation light source and the generated photocurrent, the PEC sensor has a higher sensitivity than the electrochemical sensor and has great development potential [20]. PEC sensors have been used to detect DNA [21,22], miRNA [23,24], small organic molecules [25,26], heavy metal ions [27–29], proteins [30,31], and other substances, and they have attracted much attention.

Choosing optoelectronic materials with good photoelectric activity is one of the keys to improving sensor performance [32]. Graphite phase carbon nitride [33–35], bismuth-based semiconductors [36,37], and sulfides [30,38] are all considered potential materials. WO_3 is a typical n-type tungsten oxide semiconductor that has good chemical stability, strong electron transmission ability, and a proper band gap (2.5–2.8 eV). It has been widely used in photocatalysis and sensors, as well as in other fields. Shen et al. [39] constructed a WO_3-doped gold nanoparticle gas sensor that can detect ppb-level NO_2. Lu et al. [40] prepared a core-shell WO_3/CdS heterojunction, which also had excellent photocatalytic performance in the near-infrared region. Gold nanoparticles (Au NPs) have good electrical conductivity and the ability to fix identification elements. Therefore, they are often used to construct sensors and are promising intermediates [41].

Because of their characteristics of the narrow band gap, high yield, wide absorption, and adjustable size, quantum dots (QDs) have been widely used in the field of sensors [42,43]. Quantum dots are a kind of nanoscale semiconductor, such as CdS, CdSe and CdTe. In addition, QDs can be labeled with various biomolecules after chemical modification, and their good biocompatibility has unique advantages in biosensor applications. Aptamers are low cost, have high selectivity and good specificity, and are excellent substitutes for antibodies. In addition, aptamers have a variety of targets, which can be used to identify proteins, amino acids, pathogenic bacteria, cells, and viruses [44]. Exonuclease I (Exo I) can degrade single-stranded DNA in the $3' \rightarrow 5'$ direction, and it only recognizes the $3'$ end of single-stranded DNA. Compared with other cutting enzymes, its operation is simple, so it is also widely used [45].

In this study, a PEC aptamer sensor based on WO_3, CdTe QDs, and Exo I auxiliary signal amplification was constructed to detect *Listeria monocytogenes*. WO_3 was used as a photosensitive material and modified with Au NPs as intermediates for connecting the complementary DNA (cDNA) of the aptamer. Then, the quantum dot-coupled aptamer (QD–Ap) is modified on the electrode by DNA hybridization. The photocurrent is significantly enhanced because of the sensitization of quantum dots. When the target bacteria and Exo I are present, the photocurrent is significantly reduced. The constructed sensing platform provides a new strategy with a high potential for the rapid and efficient detection of *Listeria monocytogenes* in food.

2. Experimental Section

2.1. Chemicals and Reagents

Sodium tungstate dihydrate ($Na_2WO_4{\cdot}2H_2O$), ammonium oxalate monohydrate ($(NH_4)_2C_2O_4{\cdot}H_2O$), ascorbic acid (AA), cadmium chloride hemi(pentahydrate) ($CdCl_2{\cdot}2.5H_2O$), and 6-mercapto-1-hexanol (MCH) were obtained from Macklin (Shanghai, China). Potassium tellurite (K_2TeO_3), chloroauric acid ($HAuCl_4$), 1-ethyl-3(3-dimethylamin opropyl) carbodiimide hydrochloride (EDC), ethanol, Tris (2-carboxyethyl) phosphine (TCEP), and *N*-hydroxysuccinimide (NHS) were purchased from Aladdin (Shanghai, China). Sodium borohydride ($NaBH_4$) and thioglycolic acid (TGA) were obtained from Sinopharm Chemical Reagent Co., Ltd. (Shanghai, China). Tryptone, yeast extract powder, and agar (bacteriological) were obtained from Qingdao Hope Bio-Technology Co., Ltd. (Qingdao, China). Tris HCl solution, Exonuclease I (Exo I), and phosphate-buffered saline (PBS) were purchased from Sangon Biotech Co., Ltd. (Shanghai, China). Raw chicken was purchased from a local supermarket. *Listeria monocytogenes* (ATCC 19115), *Staphylococcus aureus*, *Escherichia coli O157:H7*, and *Salmonella typhimurium* were obtained from China General Microbiological Culture Collection Center (Beijing, China). The DNA strands used were ordered from Sangon Biotech Co., Ltd. (Shanghai, China). All DNA sequences are summarized as follows: aptamer (Ap), 5′-NH_2-$(CH_2)_6$-ATA CCA GCT TAT TCA ATT CCA AAA GCG CAC CCA TAT ATG TTC TAT GTC CCC CAC CTC GAG ATT GCA CTT ACT ATC T-3′ [46]; and aptamer-complementary DNA (cDNA), 5′-GTG CGC TTT TGG AAT TGA ATA AGC TGG TAT TTT TTT-$(CH_2)_6$-SH-3′.

2.2. Apparatus

The X-ray diffraction (XRD) patterns of samples were measured using a 7000S device (Shimadzu, Japan). The test conditions were set with a Cu-Ka XRD source, a scan rate of 5°/min, and a scan range of 10–70 degrees at room temperature. Surface morphologies of the prepared materials were observed using field emission scanning electron microscopy (JSM-7800F, Tokyo, Japan). Transmission electron microscopy (TEM) was performed using a JEOL JEM-2100 UHR (Tokyo, Japan). The UV–visible (UV–vis) absorption spectra were recorded using a TU-1901 UV–visible spectrophotometer (Beijing, China). Photoluminescence (PL) spectroscopy was characterized by an F-7000 fluorescence spectrometer (Hitachi, Japan) at room temperature. A xenon lamp was applied as the excitation source for the photoelectrochemical tests. Photocurrent and electrochemical characterizations were conducted with a CHI 660C electrochemical workstation (Shanghai, China) with a typical three-electrode system.

2.3. Preparation of WO_3 Nanoplate

First, 0.4 g of $Na_2WO_4{\cdot}2H_2O$ and 0.17 g of $(NH_4)_2C_2O_4{\cdot}H_2O$ was dissolved in 33 mL deionized water. After stirring for 10 min, 9 mL of HCl solution (37%) was added. After stirring for another 10 min, 8 mL of H_2O_2 (30%) was added. Stirring was continued for 20 min. Then, 30 mL of absolute ethanol was added, and the solution was stirred for 30 min. The conductive surface of the pretreated SnO_2 transparent conductive glass doped with fluorine conductive glass (FTO) was facing downward, inclined at 45° close to the beaker wall, placed in the above solution. This was placed in a constant temperature bath at 85 °C for 200 min and was then washed with deionized water and dried at 60 °C. Finally, the WO_3/FTO electrode was prepared by calcining at 500 °C for two hours in a muffle furnace, cooling to room temperature, washing with deionized water, and drying.

2.4. Preparation of Au NPs

Two milliliters of 50 mmol/L $HAuCl_4$ solutions were added into a double-necked flask with 98 mL deionized water. Stirring and refluxing were performed in an oil bath at 120 °C. When reflux began, the sodium citrate solution was added (10 mL, 38.8 Mm). When the solution turned wine-red, after refluxing for 20 min, the solution was cooled to room temperature and stored in a refrigerator.

2.5. Preparation of CdTe QDs

$CdCl_2{\cdot}2.5H_2O$ (0.0457 g) was dissolved in 50 mL of deionized water; then, 18 µL of TGA (thioglycolic acid) was added, and the solution was stirred for 10 min. The pH was then adjusted to 10.7 with 1 M NaOH. K_2TeO_3 (0.010 g) was dissolved in deionized water (50 mL) and added to the above solution after fully stirring. After 5 min, 80 mg $NaBH_4$ was added. Then, the flask was connected to the condenser and condensed at 100 °C for 3 h. After cooling to room temperature, the TGA-coated CdTe QDs were centrifugally washed with ethanol. Then, the same amount of deionized water was added, and the solution was stored in a refrigerator at 4 °C.

2.6. Preparation of the CdTe QD–Ap Conjugates

Quantum dots (QDs) were coupled with aptamers (Ap) through an EDC–NHS coupling reaction. CdTe QDs (400 μL) were activated by 30 μL of 40 mM EDC and 30 μL of 15 mM NHS for 1 h. Then, 40 μL of 2 μM Ap was added, and the solution was placed in 4 °C shakers for 12 h. Finally, the solution was purified to obtain the QD–Ap conjugate.

2.7. The Culture Process of Listeria Monocytogenes

First, 1 g tryptone, 0.5 g yeast extract powder, and 1 g NaCl were added to 100 mL of deionized water, and the pH was adjusted to 7.5 with NaOH. After sterilization, the configuration of the LB liquid medium was completed. The solid medium was prepared with 1 g tryptone, 0.5 g yeast extract powder, 1 g NaCl, 1.5 g agar powder, and 100 mL deionized water. The pH was adjusted to 7.5, and the medium was sterilized with high-pressure steam. The purchased *Listeria monocytogenes* lyophilized powder was stored with glycerin. Next, 20% glycerin was added to the bacterial solution, and the mixture was evenly mixed, divided into cryogenic tubes, and stored in a refrigerator at −80 °C. Before the experiment, an ultra-clean worktable was sterilized by ultraviolet irradiation for 30 min, and the *Listeria monocytogenes* frozen solution was thawed naturally at room temperature. After sterilization, 100 μL of thawing solution was added to the liquid medium, and then it was shaken in a shaking table at 37 °C for 15 h activation. The configured solid medium was poured into a Petri dish plate next to an alcohol lamp on the ultra-clean workbench to allow it to solidify naturally. After the plate was completely solidified, the secondary activation solution was picked up with the inoculation ring and streaked on the plate. Then, the plate was placed in a 37 °C incubator for 12 h. A single *Listeria monocytogenes* colony was added into the liquid medium, which was shaken on a shaking table at 37 °C for 15 h. Thus, the *Listeria monocytogenes* stock solution culture was completed. A series of 10 times gradient dilutions of *Listeria monocytogenes* stock solution was carried out.

2.8. Fabrication of the Aptamer Sensor

The working electrode was constructed as illustrated in Figure 1. First, 30 μL of Au NPs was dropped onto the surface of the WO_3/FTO electrode. Then, 30 μL of cDNA was mixed with TCEP (0.6 μL, 10 mM) and dropped onto the surface of the Au/WO_3/FTO electrode and incubated. After that, the electrode was washed with Tris HCl to remove the unconnected cDNA. Next, 30 μL of 1 mM 6-mercapto-1-hexanol (MCH) was added dropwise to the electrode, sealed for 1 h, and washed with Tris HCl to obtain an MCH/cDNA/Au/WO_3/FTO electrode. Next, 30 μL of CdTe Quantum dot–aptamers (QD–Aps) was added to the MCH/cDNA/Au/WO_3/FTO electrode and incubated at 4 °C for 1 h to hybridize the cDNA with Ap. When the QD–Ap/MCH/cDNA/Au/WO_3/FTO electrode was obtained, photocurrent detection was carried out, and the detection value was recorded as a. After rinsing with Tris HCl, 30 μL of pathogenic bacteria solutions of different concentrations containing 20 U of Exonuclease I (Exo I) was dropped onto the QD–AP/MCH/cDNA/Au/WO_3/FTO electrode and incubated for 60 min. After rinsing with Tris HCl, the electrode was placed in a refrigerator at 4 °C for photocurrent detection in the next step, and the photocurrent was recorded as b. The current change, detected twice, was recorded as ΔI = a–b. Finally, the relationship between ΔI and the concentration of *Listeria monocytogenes* was calculated and analyzed.

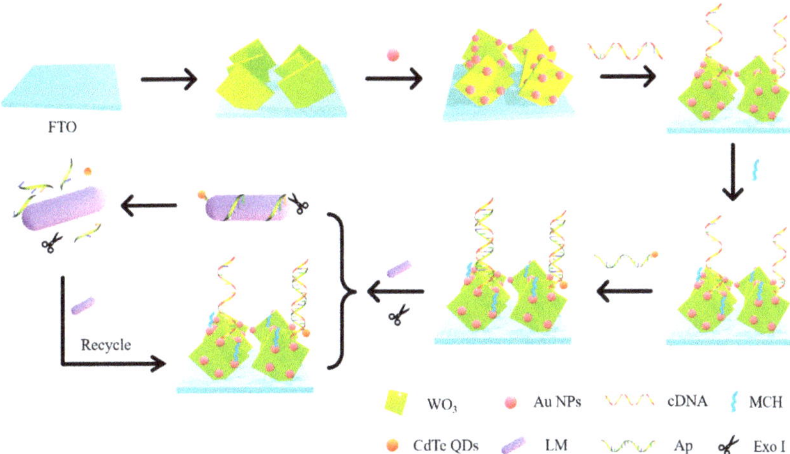

Figure 1. Construction of working electrode and working mechanism of the sensor.

2.9. Recovery Experiments

A 25 g chicken sample was placed in a sterile homogenization bag containing 225 mL of LB broth without additives and homogenized continuously on a flapping homogenizer for 2 min, without incubation of the chicken homogenate after contamination. A quantity of 1 mL of 1.3×10^1 CFU/mL, 1.3×10^4 CFU/mL, or 1.3×10^7 CFU/mL *Listeria monocytogenes* liquid was added to 9 mL of sample homogenate, and the test tube was shaken to mix it evenly. The sample homogenate containing different concentrations of pathogenic bacteria was centrifuged for 5 min under low-speed centrifugation at 2000 r/min. The supernatant was absorbed and added into the sterilization centrifuge tube and centrifuged at a speed of 12,000 r/min for 3 min. After that, the supernatant was discarded, and 1 mL of normal saline was added for full mixing. Then the *Listeria monocytogenes* pollution solution of the adult model was configured, and the prepared sensor was used for analysis and detection.

3. Results and Discussion

3.1. Photoelectrochemical Reaction Mechanism of the Sensor

The charge transfer of the working electrode is displayed in Figure 2. WO_3 has a high electron-hole recombination rate and a low photoelectric conversion efficiency. Therefore, sensitizing the CdTe quantum dots with a small band gap can broaden the spectral response range, improve the photoelectric conversion efficiency, and increase the photocurrent intensity. Au NPs mainly connect aptamers through Au-S bonds and can also increase the absorption of light by quantum dots. AA is used as the electron donor in the electrolyte solution.

Under illumination, the electrons generated by the CdTe quantum dots' transition from the valence band to the conduction band flow through the Au NPs and are injected into the conduction band of WO_3. Then, together with the photogenerated electrons generated inside WO_3, they are finally transferred to the FTO electrode to generate the current signal. The WO_3 and CdTe valence bands leave a large number of photogenerated holes. At this time, the AA in the solution provides electrons and is oxidized by the holes.

When there were no *Listeria monocytogenes*, the CdTe QDs were close to the electrode surface and produced sensitization, and the photocurrent increased. In the presence of *Listeria monocytogenes*, Ap specifically combined with *Listeria monocytogenes*, causing the CdTe QDs to leave the electrode surface and reduce the photocurrent intensity. Meanwhile, Exo I could recognize and cut Ap. *Listeria monocytogenes* bound to Ap continued to bind to Ap on the electrode, and the QDs were far away from the electrode. In this cycle, the QDs on the electrode gradually decreased, and the photocurrent decreased significantly. The sensitive detection of *Listeria monocytogenes* was achieved.

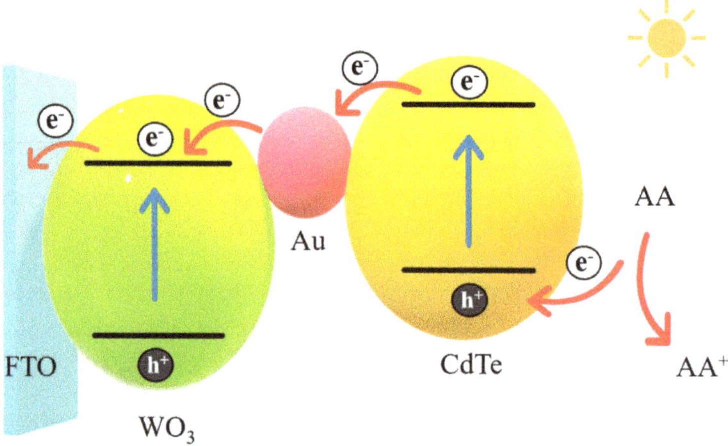

Figure 2. Photogenerated electron transfer mechanism.

3.2. Characterization of Synthesized Materials

Figure 3a shows the XRD patterns of FTO, WO_3/FTO, and Au/WO_3/FTO. The WO_3/FTO diffraction peaks (002), (200), (020), (-112), (202), (222), (140), (240), and (420) are obvious in the figure, which proves the formation of monoclinic crystal WO_3 (JCPDS no. 43-1035). When the Au NPs were modified, no new peaks appeared, but the intensity of the WO_3 peaks was significantly reduced. This may be because of the Au NP layer covering the surface of WO_3, but the content of Au NPs was too small. Figure 3b shows the XRD pattern of CdTe QDs. The 2θ value corresponds to the three crystal planes (111), (220) and (311) of the cubic crystal CdTe on the standard card no. 65-1046. The above results indicate that the sensor electrode material was successfully prepared.

Figure 4 shows the SEM images of WO_3 and Au/WO_3 modified on FTO conductive glass. It can be seen that WO_3 nanosheets were arranged vertically on the surface of the FTO glass (Figure 4a) with regular shapes and a thickness of 70.8–74.7 nm. Figure 4b shows a cross-sectional SEM image of WO_3/FTO. The WO_3 layer deposited on the surface of FTO was relatively uniform, with a thickness of about 422 nm. The vertically arranged structure increased the surface area of the WO_3 layer and could improve the utilization rate of light. After modification with Au NPs, as shown in Figure 4c, the basic morphology of the WO_3 layer did not change, and the surface of the smooth nanosheet became rough, indicating that the Au NPs were successfully modified. There was no change in the cross-sectional thickness after the modification of Au NPs.

The TEM image (Figure 5a) of Au NPs shows that the prepared Au NPs were spherical and uniform in size, and the particle size was between 10 and 15 nm. Obvious lattice fringes can be seen in Figure 4b, and the lattice spacing of the Au (111) lattice plane was about 0.235 nm after software analysis.

To further confirm the surface element composition, we performed an XPS test on WO_3/Au. Figure 6a shows the full spectrum of WO_3/Au. W, Au, O, and C were observed. Figure 6b shows the XPS spectrum of W 4f. It shows that the characteristic peaks of the binding energy of W $4f_{7/2}$ and W $4f_{5/2}$ were 35.7 eV and 37.7 eV, consistent with W^{6+}. The XPS spectrum (Figure 6c) of O 1s showed two characteristic peaks of the binding energies of 530.3 eV and 531.9 eV. The peak at 530.3 eV aligns with the O^{2-} ions characteristic of the WO_3 phase, corresponding to lattice oxygen. The peak at 531.9 eV corresponds with the chemically adsorbed oxygen species at the oxygen vacancy in WO_3 [41,47,48]. The XPS spectrum of Au 4f consisted of two parts, as shown in Figure 6d. The characteristic peaks at 83.8 eV and 87.5 eV can be ascribed to Au $4f_{7/2}$ and Au $4f_{5/2}$, respectively.

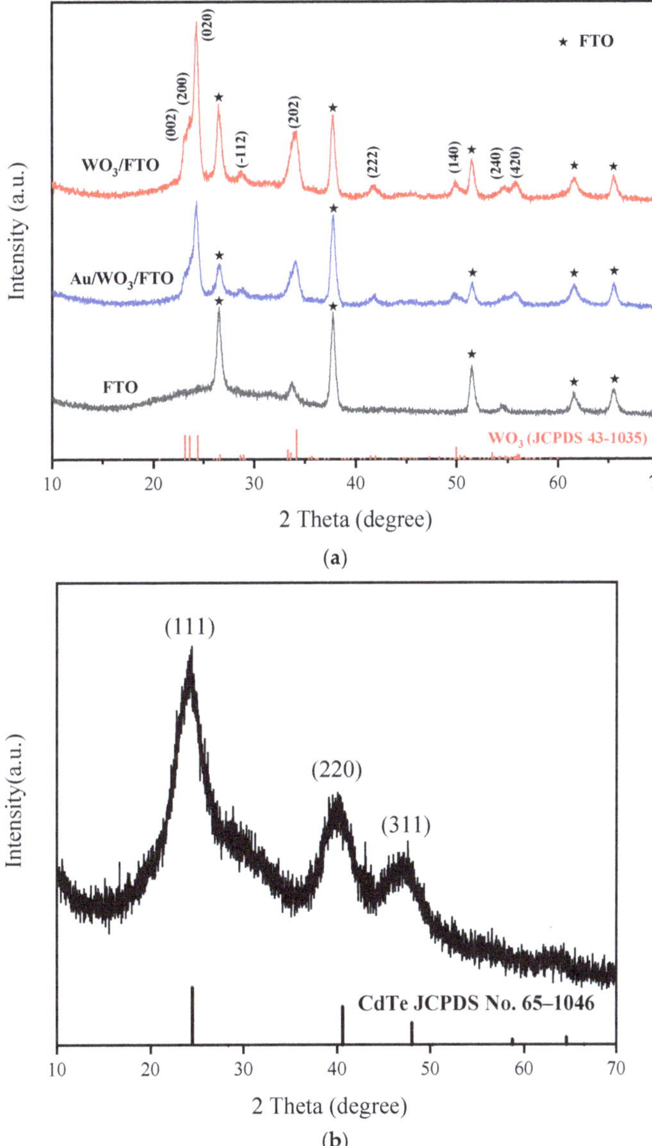

Figure 3. XRD pattern of FTO, WO_3/FTO, and Au/WO3/FTO (**a**) and XRD pattern of CdTe QDs (**b**).

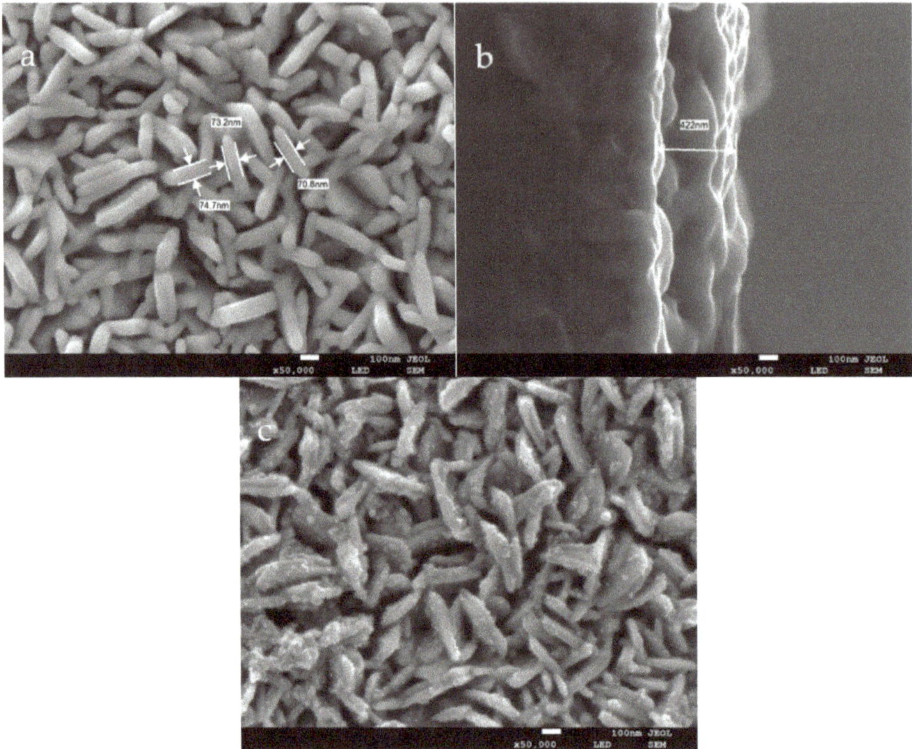

Figure 4. SEM images of the WO3/FTO front (**a**) and cross-section (**b**) and Au/WO3/FTO (**c**).

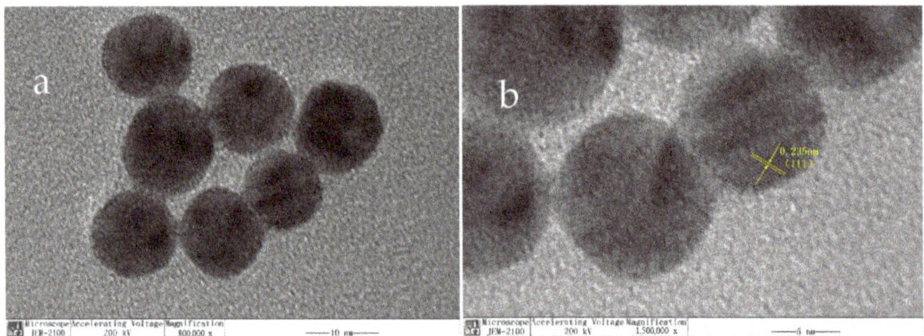

Figure 5. TEM image (**a**) and HRTEM image (**b**) of Au NPs.

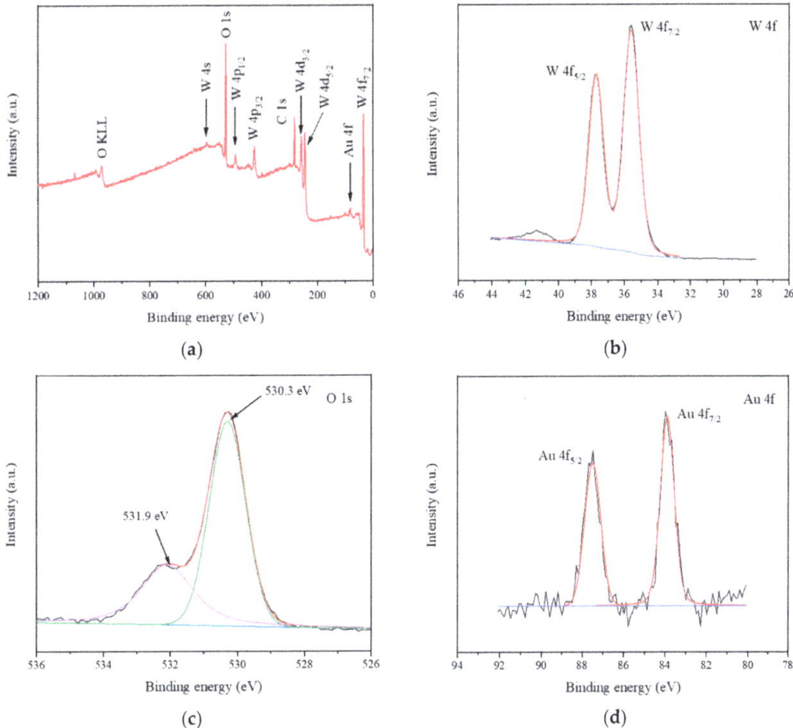

(a) (b) (c) (d)

Figure 6. XPS spectra of WO_3/Au: survey spectrum (**a**), high-resolution XPS of W 4f (**b**), O 1s (**c**), Au 4f (**d**).

Figure 7 shows the UV–vis (a) and fluorescence spectra (b) of CdTe QDs. The curves a, b and c in the figure represent quantum dots prepared through reactions for 30 min, 1 h and 3 h, respectively. With the extension of the reaction time, the peaks of the ultraviolet–visible absorption spectrum and the fluorescence spectrum gradually shifted to the long-wavelength direction, which enhanced the absorption range of visible light. This may be due to the quantum confinement effect, through which the emission wavelength was red-shifted. The stronger the peak in the fluorescence spectrum, the higher the electron-hole recombination rate. After optimization experiments, CdTe QDs with a reflow time of 3 h were selected to construct the sensor. The exciting absorption peak was 538 nm, and its maximum fluorescence emission wavelength was 582 nm.

3.3. Photocurrent Characterization of the Aptamer Sensor

In order to prove the successful preparation of the working electrode of the aptamer sensor, a photocurrent response test was performed in a 10 × PBS buffer containing 0.12 mol/L of AA with a pH of 7.4. As shown in Figure 8, the light source was turned on and off at 20 s and 40 s, and photocurrent changes within 60 s were observed. The blank FTO had no change in photocurrent (curve a), and when modified with WO_3, the photocurrent significantly increased (curve b). This is because WO_3 had good photoelectric activity and generated current under light conditions. When Au NPs were modified on the electrode surface, the photocurrent decreased (curve c). Since the Fermi level of gold nanoparticles is lower than that of WO_3, the work function of gold nanoparticles is greater than that of WO_3. To balance the two Fermis levels, part of the photogenerated electrons on the conduction band of WO_3 is transferred to the gold nanoparticles, which causes a decrease in the photocurrent. When cDNA and MCH were modified, the photocurrent had a slight increase (curve d), which may have been due to the modification of cDNA and MCH, weakening the balance trend of the Fermi level [49]. Therefore, the tendency of photogenerated electrons on the conduction band of WO_3 to transfer to gold nanoparticles was weakened. After the QD–Ap conjugate was modified on the electrode, the photocurrent was significantly enhanced because of the sensitization of the quantum dots (curve e). When 30 μL of 1.3 × 10^6 CFU/mL *Listeria monocytogenes* (LM) solution containing 20 U of Exo I was added dropwise and incubated for 60 min, the photocurrent decreased significantly (curve f). This is because the specific binding of *Listeria monocytogenes* and Ap caused the QD–Ap conjugate to detach from the electrode surface, and the quantum dot sensitization was weakened. The shearing effect of Exo I also released the *Listeria monocytogenes* that had been bound to Ap, which re-attached to the electrode surface. The binding of Ap further weakened the sensitization effect. To verify the shear cycle amplification effect of Exo I, a control experiment was carried out. As shown in Figure 9, the photocurrent of the aptamer electrode (curve b) with 30 μL of 1.3 × 10^6 CFU/mL *Listeria monocytogenes* containing Exo I was 26% lower than that of the electrode without Exo I (curve a). This shows that Exo I has a significant signal-amplifying effect on the photocurrent detection process. In summary, this shows that the aptamer sensor was successfully constructed and can be used for *Listeria monocytogenes* detection.

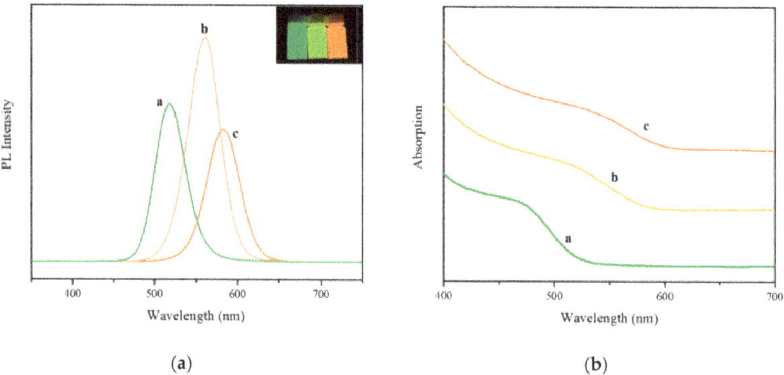

Figure 7. UV-vis (**a**) and fluorescence spectra (**b**) of CdTe QDs, a: 30 min, b: 1 h, c: 3 h. Inset in Figure (**a**) shows the image of CdTe QDs under UV lamps with different reflow times.

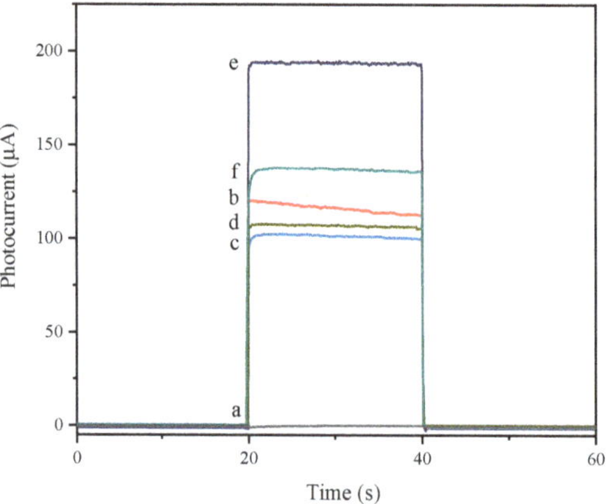

Figure 8. Photocurrent graphs of working electrodes modified with different materials; a: FTO, b: WO_3/FTO, c: Au/WO_3/FTO, d: MCH/cDNA/Au/WO_3/FTO, e: QD–Ap/MCH/cDNA/Au/WO_3/FTO, f: LM-Exo I/QD–Ap/MCH/cDNA/Au/WO_3/FTO.

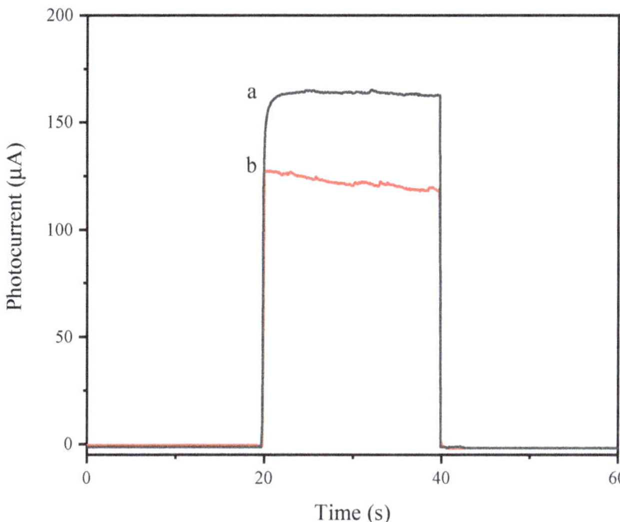

Figure 9. Photocurrent graph of aptamer electrode with or without Exo I modification, a: LM/QD–Ap/MCH/cDNA/Au/WO_3/FTO, b: LM-Exo I/QD–Ap/MCH/cDNA/Au/WO_3/FTO.

3.4. EIS and CV Characterization of Aptamer Sensors

Electrochemical impedance spectroscopy (EIS) was used to further prove the successful construction of the aptamer sensor. EIS was used to a different frequency AC signal to the system; analyze the change in impedance with frequency; analyze the electrode dynamics, diffusion, and electric double layer; and study the mechanism of the solid electrolyte and corrosion protection electrode materials. Electrochemical impedance analysis was performed on working electrodes with different modification processes in solutions. A typical EIS spectrum is a curve with a semicircle and a "tail", which correspond to the high-frequency region and the low-frequency region, respectively. The high-frequency area is dominated by charge transfer, and the low-frequency area is dominated by mass transfer. Among them, the diameter of the high-frequency region circle is equal to the charge transfer resistance (Rct). The larger the diameter, the greater the obstruction of the oxidation–reduction probe on the electrode surface. In Figure 10, FTO has the smallest diameter (curve a), indicating that FTO without any modification can transfer electrons more effectively. After the deposition of WO_3, the impedance increased (curve b). Because of the excellent conductivity of gold nanoparticles, the impedance decreased after modification with Au NPs (curve c). The impedance value increased after cDNA and MCH were modified (curve d). This is because the oligonucleotide was negatively charged, and the oxidation–reduction probe Fe $(CN)_6{}^{3-/4-}$ was also negatively charged; the repulsive force between the two caused an increase in the resistance of electron transport, and MCH is non-conductive, which led to an increase in impedance. When the QD–Ap conjugate was modified on the electrode, the impedance was further increased because of the weak conductivity of QDs and the increase in oligonucleotides. The impedance decreased after incubating Exo I and *Listeria monocytogenes*. This is because many oligonucleotides and QDs left the electrode surface after the specific binding of *Listeria monocytogenes* and Ap and Exo I shearing. Therefore, the change of impedance value proves that the aptamer sensor was constructed successfully.

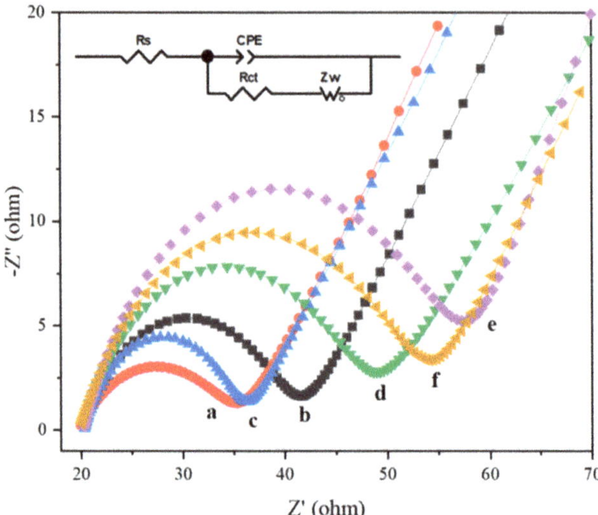

Figure 10. Electrochemical impedance spectroscopy, a: FTO, b: WO_3/FTO, c: Au/WO_3/FTO, d: MCH/cDNA/Au/WO_3/FTO, e: QDs-Ap/MCH/cDNA/Au/WO_3/FTO, f: LM-Exo I/QDs-Ap/MCH/cDNA/Au/WO_3/FTO. Inset is the equivalent circuit model.

Cyclic voltammetry is also a commonly used electrochemical analysis method, as shown in Figure 11. On the bare FTO electrode (curve a), a pair of obvious redox anode and cathode peaks can be observed at −0.15 V and 0.5 V, and the two peaks are symmetrical, which proves that the reaction is reversible. After being modified with WO_3, the redox peak current decreased (curve b). When Au NPs were modified, the current value increased (curve c), indicating that Au NPs can accelerate electron transfer. When cDNA, MCH (curve d), and QD–Ap conjugate (curve e) were modified sequentially, the redox peak current gradually decreased, and the current value increased after incubating Exo I and *Listeria monocytogenes* (curve f). The above test results were consistent with the EIS results, providing further evidence for the successful construction of the sensor.

3.5. Optimization of Experimental Parameters

The test conditions of the sensor were optimized, and the concentration of AA, the pH value of the electrolyte, the reflow time of the quantum dots, and the incubation time of the electrode with *Listeria monocytogenes* and Exo I were investigated.

AA acts as an electron donor and plays a significant role in increasing photocurrent. As shown in Figure 12a, when there was no AA in the electrolyte, the photocurrent of WO_3/FTO was the smallest. With the gradual increase in AA concentration, the photocurrent reached its maximum value at 0.10 mol/L. When the AA concentration exceeded 0.10 mol/L, the current value gradually decreased. It may be that the excessive AA concentration increases the absorbance of the electrolyte and reduces the light intensity on the electrode surface. Therefore, 0.10 mol/L was chosen as the optimum concentration of AA.

In addition, aptamers can only remain active in relatively neutral solutions. Figure 12b shows that the WO_3/FTO electrode photocurrent reached the maximum when the electrolyte pH was 7.4, so 7.4 was selected as the optimal pH.

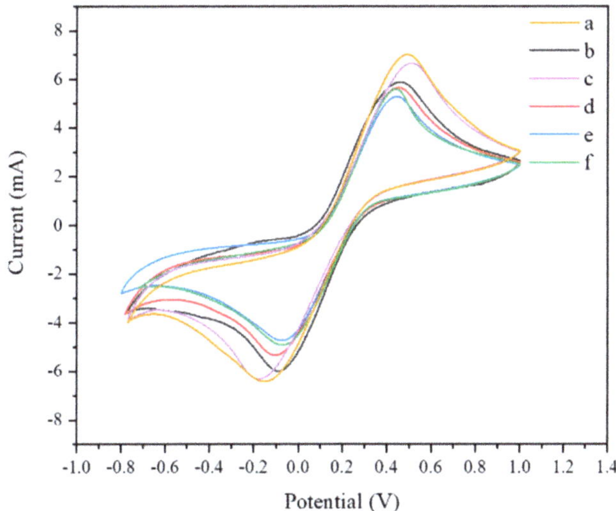

Figure 11. Cyclic voltammogram, a: FTO, b: WO_3/FTO, c: Au/WO_3/FTO, d: MCH/cDNA/Au/WO_3/FTO, e: QD–Ap/MCH/cDNA/Au/WO_3/FTO, f: LM-Exo I/QD–Ap/MCH/cDNA/Au/WO_3/FTO.

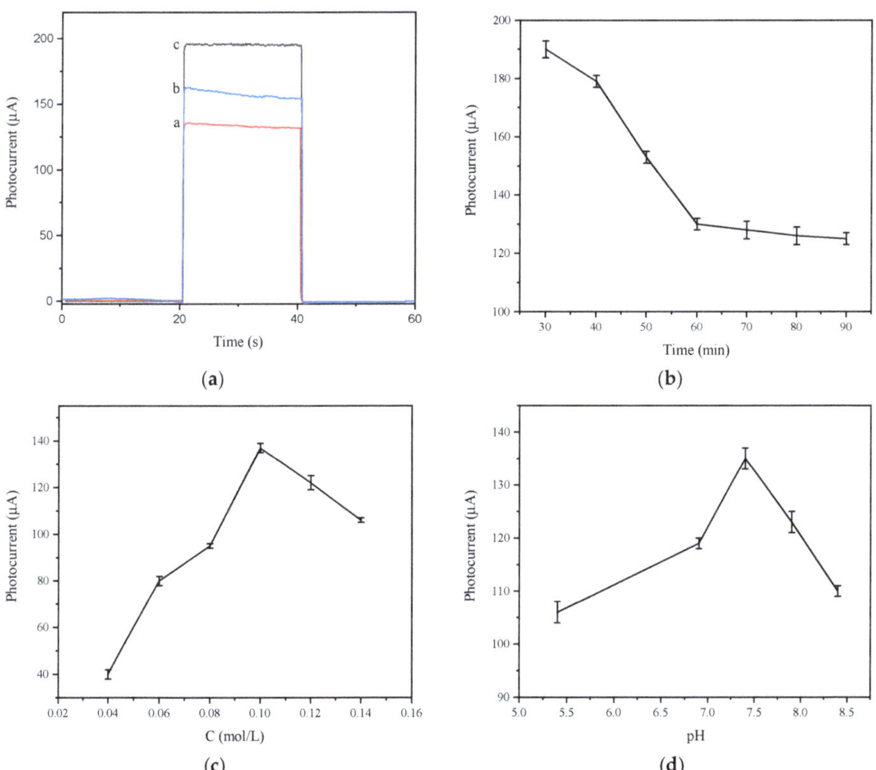

Figure 12. Photocurrent graphs of different AA concentrations (**a**); pH (**b**); quantum dot reflux times (**c**), a: 30 min, b: 1 h, c: 3 h; and incubation time (**d**).

Three CdTe quantum dots with different reflow times were prepared. It can be seen from the above that the positions of the strongest peaks of the ultraviolet–visible absorption spectra and fluorescence spectra of CdTe quantum dots with different reflow times are different. As time increased, the positions of the strongest peaks moved toward the long-wavelength direction. To further prove the size of the photocurrent of quantum dots at different reflow times, three different QDs were used to construct aptamer electrodes and perform photocurrent detection. Figure 12c shows that the photocurrent was the largest when the reflux time was 3 h. Therefore, CdTe QDs with a reflow time of 3 h were chosen to construct an aptamer sensor.

The incubation time of Exo I and *Listeria monocytogenes* also affects the change in photocurrent. Proper incubation time not only reduces the time used in the preparation process but also enables the aptamer sensor to achieve optimal performance. As can be seen from Figure 12d, the photocurrent gradually decreased with an increase in time before the incubation time of 60 min. After 60 min, the photocurrent decreased, but it gradually stabilized. Therefore, the optimal incubation time of Exo I and *Listeria monocytogenes* was 60 min.

3.6. Analytical Performance of Aptamer Sensor

The prepared aptamer sensor relies on the changes in the photocurrent of the electrodes before and after the addition of pathogenic bacteria to quantitatively and qualitatively detect the pathogenic bacteria. The change in sensor photocurrent is directly related to the concentration of pathogenic bacteria. Figure 13a shows the photocurrent change curve of different concentrations of *Listeria monocytogenes*. It shows that as the *Listeria monocytogenes* concentration increased, the photocurrent gradually decreased. This is because of the combination of a large amount of *Listeria monocytogenes* and Ap on the electrode surface, which weakened the sensitization effect of QDs and reduced the photocurrent. Figure 13b shows that from 1.3 $\times$ 10 CFU/mL to 1.3 $\times$ 10^7 CFU/mL, the change in photocurrent had a very good linear relationship with the concentration of pathogenic bacteria. The linear equation obtained was $\Delta I = 9.76\log C_{LM} - 4.44$ ($R^2 = 0.9980$), and the detection limit was 45 CFU/mL. This may be due to the sensitization of quantum dots and the auxiliary amplification effect of Exo I, allowing the aptamer sensor to have a wider detection range and a smaller detection limit.

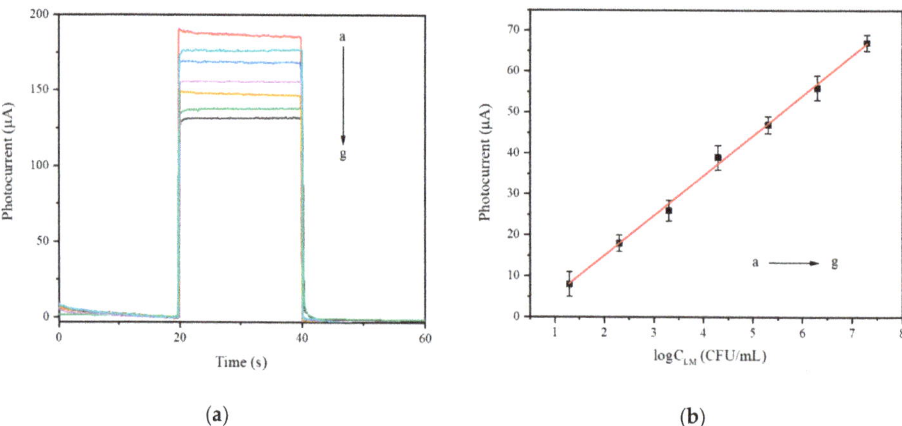

Figure 13. Photocurrent curve (**a**) and a calibration curve (**b**) of the aptamer sensor with different concentrations of *Listeria monocytogenes* (a→g: 1.3 $\times$ 10^1→1.3 $\times$ 10^7CFU/mL).

3.7. Selectivity, Stability, and Reproducibility of the PEC Sensing Platform

Specificity is a significant indicator of the sensor. The specificity of the sensor was characterized by the change of photocurrent before and after incubation with different interferents. A quantity of 1.3×10^5 CFU/mL *Listeria monocytogenes* (A) was selected for detection; *Staphylococcus aureus* (B), *Escherichia coli O157:H7* (C), and *Salmonella typhimurium* (D) at the same concentration were selected as interferents, and physiological saline (E) was used as a blank control. As can be seen from Figure 14a, the photocurrent changed significantly before and after the incubation of *Listeria monocytogenes*. After calculation, the change value RSD was 2.94%. The photocurrent changes in the interferents were similar to those in the blank group. This result indicates that the prepared aptamer sensor has satisfactory specificity for *Listeria monocytogenes*.

To investigate the stability of the sensor, the electrode was incubated with 1.3×10^5 CFU/mL *Listeria monocytogenes*, and Exo I was repeatedly switched on and off within 400 s to observe the change in photocurrent to judge the stability of the sensor. The light source switch interval was 20 s, as shown in Figure 14b. It can be seen from the figure that there was almost no change in the photocurrent within 400 s. After the second electrode was placed in a refrigerator at 4 °C for a week, the photocurrent detection was performed again under the same conditions, and the current value dropped by about 5%, which proved that the sensor had good stability.

Five electrodes were constructed under the same conditions, 1.3×10^5 CFU/mL *Listeria monocytogenes* and Exo I were incubated with photoelectric detection, and the reproducibility of the sensor was analyzed. The experimental calculation showed that the RSD was 2.1%, proving that the prepared aptamer sensor has good accuracy and reproducibility.

3.8. Analysis of Real Samples

To verify the feasibility of the sensor to detect *Listeria monocytogenes* in actual samples, we added 1×10^2 CFU/mL, 1×10^5 CFU/mL, and 1×10^7 CFU/mL of *Listeria monocytogenes* to chicken samples that did not contain *Listeria monocytogenes* for photocurrent detection. The *Listeria monocytogenes* concentration was obtained and compared with the actual addition to calculate the recovery rate. As shown in Table 1, for three different dilutions of *Listeria monocytogenes*, the prepared aptamer sensor was used for detection, and the recovery rate met the requirements. This shows that the sensor has great potential in practical applications.

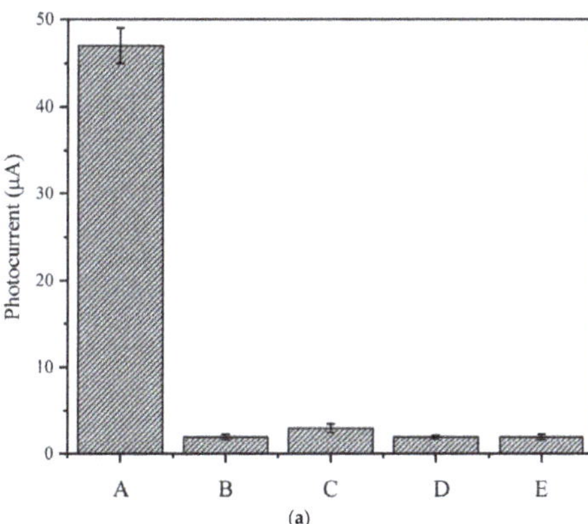

Figure 14. *Cont.*

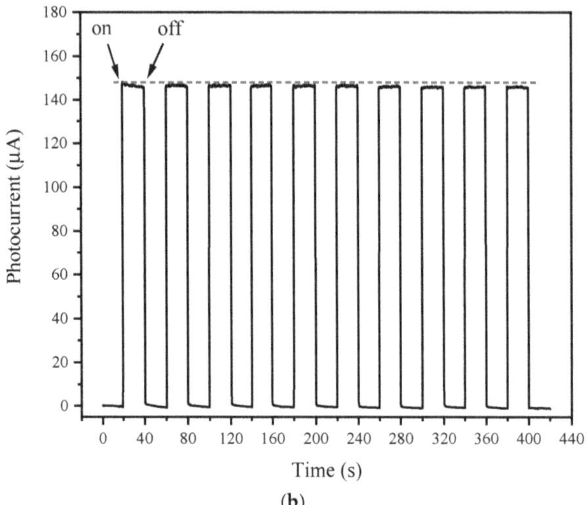

Figure 14. (**a**) Photocurrent response of the sensor to 1.3×10^5 CFU/mL of *Listeria monocytogenes* (A), *Staphylococcus aureus* (B), *Escherichia coli O157:H7* (C), *Salmonella typhimurium* (D), saline (E); (**b**) photocurrent response of the aptamer electrode when the light was turned on and off within 400 s continuously.

Table 1. Detection of *Listeria monocytogenes* in chicken samples with sensors.

Samples	Added (CFU/mL)	Found (CFU/mL)	Recovery (%)	RSD (%)
1	1×10^2	0.91×10^2	91	2.2
2	1×10^5	1.12×10^5	112	3.4
3	1×10^7	0.89×10^7	89	5.3

4. Conclusions

In the present study, a photoelectrochemical sensor based on WO_3 was constructed to detect *Listeria monocytogenes*. When pathogenic bacteria were present, the quantum dots fell off the surface of the electrode, causing the photocurrent to decrease. At the same time, combined with the shearing effect of Exo I, the photocurrent was significantly reduced, thereby amplifying the signal changes before and after the addition of pathogenic bacteria and increasing the detection limit. Finally, the performance of the sensing platform was verified through the aspects of specificity, stability, and repeatability. Under the optimal conditions, a range of 1.3×10^1 to 1.3×10^7 CFU/mL was observed, with a detection limit of 45 CFU/mL. This method may have hopeful prospects for the rapid detection of *Listeria monocytogenes*.

Author Contributions: Methodology, formal analysis, investigation, writing—original draft, L.Z.; conceptualization, supervision, writing—review and editing, H.H.; visualization, C.D.; methodology, H.G., S.J. and G.Z.; validation, J.B.; resources, S.Y.; supervision, H.H. All authors have read and agreed to the published version of the manuscript.

Funding: This research was funded by the National Key Research and Development Program of China (2019YFC1605901), the Science and Technology Program of Liaoning Provincial Education Department (J2020049), and Liaoning Province's Program for Promoting Liaoning Talents (XLYC1808034).

Institutional Review Board Statement: Not applicable.

Informed Consent Statement: Not applicable.

Data Availability Statement: The datasets generated for this study are available on request to the corresponding author.

Conflicts of Interest: The authors declare that they have no conflicts of interest.

References

1. Liu, R.; Zhang, Y.; Ali, S.; Haruna, S.A.; He, P.; Li, H.; Ouyang, Q.; Chen, Q. Development of a fluorescence aptasensor for rapid and sensitive detection of *Listeria monocytogenes* in food. *Food Control* **2021**, *122*, 107808. [CrossRef]
2. Zhang, X.; Wang, S.; Chen, X.; Qu, C. Review controlling *Listeria monocytogenes* in ready-to-eat meat and poultry products: An overview of outbreaks, current legislations, challenges, and future prospects. *Trends Food Sci. Technol.* **2021**, *116*, 24–35. [CrossRef]
3. Forauer, E.; Wu, S.T.; Etter, A.J. *Listeria monocytogenes* in the retail deli environment: A review. *Food Control* **2021**, *119*, 107443. [CrossRef]
4. Buchanan, R.L.; Gorris, L.G.; Hayman, M.M.; Jackson, T.C.; Whiting, R.C. A review of *Listeria monocytogenes*: An update on outbreaks, virulence, dose-response, ecology, and risk assessments. *Food Control* **2017**, *75*, 1–13. [CrossRef]
5. Tian, Y.; Liu, T.; Liu, C.; Xu, Q.; Fang, S.; Wu, Y.; Wu, M.; Liu, Q. An ultrasensitive and contamination-free on-site nucleic acid detection platform for *Listeria monocytogenes* based on the CRISPR-Cas12a system combined with recombinase polymerase amplification. *LWT* **2021**, *152*, 112166. [CrossRef]
6. Huang, Y.M.; Hsu, H.Y.; Hsu, C.L. Development of electrochemical method to detect bacterial count, *Listeria monocytogenes*, and somatic cell count in raw milk. *J. Taiwan Inst. Chem. Eng.* **2016**, *62*, 39–44. [CrossRef]
7. Li, Y.; Wu, L.; Wang, Z.; Tu, K.; Pan, L.; Chen, Y. A magnetic relaxation DNA biosensor for rapid detection of *Listeria monocytogenes* using phosphatase-mediated Mn (VII)/Mn (II) conversion. *Food Control* **2021**, *125*, 107959. [CrossRef]
8. Bundidamorn, D.; Supawasit, W.; Trevanich, S. Taqman® probe based multiplex RT-PCR for simultaneous detection of *Listeria monocytogenes*, *Salmonella* spp. and Shiga toxin-producing Escherichia coli in foods. *LWT* **2021**, *147*, 111696. [CrossRef]
9. Chen, Q.; Yao, C.; Yang, C.; Liu, Z.; Wan, S. Development of an in-situ signal amplified electrochemical assay for detection of *Listeria monocytogenes* with label-free strategy. *Food Chem.* **2021**, *358*, 129894. [CrossRef]
10. Etty, M.C.; d'Auria, S.; Fraschini, C.; Salmieri, S.; Lacroix, M. Effect of the optimized selective enrichment medium on the expression of the p60 protein used as *Listeria monocytogenes* antigen in specific sandwich ELISA. *Res. Microbiol.* **2019**, *17*, 182–191. [CrossRef]
11. Nguyen, T.T.; Van Giau, V.; Vo, T.K. Multiplex PCR for simultaneous identification of *E. coli* O157: H7, *Salmonella* spp. and L. *monocytogenes* in food. *3 Biotech* **2016**, *6*, 205. [PubMed]
12. Qin, H.; Shi, X.; Yu, L.; Li, K.; Wang, J.; Chen, J.; Yang, F.; Xu, H.; Xu, H. Multiplex real-time PCR coupled with sodium dodecyl sulphate and propidium monoazide for the simultaneous detection of viable *Listeria monocytogenes*, *Cronobacter sakazakii*, *Staphylococcus aureus* and *Salmonella* spp. in milk. *Int. Dairy J.* **2020**, *108*, 104739. [CrossRef]
13. Roumani, F.; Azinheiro, S.; Carvalho, J.; Prado, M.; Garrido-Maestu, A. Loop-mediated isothermal amplification combined with immunomagnetic separation and propidium monoazide for the specific detection of viable *Listeria monocytogenes* in milk products, with an internal amplification control. *Food Control* **2021**, *125*, 107975. [CrossRef]
14. Sadighbayan, D.; Sadighbayan, K.; Tohid-Kia, M.R.; Khosroushahi, A.Y.; Hasanzadeh, M. Development of electrochemical biosensors for tumor marker determination towards cancer diagnosis: Recent progress. *TrAC Trends Anal. Chem.* **2019**, *118*, 73–88. [CrossRef]
15. Mohankumar, P.; Ajayan, J.; Mohanraj, T.; Yasodharan, R. Recent developments in biosensors for healthcare and biomedical applications: A review. *Measurement* **2021**, *167*, 108293. [CrossRef]
16. Ren, R.; Cai, G.; Yu, Z.; Tang, D. Glucose-loaded liposomes for amplified colorimetric immunoassay of streptomycin based on enzyme-induced iron (II) chelation reaction with phenanthroline. *Sens. Actuators B Chem.* **2018**, *265*, 174–181. [CrossRef]
17. Zeng, R.; Su, L.; Luo, Z.; Zhang, L.; Lu, M.; Tang, D. Ultrasensitive and label-free electrochemical aptasensor of kanamycin coupling with hybridization chain reaction and strand-displacement amplification. *Anal. Chim. Acta* **2018**, *1038*, 21–28. [CrossRef] [PubMed]
18. Gao, X.; Niu, S.; Ge, J.; Luan, Q.; Jie, G. 3D DNA nanosphere-based photoelectrochemical biosensor combined with multiple enzyme-free amplification for ultrasensitive detection of cancer biomarkers. *Biosens. Bioelectron.* **2020**, *147*, 111778. [CrossRef]
19. Shin, J.; Yan, Y.; Bai, W.; Xue, Y.; Gamble, P.; Tian, L.; Kandela, I.; Haney, C.R.; Spees, W.; Lee, Y.; et al. Bioresorbable pressure sensors protected with thermally grown silicon dioxide for the monitoring of chronic diseases and healing processes. *Nat. Biomed. Eng.* **2019**, *3*, 37–46. [CrossRef]
20. Zhao, C.Q.; Ding, S.N. Perspective on signal amplification strategies and sensing protocols in photoelectrochemical immunoassay. *Coord. Chem. Rev.* **2019**, *391*, 1–14. [CrossRef]
21. Li, N.; Sun, Y.; Wang, F.; Huang, C.; Fu, C.; Zhang, L.; Liu, Y.; Ge, S.; Yu, J. Target dual-recycling-induced bipedal DNA walker and Bi_2WO_6/Bi_2S_3 cascade amplification strategy in photoelectrochemical biosensor for TP53 detection. *Sens. Actuators B Chem.* **2021**, *345*, 130386. [CrossRef]
22. Wang, H.H.; Li, M.J.; Tu, Y.P.; Wang, H.J.; Chai, Y.Q.; Li, Z.H.; Yuan, R. Fullerenol as a photoelectrochemical nanoprobe for discrimination and ultrasensitive detection of amplification-free single-stranded DNA. *Biosens. Bioelectron.* **2021**, *173*, 112802. [CrossRef]

23. Wen, Y.X.; Qing, M.; Chen, S.L.; Luo, H.Q.; Li, N.B. Y-type DNA structure stabilized p-type CuS quantum dots to quench photocurrent of ternary heterostructure for sensitive photoelectrochemical detection of miRNA. *Sens. Actuators B Chem.* **2021**, *329*, 129257. [CrossRef]
24. Song, W.; Zhang, F.; Song, P.; Zhang, Z.; He, P.; Li, Y.; Zhang, X. Untrasensitive photoelectrochemical sensor for microRNA detection with DNA walker amplification and cation exchange reaction. *Sens. Actuators B Chem.* **2021**, *327*, 128900. [CrossRef]
25. Kang, Q.; Zhang, Q.; Zang, L.; Zhao, M.; Chen, X.; Shen, D. Enhancement anti-interference ability of photoelectrochemical sensor via differential molecularly imprinting technique demonstrated by dopamine determination. *Anal. Chim. Acta* **2020**, *1125*, 201–209. [CrossRef] [PubMed]
26. Dashtian, K.; Hajati, S.; Ghaedi, M. Ti-based solid-state imprinted-Cu_2O/$CuInSe_2$ heterojunction photoelectrochemical platform for highly selective dopamine monitoring. *Sens. Actuators B Chem.* **2021**, *326*, 128824. [CrossRef]
27. Liu, Q.; Kim, J.; Cui, T. A highly sensitive photoelectrochemical sensor with polarity-switchable photocurrent for detection of trace hexavalent chromium. *Sens. Actuators B Chem.* **2020**, *317*, 128181. [CrossRef]
28. Zhang, L.; Feng, L.; Li, P.; Chen, X.; Jiang, J.; Zhang, S.; Zhang, C.; Zhang, A.; Chen, G.; Wang, H. Direct Z-scheme photocatalyst of hollow CoS_x@ CdS polyhedron constructed by ZIF-67-templated one-pot solvothermal route: A signal-on photoelectrochemical sensor for mercury (II). *Chem. Eng. J.* **2020**, *395*, 125072. [CrossRef]
29. Jiang, Q.; Wang, H.; Wei, X.; Wu, Y.; Gu, W.; Hu, L.; Zhu, C. Efficient $BiVO_4$ photoanode decorated with $Ti_3C_2T_X$ MXene for enhanced photoelectrochemical sensing of Hg (II) ion. *Anal. Chim. Acta* **2020**, *1119*, 11–17. [CrossRef]
30. Ma, Y.; Dong, Y.X.; Wang, B.; Ren, S.W.; Cao, J.T.; Liu, Y.M. CdS: Mn-sensitized 2D/2D heterostructured g-C_3N_4-MoS_2 with excellent photoelectrochemical performance for ultrasensitive immunosensing platform. *Talanta* **2020**, *207*, 120288. [CrossRef] [PubMed]
31. Shang, M.; Gao, Y.; Zhang, J.; Yan, J.; Song, W. Signal-on cathodic photoelectrochemical aptasensing of insulin: Plasmonic Au activated amorphous MoS_x photocathode coupled with target-induced sensitization effect. *Biosens. Bioelectron.* **2020**, *165*, 112359. [CrossRef]
32. Yan, P.; Jiang, D.; Tian, Y.; Xu, L.; Qian, J.; Li, H.; Xia, J.; Li, H. A sensitive signal-on photoelectrochemical sensor for tetracycline determination using visible-light-driven flower-like CN/BiOBr composites. *Biosens. Bioelectron.* **2018**, *111*, 74–81. [CrossRef]
33. Zhang, K.; Lv, S.; Zhou, Q.; Tang, D. CoOOH nanosheets-coated g-C_3N_4/$CuInS_2$ nanohybrids for photoelectrochemical biosensor of carcinoembryonic antigen coupling hybridization chain reaction with etching reaction. *Sens. Actuators B Chem.* **2020**, *307*, 127631. [CrossRef]
34. Li, Y.; Bu, Y.; Jiang, F.; Dai, X.; Ao, J.P. Fabrication of ultra-sensitive photoelectrochemical aptamer biosensor: Based on semiconductor/DNA interfacial multifunctional reconciliation via 2D-C_3N_4. *Biosens. Bioelectron.* **2020**, *150*, 111903. [CrossRef]
35. Wang, Y.; Zhou, Y.; Xu, L.; Han, Z.; Yin, H.; Ai, S. Photoelectrochemical apta-biosensor for zeatin detection based on graphene quantum dots improved photoactivity of graphite-like carbon nitride and streptavidin induced signal inhibition. *Sens. Actuators B Chem.* **2018**, *257*, 237–244. [CrossRef]
36. Zargazi, M.; Entezari, M.H. Ultrasound assisted deposition of highly stable self-assembled Bi_2MoO_6 nanoplates with selective crystal facet engineering as photoanode. *Ultrason. Sonochem.* **2020**, *67*, 105145. [CrossRef]
37. Zhou, T.; Zhang, H.; Ma, X.; Zhang, X.; Zhu, Y.; Zhang, A.; Cao, Y.; Yang, P. Construction of AgI/Bi_2MoO_6/AgBi $(MoO_4)_2$ multi-heterostructure composite nanosheets for visible-light photocatalysis. *Mater. Today Commun.* **2020**, *23*, 100903. [CrossRef]
38. Miodyńska, M.; Mikolajczyk, A.; Bajorowicz, B.; Zwara, J.; Klimczuk, T.; Lisowski, W.; Trykowski, G.; Pinto, H.P.; Zaleska-Medynska, A. Urchin-like TiO_2 structures decorated with lanthanide-doped Bi_2S_3 quantum dots to boost hydrogen photogeneration performance. *Appl. Catal. B Environ.* **2020**, *272*, 118962. [CrossRef]
39. Shen, Y.; Li, T.; Zhong, X.; Li, G.; Li, A.; Wei, D.; Zhang, Y.; Wei, K. Ppb-level NO_2 sensing properties of Au-doped WO_3 nanosheets synthesized from a low-grade scheelite concentrate. *Vacuum* **2020**, *172*, 109036. [CrossRef]
40. Lu, Y.; Li, Y.; Wang, Y.; Zhang, J. Two-photon induced NIR active core-shell structured WO_3/CdS for enhanced solar light photocatalytic performance. *Appl. Catal. B Environ.* **2020**, *272*, 118979. [CrossRef]
41. Pei, F.; Feng, S.; Wu, Y.; Lv, X.; Wang, H.; Chen, S.-M.; Hao, Q.; Cao, Y.; Lei, W.; Tong, Z. Label-free photoelectrochemical immunosensor for aflatoxin B1 detection based on the Z-scheme heterojunction of g-C_3N_4/Au/WO_3. *Biosens. Bioelectron.* **2021**, *189*, 113373. [CrossRef] [PubMed]
42. Chen, P.; Liu, Z.; Liu, J.; Liu, H.; Bian, W.; Tian, D.; Xia, F.; Zhou, C. A novel electrochemiluminescence aptasensor based CdTe QDs@ NH_2-MIL-88 (Fe) for signal amplification. *Electrochim. Acta* **2020**, *354*, 136644. [CrossRef]
43. Shahdost-Fard, F.; Fahimi-Kashani, N.; Hormozi-Nezhad, M.R. A ratiometric fluorescence nanoprobe using CdTe QDs for fast detection of carbaryl insecticide in apple. *Talanta* **2021**, *221*, 121467. [CrossRef] [PubMed]
44. Duan, N.; Wu, S.; Dai, S.; Gu, H.; Hao, L.; Ye, H.; Wang, Z. Advances in aptasensors for the detection of food contaminants. *Analyst* **2016**, *141*, 3942–3961. [CrossRef]
45. Cong, X.; Fan, G.C.; Wang, X.; Abdel-Halim, E.S.; Zhu, J.J. Enhanced photoelectrochemical aptasensing platform amplified through the sensitization effect of CdTe@ CdS core–shell quantum dots coupled with exonuclease-I assisted target recycling. *J. Mater. Chem. B* **2016**, *4*, 6117–6124. [CrossRef]
46. Lee, S.H.; Ahn, J.Y.; Lee, K.A.; Um, H.J.; Sekhon, S.S.; Park, T.S.; Kim, Y.H. Analytical bioconjugates, aptamers, enable specific quantitative detection of *Listeria monocytogenes*. *Biosens. Bioelectron.* **2015**, *68*, 272–280. [CrossRef] [PubMed]

47. Yang, X.; Wang, Y.; Fu, H.; Wang, W.; Han, D.; An, X. Experimental and theoretical study on the excellent amine-sensing performance of Au decorated WO_3 needle-like nanocomposites. *Mater. Chem. Phys.* **2019**, *234*, 122–132. [CrossRef]
48. Xia, H.; Wang, Y.; Kong, F.; Wang, S.; Zhu, B.; Guo, X.; Zhang, J.; Wu, S. Au-doped WO_3-based sensor for NO_2 detection at low operating temperature. *Sens. Actuators B Chem.* **2008**, *134*, 133–139. [CrossRef]
49. Fan, G.C.; Zhao, M.; Zhu, H.; Shi, J.J.; Zhang, J.R.; Zhu, J.J. Signal-on photoelectrochemical aptasensor for adenosine triphosphate detection based on sensitization effect of CdS: Mn@ $Ru(bpy)_2(dcbpy)$ nanocomposites. *J. Phys. Chem. C* **2016**, *120*, 15657–15665. [CrossRef]

Article

A Highly Sensitive and Selective Fluorescent Probe Using MPA-InP/ZnS QDs for Detection of Trace Amounts of Cu^{2+} in Water

Zeyu Xu [1,2,3,4], Yizhong Wang [1], Jiaran Zhang [2,3,4,*], Ce Shi [2,3,4] and Xinting Yang [2,3,4]

1 College of Electronic Information and Automation, Tianjin University of Science and Technology, Tianjin 300222, China; xzy19950811@163.com (Z.X.); yzwang@tust.edu.cn (Y.W.)
2 Information Technology Research Center, Beijing Academy of Agriculture and Forestry Sciences, Beijing 100097, China; shice001@163.com (C.S.); yangxt@nercita.org.cn (X.Y.)
3 National Engineering Research Center for Information Technology in Agriculture, Beijing Academy of Agricultural and Forestry Sciences, Beijing 100097, China
4 National Engineering Laboratory for Agri-Product Quality Traceability, Beijing Academy of Agricultural and Forestry Sciences, Beijing 100097, China
* Correspondence: jiaran1031@126.com

Abstract: Detection of copper (II) ions (Cu^{2+}) in water is important for preventing them from entering the human body to preserve human health. Here, a highly sensitive and selective fluorescence probe that uses mercaptopropionic acid (MPA)-capped InP/ZnS quantum dots (MPA-InP/ZnS QDs) was proposed for the detection of trace amounts of Cu^{2+} in water. The fluorescence of MPA-InP/ZnS QDs can be quenched significantly in the presence of Cu^{2+}, and the fluorescence intensity shows excellent linearity when the concentration of Cu^{2+} varies from 0–1000 nM; this probe also exhibits an extremely low limit of detection of 0.22 nM. Furthermore, a possible fluorescence-quenching mechanism was proposed. The MPA-InP/ZnS QDs probes were further applied to the detection of trace Cu^{2+} in real water samples and drink samples, showing good feasibility.

Keywords: fluorescence; quantum dots; copper (II) ions; water

Citation: Xu, Z.; Wang, Y.; Zhang, J.; Shi, C.; Yang, X. A Highly Sensitive and Selective Fluorescent Probe Using MPA-InP/ZnS QDs for Detection of Trace Amounts of Cu^{2+} in Water. *Foods* **2021**, *10*, 2777. https://doi.org/10.3390/foods10112777

Academic Editors: Hong Wu and Hui Zhang

Received: 13 October 2021
Accepted: 8 November 2021
Published: 11 November 2021

1. Introduction

As one of the essential trace elements for humans, Cu^{2+} plays an important role in many physiological processes, such as hemoglobin regulation, bone formation, and cell metabolism [1]. However, high concentrations of Cu^{2+} can affect the function of the kidney and liver and may cause brain diseases, such as Alzheimer's disease and Parkinson's disease [2]. Even short-term exposure to high concentrations of Cu^{2+} can cause stomach and intestinal discomfort [3]. Cu^{2+} is taken up by humans through contaminated water and food and accumulates in the human body. To this end, the maximum residue levels of Cu^{2+} in water are clearly defined. The United States Environmental Protection Agency (USEPA) stipulated that the maximum allowable amounts of Cu^{2+} in drinking water and in industrial effluents are 20 μM [4] and 1.3 mg/L [5], respectively. Therefore, detecting Cu^{2+} in water is of great significance for preventing Cu^{2+} from entering the human body and ensuring human health.

There are many methods for detecting Cu^{2+}, such as atomic absorption spectrometry [6], inductively coupled plasma optical emission spectrometry (ICP-OES) [7], inductively coupled plasma mass spectrometry (ICP-MS) [8], electrochemical methods [9], and fluorescence methods [10]. Atomic absorption spectrometry, ICP-OES, and ICP-MS are highly accurate for the detection of Cu^{2+}; however, the disadvantages of expensive equipment, cumbersome operation, and time consumption [11,12] make them unable to detect mobile Cu^{2+} in real time. Electrochemical methods have the advantages of low cost, simple operation, and portability; however, the electrodes have to be maintained due to the polarization effect [13]. Fluorescence methods have attracted increasing attention in the field of Cu^{2+} detection in recent years due to their high sensitivity, anti-interference ability, and fast response. In particular, with the development of fluorescent nanomaterials, Cu^{2+} fluorescent strategies with simple and environmentally friendly designs have attracted increasing interest [14,15].

Quantum dots (QDs) are widely used in the detection of Cu^{2+} in water due to their advantages of simple preparation, stable optical performance, adjustable emission, and surface modification [16]. Sadeghi et al. [17] developed a fluorescent probe based on CdSe QDs capped with deep eutectic solvent (DES-CdSe QDs) to determine the levels of Cu^{2+} in various drinks based on aggregation-induced emission (AIE), but CdSe QDs are poisonous and environmentally unfriendly, which limits the application of the method. Carbon dots (CDs) are some of most popular QDs in the detection of Cu^{2+} [18]. The fluorescence of CDs can be quenched in the presence of Cu^{2+} based on the complexation reaction of Cu^{2+} with the oxygen-containing functional groups on the surface of the CDs. Although CDs have the advantages of simple synthesis, low price, and nontoxicity, some metal ions that could be present in the water, such as iron ions and mercury ions, can also quench the fluorescence of the CDs [19,20], which can interfere with the detection of Cu^{2+}. Therefore, in real environmental samples, eliminating the influence of potential interfering substances to detect Cu^{2+}, even in trace amounts and with highly sensitive methods, is one of the main challenges.

In this work, we synthesized 3-mercaptopropionic acid (MPA)-capped InP/ZnS QDs (MPA-capped InP/ZnS QDs) for the first time and developed a fluorescent probe based on MPA-InP/ZnS QDs for the detection of trace Cu^{2+} in water. With proper capping of MPA, the MPA-InP/ZnS QDs exhibit excellent monodispersity in aqueous media and are suitable for the highly selective detection of trace Cu^{2+}. In the presence of Cu^{2+}, the fluorescence of MPA-InP/ZnS QDs could be significantly quenched, the sensing performance was studied, and the possible mechanism was proposed. Furthermore, the probe was applied to the detection of Cu^{2+} in environmental water and drink samples. To the best of our knowledge, there are few reports on the use of MPA-InP/ZnS QDs to detect Cu^{2+} in water.

2. Materials and Methods

2.1. Materials and Instruments

Indium (III) iodide (InI_3) (99.998%), zinc (II) chloride ($ZnCl_2$) ($\geq$98%), tris(diethylamino)-phosphine ($C_{12}H_{30}N_3P$) (97%), 100 mesh selenium powder (99.99%), hydrochloric acid (HCI) (mass fraction, 36.46%), zinc stearate ($C_{36}H_{70}O_4Zn$) (technical grade, 65%), tri-octylphosphine ($C_{24}H_{51}P$) (>97%), octadecene ($C_{18}H_{36}$) (technical grade, 90%), chloroform ($CHCI_3$) ($\geq$99%), 3-mercaptopropionic acid ($C_3H_6O_2S$) (MPA) ($\geq$99%), tetramethylam-monium hydroxide pentahydrate ($(CH_3)_4N\text{-}OH{\cdot}5H_2O$) (TMAH) ($\geq$97%), sulfur powder, oleylamine ($C_{18}H_{37}N$) (mass fraction, 80–90%), citric acid ($C_6H_8O_7$) ($\geq$99.5%), disodium hydrogen phosphate (Na_2HPO_4) ($\geq$99%), and all of the soluble metal salts (NaCl, $MgCl_2$, $AlCl_3$, KCl, $CaCl_2$, $MnCl_2$, $FeCl_3$, $CoCl_2$, $CuCl_2$, $PbCl_2$, $CdCl_2$, $BaCl_2$, and $AgNO_3$) were analytically pure and purchased from Sigma–Aldrich (St. Louis, MO, USA). All chemical reagents were directly used without further purification. The distilled water used in all experiments had a conductivity of 1.37 (μs/cm) and a resistivity of 0.73 (MΩ/cm).

Transmission electron microscopy (TEM) observations were carried out by using a transmission electron microscope (JEOL, JEM-2100, Akishima, Japan). Fourier-transform infrared spectra (FT-IR) were obtained by FT-IR spectroscopy (Tianjin Jiangdong Company, FT-IR-650, Tianjin, China). X-ray photoelectron spectroscopy (XPS) was recorded by using an XPS spectrometer (Japan Electronics, D/MAX, Tokyo, Japan). Additionally, the quantum yield (QY) was measured by using a fluorescence spectrometer (Horiba Jobin Yvon, Nanolog FL3-2 IHR, Pairs, France). The fluorescence spectra of the probes were recorded by using a fluorescence spectrometer (Edinburgh Instruments Ltd., Edinburgh FS5, Livingston, UK) at an excitation wavelength (λex) of 250 nm and an emission wavelength (λem) of 520 nm with an integration time of 0.1 s. Both the excitation and emission slits were set to a width of 5 nm. A time-resolved fluorescence spectrometer (Horiba Jobinyvon IBH Inc, Deltaflex 77-500K, Pairs, France) was used to measure the fluorescence lifetime of the probes with λex = 250 nm and λem = 520 nm. The ultraviolet-visible (UV-Vis) spectrum was scanned in the wavelength range of 190–1100 nm with a UV-Vis spectrophotometer (Hach Company, DR6000, Loveland, USA) with a step size of 5 nm. A refrigerated centrifuge (Sigma, 3k15, Landkreis Osterode, Germany) was used for the pretreatment of samples. In addition, the pH was adjusted by an automatic acid-base titration apparatus (Mettler Toledo, Titration Excellence T9, Zurich, Switzerland), and the sample was sonicated by an ultrasonic cleaner (Kunshan Shumei, KQ-250DE, Kunshan, China).

2.2. Synthesis of MPA-InP/ZnS QDs

The MPA-InP/ZnS QDs were synthesized according to a previously reported procedure [21]. The solvothermal method was used to proceed with the MPA-InP/ZnS QDs. First, 100 mg of InI_3 (0.45 mM) and 300 mg of $ZnCl_2$ (2.2 mM) were added to 5.0 mL (15 mM) of technical oleylamine, which is a coordinating solvent at 120 °C, and stirred and degassed to react for one hour; then, 0.45 mL (1.6 mM) of tris(diethylamino)-phosphine (phosphorus:indium ratio = 3.6:1) was injected into the above mixture at 180 °C under inert atmosphere for 20 min to synthesize the InP nanocrystals. Subsequently, 1 mL of saturated TOP-S (2.2 M) was slowly injected into the solution of InP nanocrystals, and the temperature was raised to 200 °C after 40 min; 4 mL of ODE with 1 g $Zn(stearate)_2$ was injected into the solution, and the temperature was increased to 220 °C after 60 min; 0.7 mL saturated TOP-S (2.2 M) was slowly injected into the solution, and the temperature was increased to 240 °C, and 2 mL of ODE with 0.5 g $Zn(stearate)_2$ was injected into the solution in turn, and temperature was increased to 260 °C after 30 min and 60 min, respectively. At the end, the solution reacted at 260 °C for 30 min to form trioctyl phosphine oxide (TOPO)-capped InP/ZnS QDs.

To improve the monodispersion and optical properties in aqueous media, the InP/ZnS QDs were capped by MPA. One hundred milligrams of the organic base TMAH were mixed well vigorous shaking with 50 μL MPA in 1 mL of chloroform and allowed to stand for 1 h for MPA deprotonation. Then, the bottom organic phase containing deprotonated MPA was transferred into a polypropylene tube. Then, 100 μL of the TOPO-capped InP/ZnS QDs (0.1 μM in chloroform) was added to the polypropylene tube and mixed at room temperature for 40 h for the ligand-exchange reaction between TOPO and MPA. Finally, the product was washed with chloroform and ethanol 3 times to obtain MPA-InP/ZnS QDs.

2.3. Detection of Cu^{2+} Based on MPA-InP/ZnS QDs

Next, 100 μL of different concentrations of Cu^{2+} (0 nM, 3 nM, 5 nM, 10 nM, 15 nM, 20 nM, 30 nM, 50 nM, 100 nM, 150 nM, 200 nM, 250 nM, 300 nM, 400 nM, 600 nM, and 1000 nM) were added to 2 mL of MPA-InP/ZnS QDs PBS solution (14 nM, pH = 8.0) for 12 min incubation at room temperature, and the fluorescence spectra were recorded (λex = 250 nm). The calibration curves of the fluorescence intensity at the emission peak of 520 nm vs. the concentrations of Cu^{2+} were plotted. The stability and selectivity of MPA-InP/ZnS QDs were investigated. The fluorescence intensity was recorded at indoor environment every day for 7 days at room temperature; different metal ions, including Na^+, Mg^{2+}, Al^{3+}, K^+, Ca^{2+}, Co^{2+}, Mn^{2+}, Fe^{3+}, Ba^{2+}, Cd^{2+}, Pb^{2+}, and Ag^+ (500 nM for each), were added to the MPA-InP/ZnS QDs solution, and the fluorescence intensity of each solution was recorded.

The probes were used for the detection of Cu^{2+} in the environmental water samples and drinking samples. In the spiked and real samples detection, 100 μL of samples were added to 2 mL of MPA-InP/ZnS QDs PBS solution (14 nM, pH = 8.0) for 12 min incubation at room temperature, and the fluorescence spectra were recorded (λex = 250 nm).

Samples 1–5 were mineral water of different brands, including Sample 1 (Uni-President; Tainan, China), Sample 2 (Jingtian Food & Beverage Co., Ltd.; Shenzhen, China), Sample 3 (Evergrande Mineral Water Group Co., Ltd.; Guangzhou, China), Sample 4 (Voss Beverage Co., Ltd.; Voss, Norway), and Sample 5 (Danone Co., Ltd.; Pairs, France); the carbonated drinks included Sample 6 (Coca-Cola; Atlanta, GA, USA) and Sample 7 (Pepsi; New York, NY, USA), and the drinking water Sample 8 (Beijing Xiangshan Tianquan Beverage Co., Ltd.; Beijing, China), all of which were purchased from the local market in Haidian District of Beijing. The river water samples (Sample 9) were obtained from the Jingmi Diversion Canal in Beijing. Tap water (Sample 10) was obtained in Haidian District of Beijing. The river water samples were centrifuged at 5000 rpm for 3 min to obtain the supernatant, which was filtered through a 0.45-μm filter membrane. The carbonated drink samples were pretreated by ultrasonic oscillation for 5 min. The Cu^{2+} of the samples were detected by ICP-MS method as comparisons [22] and all experiments were performed in triplicate. Unless otherwise specified, all experiments in this article were performed at room temperature.

3. Results and Discussion

3.1. Characterization of MPA-InP/ZnS QDs

As shown in Figure 1, the synthesized MPA-InP/ZnS QDs were spherical and well dispersed in water solution. The particle size of MPA-InP/ZnS QDs was distributed in a relatively narrow range of 1.87–5.86 nm, with an average size of 3.21 nm (inset in Figure 1a). The MPA-InP/ZnS QDs had an obvious lattice with a lattice spacing of 0.23 nm, as shown in Figure 1b, and the charge couple device (CCD) image revealed that InP/ZnS QDs have round shape and excellent light-emitting characteristics.

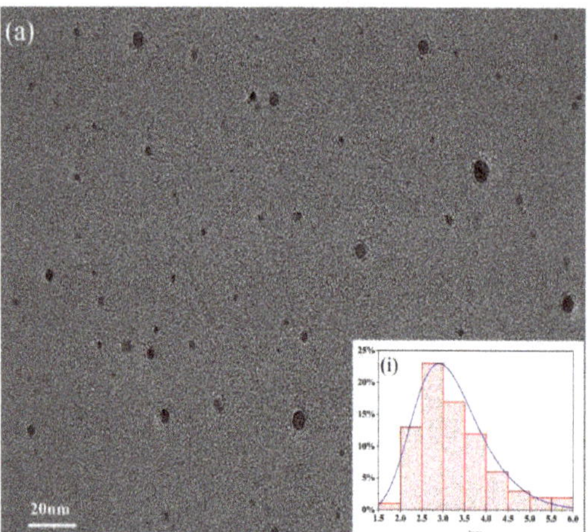

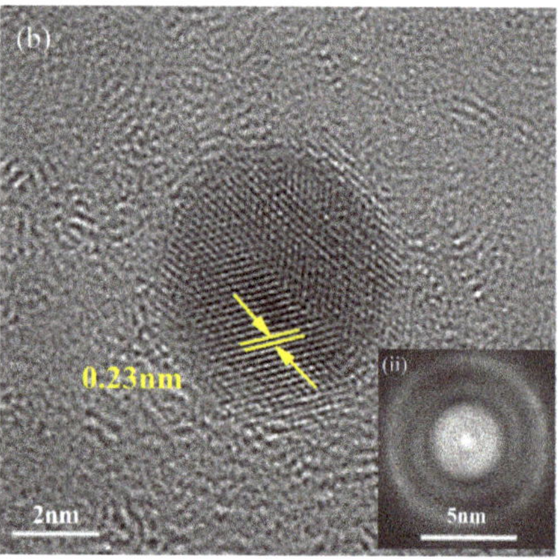

Figure 1. TEM (**a**) and HRTEM images (**b**) of MPA-InP/ZnS QDs (inset: particle size distribution (**i**) and CCD image of MPA-InP/ZnS QDs (**ii**)). TEM: transmission electron microscopy, HRTEM: high resolution transmission electron microscope, MPA-InP/ZnS QDs: mercaptopropionic acid capped InP/ZnS quantum dots, CCD: charge couple device.

FT-IR spectra were obtained to investigate the functional groups on the surface of MPA-InP/ZnS QDs. As shown in Figure 2a, the FT-IR spectra of MPA (red curve) displays that the absorption bands at 3142 cm^{-1} and 1701 cm^{-1} are attributed to the stretching vibration of the O–H and the C=O of COOH, respectively. The FT-IR spectra of MPA-InP/ZnS QDs (black curve) shows the absorption band at 3396 cm^{-1} is attributed to the stretching vibration of the O–H band of COOH. The peaks at 1640, 1568, 1486, and 1396 cm^{-1} are caused by the stretching vibrations of the C=O, C–H, –NH, and –OH bonds, respectively. The presence of carboxyl hydroxyl and amine groups was demonstrated on the surface of the MPA-InP/ZnS QDs [23], which are beneficial to the homogenous and stable dispersion of MPA-InP/ZnS QDs in water [24] and indicating the InP/ZnS QDs are successfully capped by MPA. XPS spectra were recorded to investigate the surface elements of the MPA-InP/ZnS QDs. As shown in Figure 2b, the XPS spectrum of MPA-InP/ZnS

QDs shows three peaks at 284.8, 402.4, and 532.0 eV, which correspond to the C1s, N1s, and O1s orbitals with relative atomic percentages of 73.96%, 5.01%, and 21.03%, respectively. The high-resolution C1s and O1s XPS spectra, as shown in Figure 2c,d, demonstrate the presence of C–C/C–H, C–O–H/C–O–N, O–H, and C=O, which agrees with the FT-IR spectra. The XPS spectra shows that the MPA-InP/ZnS QDs contain a large number of carboxyl groups, which further indicates that the InP/ZnS QDs are successfully capped by MPA.

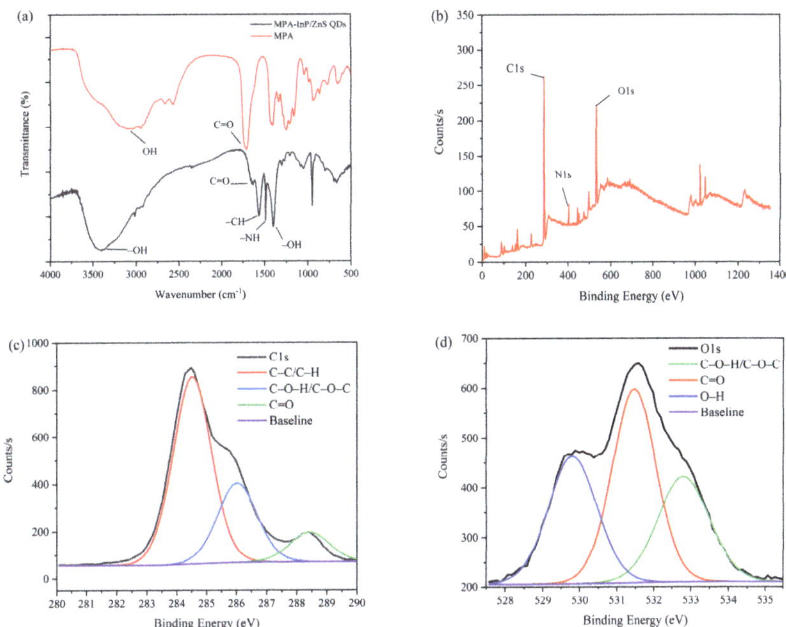

Figure 2. FT-IR spectrum (**a**), XPS patterns (**b**), C1s XPS spectrum (**c**), and O1s XPS spectrum (**d**) of MPA-InP/ZnS QDs. XPS: X-ray photoelectron spectroscopy, C1s XPS: X-ray photoelectron spectroscopy of 1s orbital electron peak of carbon atom, O1s XPS: X-ray photoelectron spectroscopy of 1s orbital electron peak of oxygen atom, MPA-InP/ZnS QDs: mercaptopropionic acid capped InP/ZnS quantum dots.

The fluorescence excitation and emission spectra of the MPA-InP/ZnS QDs were obtained to investigate the fluorescence properties. As shown in Figure 3a, the MPA-InP/ZnS QDs show an emission wavelength at 520 nm with an excitation wavelength of 250 nm. The MPA-InP/ZnS QDs are pale yellow in daylight and exhibit strong yellow color under UV light (365 nm), as shown in the insert of Figure 3a. An obvious emission peak located at approximately 520 nm with excitation wavelength scanning from 200 nm to 300 nm can be observed in Figure 3b, indicating that the emission is independent of the excitation wavelength. The quantum yield of purified MPA-InP/ZnS QDs was measured by a fluorescence spectrometer with an integrating sphere to be 12.05%.

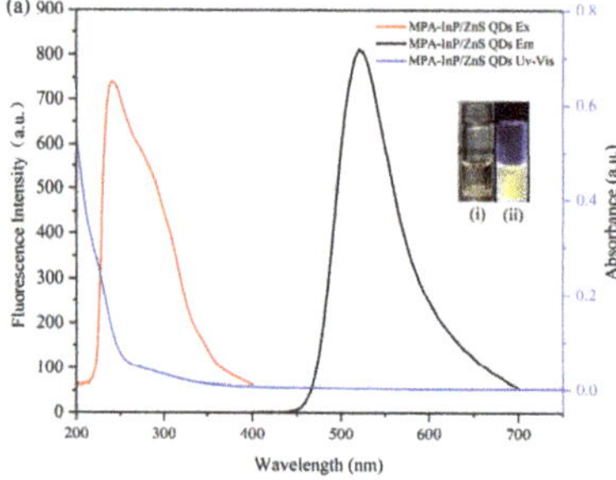
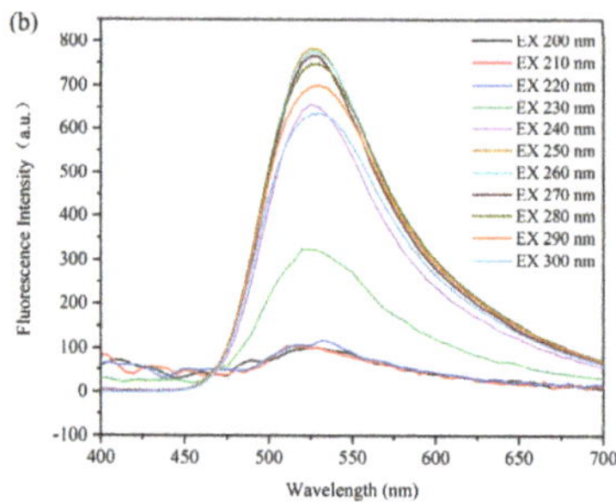

Figure 3. The absorption spectrum, excitation, and emission spectroscopy (**a**). (insert: photos of MPA-InP/ZnS QDs solutions under sunlight (**i**) and UV light (365 nm)) (**ii**) Emission spectroscopy with different excitation of MPA-InP/ZnS QDs (**b**). MPA-InP/ZnS QDs: mercaptopropionic acid capped InP/ZnS quantum dots.

3.2. Detection of Cu^{2+} by Utilizing the MPA-InP/ZnS QDs

The influence of pH, the concentration of MPA-InP/ZnS QDs, and the reaction time were investigated to optimize the conditions of reaction conditions for the detection of Cu^{2+}. As shown in Figure 4a, the ratio of the fluorescence quenching of the MPA-InP/ZnS QDs with Cu^{2+} (50 nM, 100 μL) added in PBS buffer at different pH values (2–10) was recorded, and the fluorescence was quenched most obviously at pH = 8.0. Thus, the value of 8.0 was used as the optimum pH. Figure 4b shows that the fluorescence intensity of MPA-InP/ZnS QDs at a concentration of 14 nM was the strongest, which means that the quenching degree may be the greatest in the presence of same concentration of Cu^{2+}. In order to investigate this issue further, the fluorescence intensity of MPA-InP/ZnS QDs in the concentration of 6 nM, 10 nM, 14 nM, and 18 nM in the presence of different concentrations of Cu^{2+} were measured. As shown in Figure S1 of Supplementary Materials, in the concentrations of Cu^{2+} of 0–1000 nmol/L, the MPA-InP/ZnS QDs solution with a concentration of 14 nM has the greatest degree of fluorescence quenching. The results indicated that 14 nM is the best concentration. Furthermore, the time-based fluorescence behavior of the MPA-InP/ZnS QDs with added Cu^{2+} was studied, and it is shown in Figure 4c that the fluorescence was stable after 12 min. Therefore, we chose 12 min as the incubation time for the complete reaction of MPA-InP/ZnS QDs and Cu^{2+}. All the following experiments were performed under the optimum conditions.

The fluorescence intensity of MPA-InP/ZnS QDs solutions with different concentrations of Cu^{2+} was measured at λex = 250 nm to explore the sensitivity in terms of detecting Cu^{2+} in water. Figure 5a shows a remarkable decrease in fluorescence intensity at 520 nm with increasing concentrations of Cu^{2+}. The insets of Figure 5a shows photos of MPA-InP/ZnS QD solutions with different concentrations of Cu^{2+} irradiated by a UV lamp (365 nm). The color of the MPA-InP/ZnS QDs solution under UV light gradually changed from extremely bright yellow to pale yellow until colorless with increasing Cu^{2+} concentration. The Figure 5b shows a good linear relationship (R^2 = 0.94) between the F_0/F (F_0 is the initial fluorescence intensity of MPA-InP/ZnS QDs, and F is the fluorescence intensity of MPA-InP/ZnS QDs after adding Cu^{2+}) and concentration of Cu^{2+} in the range of 0–1000 nM. Moreover, the insert of Figure 5b shows that the fluorescence intensity of the MPA-InP/ZnS QDs at low concentrations of Cu^{2+} (0–50 nM) is quenched more rapidly, which means that the Cu^{2+} detection is extremely sensitive at low concentrations. In the concentration range of 0–50 nM, the calibration curve has better linearity (R^2 = 0.99), and the limit of detection (LOD) of the prepared MPA-InP/ZnS QDs calculated by the Formula $3\sigma/x$ was 0.22 nM [18].

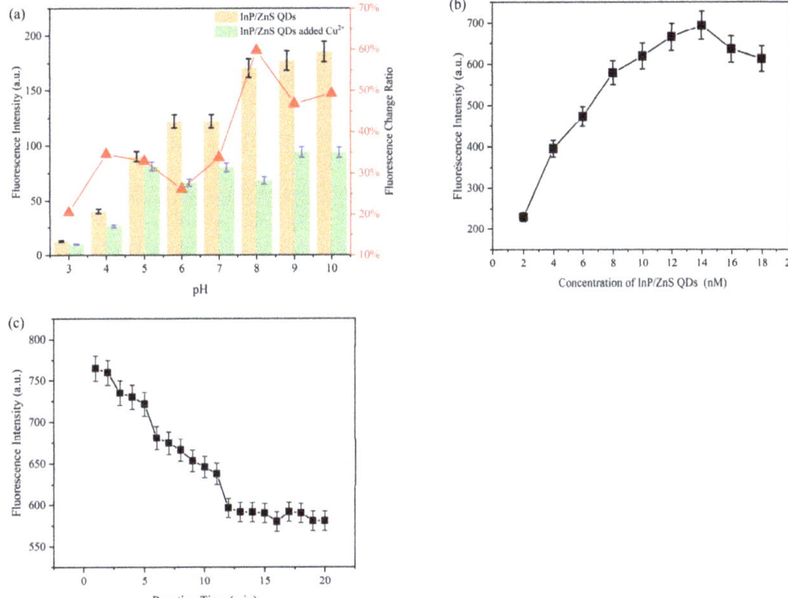

Figure 4. The fluorescence intensity of MPA-InP/ZnS QDs without and with Cu^{2+} in different pH and the fluorescence-quenching rate of MPA-InP/ZnS QDs after adding Cu^{2+} (**a**), the fluorescence intensity of MPA-InP/ZnS QDs in different concentrations (**b**), and the changes of fluorescence intensity of MPA-InP/ZnS QDs with added copper over time (**c**). MPA-InP/ZnS QDs: mercaptopropionic acid capped InP/ZnS quantum dots.

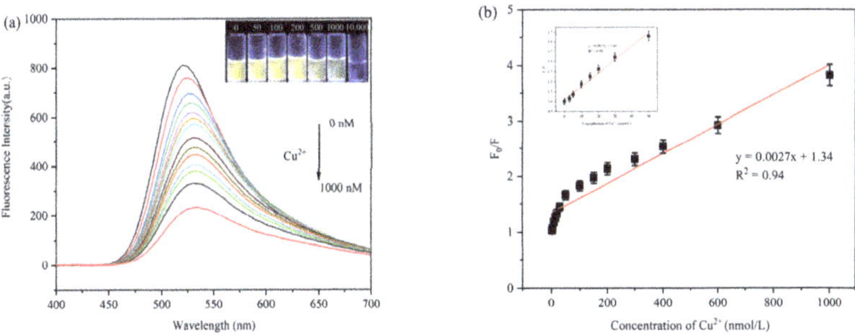

Figure 5. The fluorescence spectra of MPA-InP/ZnS QDs with different added concentrations of Cu^{2+} (0 nM, 3 nM, 5 nM, 10 nM, 15 nM, 20 nM, 30 nM, 50 nM, 100 nM, 150 nM, 200 nM, 250 nM, 300 nM, 400 nM, 600 nM, and 1000 nM) (**a**), (insert: photos of MPA-InP/ZnS QDs solutions with different concentrations of Cu^{2+} under 365 nm UV light) and the relationship between the F_0/F and the concentrations of Cu^{2+} (**b**), (insert: the relationship between the F_0/F and low concentrations of Cu^{2+} (0–50 nM)). MPA-InP/ZnS QDs: mercaptopropionic acid capped InP/ZnS quantum dots.

The stability and selectivity are important indicators for evaluating the practicability and feasibility of probes. Figure 6a shows that the changes in the fluorescence intensity of the MPA-InP/ZnS QDs solution at indoor environment are less than 8% within seven days, indicating the perfect stability of the MPA-InP/ZnS QDs probe. Furthermore, the influence of the potentially competing metal ions Na^+, Mg^{2+}, Al^{3+}, K^+, Ca^{2+}, Co^{2+}, Mn^{2+}, Fe^{3+}, Ba^{2+}, Cd^{2+}, Pb^{2+}, and Ag^+ (500 nM for each) was studied. As shown in Figure 6b, compared with

Cu^{2+}, only Ag^{+} and Fe^{3+} can quench the fluorescence of InP/ZnS QDs. The fluorescence-quenching degree of InP/ZnS QDs are 7.25% and 7.77% in the concentration of 500 nM of Ag^{+} and Fe^{3+}, respectively. While in the presence of 50 nM Cu^{2+}, the fluorescence-quenching degree of InP/ZnS QDs is 39.2%, indicating the fluorescence quenching of MPA-InP/ZnS QDs by other metal ions is almost negligible. The ion selectivity is dependent on the intrinsic affinity between the analyte and the surface ligands [25], and the MPA ligand has a high affinity constant with Cu^{2+} [26], which may be the reason of MPA-InP/ZnS QDs has higher selectivity to Cu^{2+} than other metal ions. These experimental results show the excellent sensitivity, high selectivity, and good anti-influence of the MPA-InP/ZnS QDs probe for the detection of trace Cu^{2+} in water.

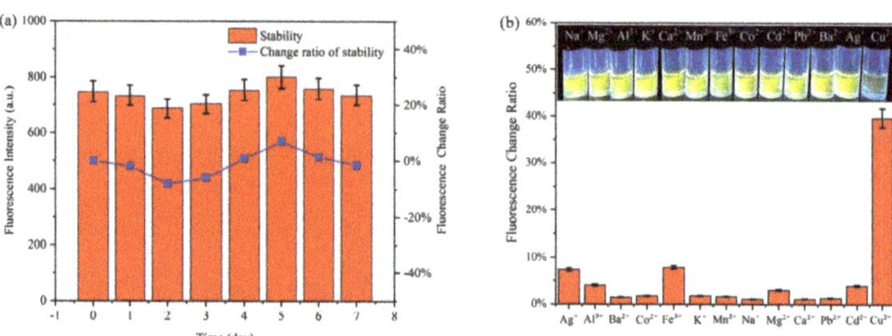

Figure 6. The stability (**a**) and the selectivity (**b**) of the fluorescent probe.

3.3. *Fluorescence-Quenching Mechanism of MPA-InP/ZnS QDs*

The fluorescence-quenching mechanism of quantum dots is complicated, generally including inner filter effect (IFE), fluorescence resonance energy transfer (FRET), photoinduced electron transfer (PET), static quenching effect (SQE), and dynamic quenching. The FRET, PET, and dynamic quenching can cause the fluorescence lifetime of the fluorophore to decay after adding a quencher, while IFE and SQE cannot [27]. As shown in Figure 7a, the fluorescence decay lifetimes of the MPA-InP/ZnS QDs without and with Cu^{2+} were 1.07 ns and 1.02 ns, respectively; these values were almost unchanged, indicating that FRET, PET, and dynamic quenching do not occur between MPA-InP/ZnS QDs and Cu^{2+}. Moreover, IFE can be confirmed by UV-Vis absorption spectrum because IFE requires a certain degree of overlap between the excitation or emission band of the fluorophore and the absorption band of the UV-Vis spectrum of the quencher [28]. As shown in Figure 7b, the absorption band of Cu^{2+} in the range of 200 nm to 270 nm with a peak at 205 nm was observed, and the absorption band of Cu^{2+} overlaps with the excitation peak of the MPA-InP/ZnS QDs at 250 nm, indicating that the IFE may exist in the quenching mechanism. To further confirm the quenching mechanism, a typical Stern–Volmer diagram was constructed:

$$F_0/F = 1 + K_{SV}[Q] = 1 + K_q\tau_0[Q] \tag{1}$$

where F_0 is the fluorescence intensity of MPA-InP/ZnS QDs, F is the fluorescence intensity observed after adding Cu^{2+}, K_{SV} is the quenching Stern–Volmer constant, and $[Q]$ is the concentration of the Cu^{2+}, K_q is the dynamic quenching Stern–Volmer constant, and τ_0 is the lifetime of the InP/ZnS QDs. The static quenching can be judged by K_{SV} in different temperatures. The value of K_{SV} decreases with the increase of temperature in the SQE process [29]. Obviously, as shown in Figure 7c, with the increasing temperature from 20 °C to 40 °C, the slope of the curves that are K_{SV} decreases, which confirms the existence of SQE. Moreover, the maximum K_q of collision quenching is 2×10^{10} L mol^{-1} s^{-1} [30]. In addition, the K_q of 20 °C, 30 °C, and 40 °C in this paper are 2.34×10^{15}, 1.31×10^{15}, and 1.21×10^{15}, respectively, which are much larger. The results further confirm the existence of SQE.

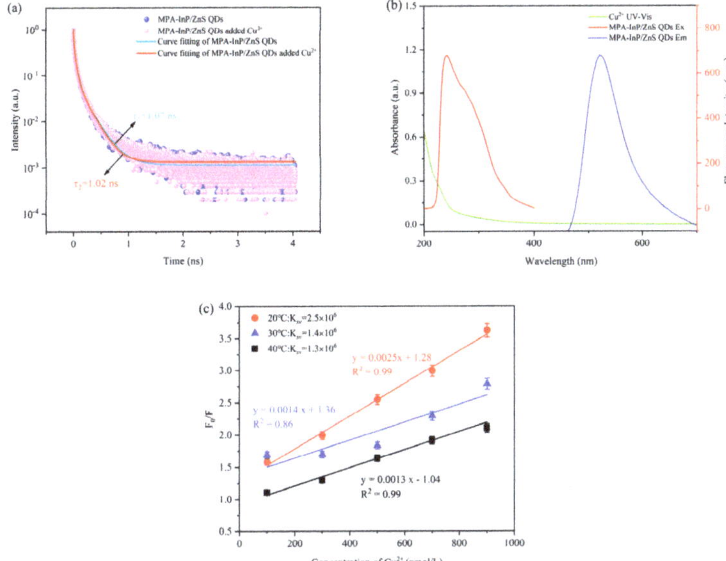

Figure 7. The fluorescence lifetime decay curves of MPA-InP/ZnS QDs without and with Cu^{2+} (**a**); the absorption spectrum of Cu^{2+}, excitation, and emission spectra of MPA-InP/ZnS QDs (**b**); the calibration curves between F_0/F and the concentrations of Cu^{2+} under different temperature (20 °C, 30 °C, and 40 °C) (**c**). MPA-InP/ZnS QDs: mercaptopropionic acid capped InP/ZnS quantum dots.

3.4. Detection of Cu^{2+} in Real Samples

The applicability and accuracy of MPA-InP/ZnS QDs for the detection of Cu^{2+} in environmental water samples and drinking samples were investigated. As shown in Table 1, spiked detections were carried out in pure water of Watsons; the recovery was 93.64–120.91%, and the relative standard deviation (RSD) (n = 3) was below 1.02%. The results of the detection of Cu^{2+} in different water samples using our prepared MPA-InP/ZnS QDs probes and ICP-MS are listed in Table 2 and Figure S2. The results of the two methods demonstrate that the MPA-InP/ZnS QDs probes are highly accurate and have the ability to detect trace amounts of Cu^{2+} in real water samples.

The performance of MPA-InP/ZnS QDs probes was also compared with several previously reported studies on the detection of Cu^{2+} in water. As shown in Table 3 and Figure S3, the detection range of fluorescent probes for Cu^{2+} can reach hundreds of micromoles; however, the LOD is difficult to reach for the nanomolar level. As the comparison, the detection range of our probe is 0–1000 nM with the LOD of 0.22 nM, which exhibited superior sensing performance, especially in the detection of trace Cu^{2+}. Our work shows promising prospects in the highly sensitive detection of trace Cu^{2+} in real water.

Table 1. Determination results of copper ions (Cu^{2+}) in pure water of Watsons samples (n = 3).

Spiked/nM	Found/nM	Recovery (%)	RSD (%)
5.00	4.97	99.33	0.58
10.00	12.09	120.91	0.54
20.00	18.73	93.64	1.02

Table 2. Detection of Cu^{2+} using this method and ICP-MS (n = 3).

Samples	This Method		ICP-MS	
	Found/nM	RSD (%)	Found/nM	RSD (%)
Sample 1	5.16	5.24	5.31	0.54
Sample 2	4.29	2.42	3.75	1.04
Sample 3	14.03	1.96	12.50	0.67
Sample 4	7.97	1.09	6.56	0.47
Sample 5	13.59	1.79	14.69	0.64
Sample 6	4.68	5.66	3.38	1.23
Sample 7	5.07	7.25	3.69	0.78
Sample 8	3.81	2.96	3.75	0.97
Sample 9	17.47	4.47	15.09	1.06
Sample 10	31.06	5.13	33.13	0.54

Table 3. Comparison of the performance of fluorescent MPA-InP/ZnS QDs probes for the detection of Cu^{2+}.

Materials	Principle	Detection Range	LOD	Reaction Time	Applications	Ref.
Mercaptoacetic acid-CdTe QDs	Dynamic quenching	40–600 nM	35.0 nM	5 min	Urine	[31]
Polyamine@C-dots	Static quenching effect	0.07–60 μM	0.02 μM	15 min	Conduit water, tap water, and mineral water	[18]
Coumarin	Static quenching effect	0–50 μM	0.27 μM	10 min	Tap water	[32]
CdTe QDs-polyethyleneimine/ polyvinyl alcohol electrospun	Dynamic quenching	0.08–800 μM	11.1 nM	50 min	Lake water	[26]
Deep eutectic solvent-CdSe QDs	Aggregation-induced emission	10–600 nM	5.3 nM	1 min	Tap water, mineral water, pineapple fruit juice, milk, and cola	[17]
MPA-InP/ZnS QDs	Static quenching effect	0–1000 nM	0.22 nM	12 min	River water, tap water, purified water, mineral water, drinking water, and beverages	This work

MPA-InP/ZnS QDs: mercaptopropionic acid capped InP/ZnS quantum dots.

4. Conclusions

In summary, MPA-capped MPA-InP/ZnS QDs were synthesized by a solvothermal method; they exhibited bright yellow fluorescence and excellent monodispersity in aqueous solution. A highly sensitive and highly selective fluorescent probe for the detection of trace Cu^{2+} was developed using the synthesized MPA-InP/ZnS QDs. The fluorescence intensity of the MPA-InP/ZnS QDs can be quenched significantly in the presence of Cu^{2+} due to the SQE, which showed an excellent linear relationship with Cu^{2+} in the concentration range of 0–1000 nM, with a detection limit of 0.22 nM. Furthermore, the probe was applied to the detection of Cu^{2+} in environmental water and drink samples, indicating that MPA-InP/ZnS QDs could be used as a fluorescence probe in the application of highly sensitive and rapid detection of trace Cu^{2+} in real water.

Supplementary Materials: The following are available online at https://www.mdpi.com/article/10.3390/foods10112777/s1, Figure S1: The response curves of InP/ZnS QDs solutions with concentrations of 6 nM, 10 nM, 14 nM and 18 nM to different concentrations of Cu^{2+}, Figure S2: The data comparison image of the detection results of the two methods, Figure S3: The results comparison image of this work and literature.

Author Contributions: Conceptualization, Z.X. and J.Z.; experimental design, Z.X.; data curation, Z.X.; formal analysis, Z.X.; investigation, Z.X.; methodology, Z.X. and J.Z.; writing—original draft, Z.X.; conceptualization, J.Z.; writing—review and editing, J.Z.; formal analysis, C.S.; supervision, C.S.

and Y.W.; funding acquisition, X.Y. All authors have read and agreed to the published version of the manuscript.

Funding: This research was funded by Research and Development Projects in Key Areas of Guangdong Province (grant number 2019B020225001) and Young Beijing Scholar.

Data Availability Statement: Not applicable.

Conflicts of Interest: The authors declare that they have no known competing financial interests or personal relationships that could have appeared to influence the work reported in this paper.

References

1. Gaetke, L.; Chow, C.K. Copper toxicity, oxidative stress, and antioxidant nutrients. *Toxicology* **2003**, *189*, 147–163. [CrossRef]
2. Luo, Y.; Zhang, L.; Liu, W.; Yu, Y.; Tian, Y. A Single Biosensor for Evaluating the Levels of Copper Ion and L-Cysteine in a Live Rat Brain with Alzheimer's Disease. *Angew. Chem. Int. Ed.* **2015**, *54*, 14053–14056. [CrossRef]
3. Georgopoulos, P.G.; Wang, S.W.; Georgopoulos, I.G.; Yonone-Lioy, M.J.; Lioy, P.J. Assessment of human exposure to copper: A case study using the NHEXAS database. *J. Exp. Sci. Environ. Epidemiol.* **2006**, *16*, 397–409. [CrossRef] [PubMed]
4. Kim, M.H.; Jang, H.H.; Yi, S.; Chang, S.; Han, M.S. Coumarin-derivative-based off-on catalytic chemodosimeter for Cu^{2+} ions. *Chem. Commun.* **2009**, *32*, 4838–4840. [CrossRef]
5. Al-Saydeh, S.A.; El-Naas, M.H.; Zaidi, S.J. Copper removal from industrial wastewater: A comprehensive review. *J. Ind. Eng. Chem.* **2017**, *56*, 35–44. [CrossRef]
6. Seidi, S.; Alavi, L. Novel and Rapid Deep Eutectic Solvent (DES) Homogeneous Liquid-Liquid Microextraction (HLLME) with Flame Atomic Absorption Spectrometry (FAAS) Detection for the Determination of Copper in Vegetables. *Anal. Lett.* **2019**, *52*, 2092–2106. [CrossRef]
7. Smirnova, S.V.; Ilin, D.V.; Pletnev, I.V. Extraction and ICP-OES determination of heavy metals using tetrabutylammonium bromide aqueous biphasic system and oleophilic collector. *Talanta* **2021**, *221*, 121485. [CrossRef]
8. Wang, W.; Evans, R.D.; Newman, K.; Toms, A. Automated separation and measurement of 226Ra and trace metals in freshwater, seawater and fracking water by online ion exchange chromatography coupled with ICP-MS. *Microchem. J.* **2021**, *167*, 106321. [CrossRef]
9. Ghorbani, M.; Pedramrad, T.; Aghamohammadhasan, M.; Seyedin, O.; Akhlaghi, H.; Afshar Lahoori, N. Simultaneous clean-up and determination of Cu(II), Pb(II) and Cr(III) in real water and food samples using a magnetic dispersive solid phase microextraction and differential pulse voltammetry with a green and novel modified glassy carbon electrode. *Microchem. J.* **2019**, *147*, 545–554. [CrossRef]
10. Yang, Y.Z.; Xiao, N.; Cen, Y.Y.; Chen, J.R.; Liu, S.G.; Shi, Y.; Fan, Y.Z.; Li, N.B.; Luo, H.Q. Dual-emission ratiometric nanoprobe for visual detection of Cu(II) and intracellular fluorescence imaging. *Spectrochim. Acta A Mol. Biomol.* **2019**, *223*, 117300. [CrossRef]
11. Almeida, J.S.; Souza, O.C.C.O.; Teixeira, L.S.G. Determination of Pb, Cu and Fe in ethanol fuel samples by high-resolution continuum source electrothermal atomic absorption spectrometry by exploring a combination of sequential and simultaneous strategies. *Microchem. J.* **2018**, *137*, 22–26. [CrossRef]
12. Arı, B.; Bakırdere, S. A primary reference method for the characterization of Cd, Cr, Cu, Ni, Pb and Zn in a candidate certified reference seawater material: TEA/$Mg(OH)_2$ assisted ID3MS by triple quadrupole ICP-MS/MS. *Anal. Chim. Acta* **2020**, *1140*, 178–189. [CrossRef]
13. Pizarro, J.; Flores, E.; Jimenez, V.; Maldonado, T.; Saitz, C.; Vega, A.; Godoy, F.; Segura, R. Synthesis and characterization of the first cyrhetrenyl-appended calix[4]arene macrocycle and its application as an electrochemical sensor for the determination of Cu(II) in bivalve mollusks using square wave anodic stripping voltammetry. *Sens. Actuators B Chem.* **2019**, *281*, 115–122. [CrossRef]
14. Chakraborty, G.; Katiyar, V.; Pugazhenthi, G. Improvisation of polylactic acid (PLA)/exfoliated graphene (GR) nanocomposite for detection of metal ions (Cu^{2+}). *Compos. Sci. Technol.* **2021**, *213*, 108877. [CrossRef]
15. Gao, B.; Chen, D.; Gu, B.; Wang, T.; Wang, Z.; Xie, F.; Yang, Y.; Guo, Q.; Wang, G. Facile and highly effective synthesis of nitrogen-doped graphene quantum dots as a fluorescent sensing probe for Cu^{2+} detection. *Curr. Appl. Phys.* **2020**, *20*, 538–544. [CrossRef]
16. Zhao, L.; Li, H.; Xu, Y.; Liu, H.; Zhou, T.; Huang, N.; Li, Y.; Ding, L. Selective detection of copper ion in complex real samples based on nitrogen-doped carbon quantum dots. *Anal. Bioanal. Chem. Res.* **2018**, *410*, 4301–4309. [CrossRef] [PubMed]
17. Sadeghi, S.; Davami, A. CdSe quantum dots capped with a deep eutectic solvent as a fluorescent probe for copper(II) determination in various drinks. *Mikrochim. Acta* **2020**, *187*, 147. [CrossRef]
18. Ali, H.R.H.; Hassan, A.I.; Hassan, Y.F.; El-Wekil, M.M. Development of dual function polyamine-functionalized carbon dots derived from one step green synthesis for quantitation of Cu^{2+} and S^{2-} ions in complicated matrices with high selectivity. *Anal. Bioanal. Chem.* **2020**, *412*, 1353–1363. [CrossRef] [PubMed]
19. Zhang, S.; Li, J.; Zeng, M.; Xu, J.; Wang, X.; Hu, W. Polymer nanodots of graphitic carbon nitride as effective fluorescent probes for the detection of Fe^{3+} and Cu^{2+} ions. *Nanoscale* **2014**, *6*, 4157. [CrossRef]

20. Xu, Y.; Fan, Y.; Zhang, L.; Wang, Q.; Fu, H.; She, Y. A novel enhanced fluorescence method based on multifunctional carbon dots for specific detection of Hg^{2+} in complex samples. *Spectrochim. Acta A Mol. Biomol.* **2019**, *220*, 117109. [CrossRef] [PubMed]
21. Tessier, M.D.; Dupont, D.; De Nolf, K.; De Roo, J.; Hens, Z. Economic and Size-Tunable Synthesis of InP/ZnE (E = S, Se) Colloidal Quantum Dots. *Chem. Mater.* **2015**, *27*, 4893–4898. [CrossRef]
22. Leyden, E.; Farkas, J.; Gilbert, S.; Hutson, J.; Mosley, L.M. A simple and rapid ICP-MS/MS determination of sulfur isotope ratios (34S/32S) in complex natural waters: A new tool for tracing seawater intrusion in coastal systems. *Talanta* **2021**, *235*, 122708. [CrossRef]
23. Jo, J.; Kim, M.; Han, C.; Jang, E.; Do, Y.R.; Yang, H. Effective surface passivation of multi-shelled InP quantum dots through a simple complexing with titanium species. *Appl. Surf. Sci.* **2018**, *428*, 906–911. [CrossRef]
24. Han, Z.; Nan, D.; Yang, H.; Sun, Q.; Pan, S.; Liu, H.; Hu, X. Carbon quantum dots based ratiometric fluorescence probe for sensitive and selective detection of Cu^{2+} and glutathione. *Sens. Actuators B Chem.* **2019**, *298*, 126842. [CrossRef]
25. Lou, Y.; Zhao, Y.; Chen, J.; Zhu, J. Metal ions optical sensing by semiconductor quantum dots. *J. Mater. Chem. C* **2014**, *2*, 595–613. [CrossRef]
26. Li, X.; Yang, T.; Wang, J.; Huang, C. CdTe Quantum Dots-Electrospun Nanofibers Assembly for Visual and Portable Detection of Cu^{2+}. *Chin. J. Anal. Chem.* **2021**, *49*, 207–215. [CrossRef]
27. Meng, Y.; Jiao, Y.; Zhang, Y.; Li, Y.; Gao, Y.; Lu, W.; Liu, Y.; Shuang, S.; Dong, C. Multi-sensing function integrated nitrogen-doped fluorescent carbon dots as the platform toward multi-mode detection and bioimaging. *Talanta* **2020**, *210*, 120653. [CrossRef]
28. Tang, X.; Yu, H.; Bui, B.; Wang, L.; Xing, C.; Wang, S.; Chen, M.; Hu, Z.; Chen, W. Nitrogen-doped fluorescence carbon dots as multi-mechanism detection for iodide and curcumin in biological and food samples. *Bioact. Mater.* **2021**, *6*, 1541–1554. [CrossRef]
29. Gan, X.; Liu, S.; Liu, Z.; Hu, X. Determination of Tetracaine Hydrochloride by Fluorescence Quenching Method with Some Aromatic Amino Acids as Probes. *J. Fluoresc.* **2012**, *22*, 129–135. [CrossRef]
30. Cui, M.; Song, G.; Wang, C.; Song, Q. Synthesis of cysteine-functionalized water-soluble luminescent copper nanoclusters and their application to the determination of chromium(VI). *Mikrochim. Acta* **2015**, *182*, 1371–1377. [CrossRef]
31. Xie, Y.F.; Jiang, Y.J.; Zou, H.Y.; Wang, J.; Huang, C.Z. Discrimination of copper and silver ions based on the label-free quantum dots. *Talanta* **2020**, *220*, 121430. [CrossRef] [PubMed]
32. Wu, W.; Min, S.; Tong, Q.; Wang, J.; Hu, J.; Dhamsaniya, A.; Shah, A.K.; Mehta, V.P.; Dong, B.; Song, B. Highly sensitive and selective "turn-off" fluorescent probes based on coumarin for detection of Cu^{2+}. *Colloids Interface Sci. Commun.* **2021**, *43*, 100451. [CrossRef]

Article

Preparation of Specific Nanobodies and Their Application in the Rapid Detection of Nodularin-R in Water Samples

Jinyi Yang [1,†], Rui Si [1,†], Guangpei Wu [1], Yu Wang [2], Ruyu Fang [1], Fei Liu [1], Feng Wang [1], Hongtao Lei [1], Yudong Shen [1], Qi Zhang [3] and Hong Wang [1,*]

[1] Guangdong Provincial Key Laboratory of Food Quality and Safety, National-Local Joint Engineering Research Center for Processing and Safety Control of Livestock and Poultry Products, College of Food Science, South China Agricultural University, Guangzhou 510642, China; yjy361@163.com (J.Y.); huanongxiaosi@163.com (R.S.); wgphavoc@163.com (G.W.); 18819259799@163.com (R.F.); m13476079515@163.com (F.L.); wangfeng_sp@163.com (F.W.); hongtao@scau.edu.cn (H.L.); shenyudong@scau.edu.cn (Y.S.)

[2] Guangzhou Institute of Food Inspection, Guangzhou 510080, China; xxwangyu@163.com

[3] Oil Crops Research Institute, Chinese Academy of Agricultural Sciences, Xudong 2nd Road No. 2, Wuchang District, Wuhan 430062, China; zhangqi01@caas.cn

* Correspondence: gzwhongd@163.com; Tel.: +86-20-85283448; Fax: +86-20-85280270

† These authors contribute equally to this work.

Citation: Yang, J.; Si, R.; Wu, G.; Wang, Y.; Fang, R.; Liu, F.; Wang, F.; Lei, H.; Shen, Y.; Zhang, Q.; et al. Preparation of Specific Nanobodies and Their Application in the Rapid Detection of Nodularin-R in Water Samples. *Foods* **2021**, *10*, 2758. https://doi.org/10.3390/foods10112758

Academic Editor: Juliana Moura Luna

Received: 28 September 2021
Accepted: 5 November 2021
Published: 10 November 2021

Abstract: Nanobodies have several advantages, including great stability, sensibility, and ease of production; therefore, they have become important tools in immunoassays for chemical contaminants. In this manuscript, nanobodies for the detection of the toxin Nodularin-r (NOD-R), a secondary metabolite of cyanobacteria that could cause a safety risk for drinks and food for its strong hepatotoxicity, were for the first time selected from an immunized Bactrian camel VHH phage display library. Then, a sensitive indirect competitive enzyme-linked immunosorbent assay (ic-ELISA) for NOD-R, based on the nanobody N56 with great thermostability and organic solvent tolerance, was established under optimized conditions. The results showed that the limit of detection for NOD-R was 0.67 μg/L, and the average spike recovery rate was between 84.0 and 118.3%. Moreover, the ic-ELISA method was validated with spiked water sample and confirmed by UPLC–MS/MS, which indicated that the ic-ELISA established in this work is a reproducible detection assay for nodularin residues in water samples.

Keywords: nodularin; nanobody; detection; ic-ELISA

1. Introduction

Nodularins (NODs) are harmful cyanobacteria toxins produced by *Nodularia spumigena*, especially when blooms brake out in lakes or rivers [1]. So far, different forms of NOD variants have been found. Among these variants, NOD-R is the most abundant and the only one commercially available [2]. NOD-R is a characteristic cyclic peptide composed of five amino acids (-D-MeAsp–L-Y–Adda–D-Glu–Mdhb), in which the Adda structural region ((2S,3S,8S,9S)-3-amino- 9-methoxy-2,6,8-trimethyl-10-phenyl deca-4,6-dienoic acid) is also found in microcystins (MCs), another group of cyanobacteria toxins composed of heptapeptides. Although NOD-R is similar to MCs in structure (Figure 1), it penetrates more easily into hepatocytes than MCs and exhibits stronger acute toxicity [3]. It has been found that NOD-R can inhibit the catalytic subunit of seine/threonine-specific protein phosphatases (PPs) 1 and 2A (PP1 and PP2A) and lead to severe hepatotoxicity [4–6]. Meanwhile, NOD-R cannot be completely eliminated via routine water treatment processes because of its water solubility and heat stability; therefore, contaminated drinking water and polluted recreational water might increase the risk of human exposure through intake or skin contact. A guideline value specific for NOD-R in drinking water and recreational water has not yet been established by the World Health Organization (WHO); however, many countries commonly take 1.0 μg/L, the maximum residue limit (MRL) for MC-LR, as a reference [7]. Rapid and sensitive detection methods for NOD-R are required to continuously monitor the level of NOD-R in the environment as well as in drinking water or foods [8].

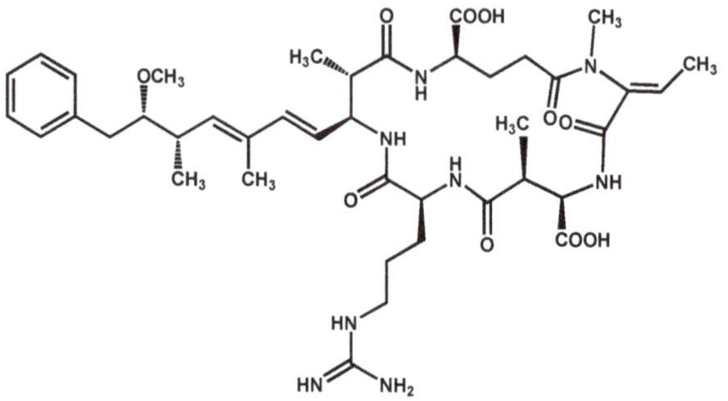

Figure 1. Structure of NOD-R.

Recently, several methods for the detection of NOD-R residues have been developed, including high-performance liquid chromatography (HPLC) [9], mass spectrometry (MS) [10], liquid chromatography–mass spectrometry (LC–MS) [11], as well as some rapid detection methods such as immunoassays based on specific antibodies or biosensor assays based on aptamers [12]. Compared with instrumental assay methods, immunoassay technology has the obvious advantages of rapidness, simplicity, and lower cost [13]. For example, only several cents to dollars are commonly spent on each sample when using common ELISA kits, whereas at least tens of dollars are necessary when using UPLC–MS/MS. In addition, antibodies are routinely stable in conformation and common in practical utilization; thus, immunodetection of NOD-R has attracted attention. Both polyclonal antibodies (PAbs) and monoclonal antibodies (MAbs) against NOD-R were obtained, and several detection methods, e.g., indirect competitive enzymatic-linked immunoassay (ic-ELISA), fluorescence polarization immunoassay (FPIA), and optical surface plasmon resonance (SPR) immunosensor detection, were developed, obtaining a limitation of detection (LOD) for NOD-R ranging from 0.051 μg/L to 0.95 μg/L [14,15].

Actually, except for a few examples, most of the antibodies reported above are broad-specificity antibodies originally prepared for MC-LR, but they can recognize also other MCs as well as NOD-R simultaneously. Previously, with MC-LR-KLH as a common immunogen, multi-residue-recognizing PAbs [16] and MAbs [data not shown] were generated in our lab with IC_{50} of 0.29 μg/L and 18 μg/L for NOD-R, respectively.

Obviously, antibodies, as molecular recognition elements, are the core components of immunoassays. In recent years, single-domain antibodies derived from heavy-chain-only antibodies, namely VHH, have been found in camelids, including camels and llamas, as well as VNAR in cartilaginous fishes, including sharks [17]. These antibodies, also called nanobodies (Nbs), have attracted attention for their application in in immunoassays and have shown superiority over traditional PAbs and MAbs because of their small size (−15 kDa) and high solubility and stability [18]. In addition, Nbs can be efficiently generated by the phage display method and can be genetically encoded and expressed in *Escherichia coli* host cells with high yields [19,20]. Until now, several specific Nbs for some chemical contaminants such as parathion [21], aflatoxin B1 [22] and triazophos [23] have been successfully prepared, and related immunoassays have been established.

In this study, an immunized Bactrian camel phage display library was constructed, and then Nbs against NOD-R were selected and prepared. Furthermore, based on optimized reaction conditions, an ic-ELISA for NOD-R in water samples was established. Finally, the performance of the ic-ELISA was measured and validated by UPLC–MS/MS.

2. Materials and Methods

2.1. Materials and Reagents

Nodularin-R and microcystins (MC-LR, -LA, -LY, -LW, -LF, -YR, -WR, -RR) (Enzo, USA) were used as standards. Nodularin-R-ovalbumin (NOD-R-OVA, prepared in our lab) was the coating antigen. Microcystin-LR-keyhole limpet hemocyanin (MC-LR-KLH, prepared in our lab) was used as the immunogen. An anti-MC-LR monoclonal antibody (anti-MC-LR MAb, prepared in our lab) was used for comparison with the nanobodies. pComb3XSS vector and *E. coli* BL21(DE3) available in our lab were used to construct a recombinant vector and a transformed host strain, respectively. Helper phage M13K07 (New England Biolabs, Ipswich, MA, USA) was used for the construction of a phage displayed nanobody library. The anti-VHH-HRP polyclonal antibody from rabbit (Genscript Bio. Co.Ltd, Nanjing, China) was used as a secondary antibody in ELISA. HisPur Ni-NTA resin (GE Healthcare, Beijing, China) was utilized for protein purification. The primers used in this work were synthesized by Invitrogen Biotechnology Co. (Shanghai, China). Other chemical reagents were provided by Sigma (St. Louis, MS, USA) and Thermo Fisher Scientific (Thermo, Waltham, MA, USA).

2.2. Instruments

A NanoDrop 2000C spectrophotometer (Thermo, Waltham, MA, USA) was used to measure absorbance in the ultraviolet–visible spectrum. Centrifugation of phages and bacteria was performed using SORVALL LYNX 4000 centrifuges (Thermo, Waltham, MA, USA). Biologic LP (Bio-Rad, Hercules, CA, USA) was used to purify VHH antibodies. The absorbance values in ELISA were measured using a Multiskan MK3 microplate reader (Thermo Scientific, Waltham, MA, USA).

2.3. Construction of a Phage-Displayed Nanobody Library

The procedure for constructing the nanobody library by the phage display technology was similar to that previously published [24]. In practice, a three-year-old Bactrian camel was immunized with 500 μL of MC-LR-KLH(1.0 mg/mL) emulsified with Freund's adjuvant at a ratio of 1:1 (*v/v*) every two weeks for a total of 6 times. Particularly, the first time of immunization was performed with complete adjuvant, and the boost immunizations with incomplete adjuvant. One week after the sixth immunization, 100 mL of fresh peripheral blood was collected, and lymphocytes were isolated, followed by RNA extraction. After the synthesis of cDNA by RT-PCR, the VHH genes were amplified by two-step nested PCR using the primers shown in Table 1. The VH−CH1−CH2 and VHH−CH2 regions were firstly amplified using the primers CALL001and CALL002. Next, the primers Fr4-*Sfi*I and Fr1-*Sfi*I were used to amplify the VHH genes. The VHH genes and the pComb3XSS vector were digested using the restriction endonuclease *Sfi*I, ligated, and transfected into competent *E. coli* ER2738 cells. All transformants were collected by scraping from LB–agar plates containing ampicillin and tetracycline. After infection with helper phage M13K07, a phage-displayed Nb library was constructed. The capacity and diversity of the constructed library were identified by sequence analysis of 20 randomly selected clones.

Table 1. Primers' sequences used to amplify the VHH genes (W = A or T, S = G or C, M = A or C, R = G or A).

Primer Names	Nucleotide Sequences (5′→3′)
CALL001	GTCCTGGCTGCTCTTCTACAAGG
CALL002	GGTACGTGCTGTTGAACTGTTCC
Fr4-*Sfi*I	ACTGGCCCAGGCGGCCGAGGTGCAGCTGSWGSAKTCKG
Fr1-*Sfi*I	ACTGGCCGGCCTGGCCTGAGGAGACGGTGACCWGGGTC

2.4. Biopanning and Identification of Nanobody Clones for NOD-R

Four rounds of biopanning in 96-well microtiter plates were performed, as described in references to select Nb clones for NOD-R [25]. In practice, the antigen NOD-R-OVA was coated at decreasing concentrations (10, 2.5, 0.5, 0.1 μg/mL) in four rounds of biopanning at 4 °C with overnight incubation. Meanwhile, 1 mg/mL of OVA and BSA was coated in the parallel wells under the same condition (4 °C, overnight). After shaking out the excess coating solution, a 1% gelatin solution was used to block the coated wells at 37 °C for 2 h. In each round of biopanning, 100 μL of phage library (about 1 × 1011 pfu/mL) was introduced into OVA-wells and BSA-wells in turn, for 1 h at 37 °C, to pre-absorb the non-specific phages; then, the supernatant was transferred to NOD-R-OVA-coated wells and incubated for another 1 h at 37 °C. After washing with PBST for six times, elution was conducted by incubating with decreasing concentrations of NOD-R (2, 0.5, 0.1 μg/mL) and 10 mg/mL of trypsin under gentle shaking at 37 °C for 1 h in each round. Finally, 10 μL of elution solution was used to test the phage titer, and the remainder was amplified for the next round of biopanning.

After four rounds of biopanning, 20 clones were picked out randomly and induced by 1.0 mmol/L isopropylthio-β-D-galactoside (IPTG) to express Nbs in 96-deep-well plates at 37 °C with shaking at 250 rpm for 16 h. Then, ic-ELISA was used to identify the positive clones. Furthermore, the positive clones were subjected to DNA sequencing.

2.5. Expression and Characteristics of Nanobodies

The recombinant vector pComb3XSS-vhh with the NOD-R VHH gene was extracted from the positive *E. coli* ER2738 clones and then transfected into competent cells of the host strain *E. coli* BL21(DE3) to express specific Nbs. For scaled-up expression, 10 mL of the overnight culture was added to 1 L of LB medium (with 100 mg/mL ampicillin) and incubated at 37 °C with shaking at 250 rpm. When the absorbance of the culture at a wavelength of 600 nm (A_{600nm}) reached 0.6–0.8, 1 mM IPTG was added to induce the expression of the target protein, followed by continuous incubation overnight. Then, the bacteria precipitate was harvested after centrifugation at 12,000 rpm for 15 min. The extraction solution (0.2 M Tris-HCl, pH 8.0, 0.5 mM EDTA, 0.5 M sucrose) was added, and the soluble proteins were isolated via cycles of freezing and thawing and by using the osmotic pressure method. The expressed Nbs were purified using the HisPur Ni-NTA resin, followed by 15% SDS-PAGE and Western blotting identification.

To assess the thermostability of the candidate Nbs, 1 mg/mL of purified Nbs was separately incubated at different temperatures, i.e., 25, 37, 50, 65, 80, 90 °C for 10 min, or at 90 °C for 10, 20, 30, 40, 50, and 60 min. The organic solvent tolerance of Nbs was evaluated using different concentrations (10%, 20%, 30%, 40%, 50%, 60%, 70%, 80%, and 100%) of methanol (MeOH) or acetonitrile. PBS buffer (10 mm PBS, pH = 7.0) was used as the dilution reagent (*v/v*), and Mc-MAb was used as the control.

2.6. Establishment of an Indirect Competitive ELISA Based on the Nanobody N56

The ic-ELISA for NOD-R was developed under optimized reaction conditions. The concentrations of N56 and coating antigen NOD-R-OVA were optimized by checkboard titration. Different buffers (PB, PBS, PBST, Tris-HCl, HEPES), pH values (6.5, 7.0, 7.5, 8.0), and ionic strengths of PBS (5 mM, 10 mM, 20 mM, 40 mM) were chosen to optimize the conditions based on absorbance values of approximately 1.0 in ELISA at 450 nm wavelength or on the Amax (maximal absorbance)/IC_{50} ratio. Then, the established ic-ELISA was performed as follows. In practice, 100 μL/well of NOD-R-OVA solution was used to coat a 96-well plate, which was incubated overnight at 4 °C and then blocked with 3% BSA in PBS buffer for 1 h. After five washings with PBST buffer, 50 μL/well of an N56 solution at optimal concentration and 50 μL/well of NOD-R at different concentrations were added into the microwells, which were incubated for 40 min at 37 °C with a slight oscillation. After additional five washings, 100 μL/well of rabbit anti-VHH-HRP polyclonal antibody (1:5000 dilution) was added in the microwells, and incubation was carried out at 37 °C for 30 min. After 5 more washings, 100 μL/well of TMB peroxidase substrate buffer was added, and the enzyme reaction was stopped after a 10 min incubation by the addition of 50 μL/well of 10% H_2SO_4 (*v/v*). The absorbance values of the microplates were measured using a microplate reader at 450 nm wavelength. Furthermore, a standard curve was obtained based on the established four-parameter logistic equation utilizing Origin 8.5 (Origin Lab Corp., Northampton, MA, USA). Consequently, the limit of detection (LOD) was defined as the concentration with 10% inhibitory activity and the half-maximal inhibitory concentration (IC_{50}) as the concentration with 50% inhibitory activity. The rate of cross reactivity was calculated according to a previously formula, namely, CR (%) = IC_{50} (NOD-R, μg/L)/IC_{50} (other toxins, μg/L) × 100.

2.7. Analysis of Water Samples by Ic-ELISA for NOD-R Detection

Tap water was from Guangdong Provincial Key Laboratory of Food Quality and safety, College of Food Science, South China Agricultural University (Guangzhou, China), lake water was from NingYin Lake in South China Agricultural University, and river water was from CheBei river. After filtering through a 0.22 μm nitrocellulose membrane, all water samples (5 mL) were spiked with different concentrations of NOD-R standard obtaining concentrations of 0, 0.5, 1, 2 μg/L for assessment of the recovery. Furthermore, the spiked samples were analyzed by the established ic-ELISA. According to the ic-ELISA standard curve, the values of mean recovery, SD, and CV were calculated.

The N56-based ic-ELISA was verified by UPLC–MS/MS operated at the Guangzhou Institute for Food Control, China. The conditions were established according to the literatures [16]: Accucore aQ C18 column (150 mm × 2.1 mm, 2.6 μm, Thermo Scientific, USA) was used for chromatographic separation. The mobile phase A was a MeOH solution containing 0.1% formic acid (*v/v*) and 5 mmol/L ammonium formate, while the mobile phase B was an aqueous solution with the same constituents as the mobile phase A. The gradient elution was 0–0.5 min, 80% B; 0.5–4 min, 80–25% B; 4–7 min, 25–0% B; 7–7.5 min, 0% B; 7.5–8 min, 0–80% B; 8–10 min, 80%B. The flow rate was 0.3 mL/min, and the injection volume of each sample was 5 μL. The temperature of the column oven was set at 40 °C. The mass spectra were obtained with the AB TRIPLE QUAD 4500 mass spectrometer under positive ionization mode, with the following parameters' values: curtain gas flow rate, 30 L/min; nebulizer gas flow rate, 50 L/min; auxiliary gas flow rate, 55 L/min; collision gas, 8 L/min; spray voltage, 5.5 kV.

3. Results and Discussion

3.1. Selection of Anti-NOD-R Nanobodies from the Phage Display Library

Considering that MC-LR-KLH previously prepared in our lab as a common immunogen can induce excellent immunization and has been successfully used to generate specific PAbs and MAbs which could recognize NOD-R, in this work, this molecule was chosen as the immunogen for the camel [16]. After six inoculations, the serum titer was up to 1:64,000, and inhibition of NOD-R was higher than 80%. By two-step nested PCR, VHH genes were amplified, and a bacterial library with the capacity of 8.0×10^7 cfu/mL was constructed. Then, rescued by helper phage M13K07, the titer of phage-displayed Nbs library was about 1.0×10^{12} pfu/mL. Furthermore, after four rounds of biopanning, three clones, namely, N4, N56, and N159, were identified as positive ones and exhibited recognition activity for NOD-R (Figure 2). The inhibition rates of these three clones identified by ic-ELISA were all higher than 80%. By sequence alignment analysis, these three clones, composed of 131 amino acid residues, were confirmed to possess the characteristics of Nbs and to have high similarity, except for a few residues in frame regions (FRs) (Figure 3). Therefore, they were considered as the same category of Nbs, and N56 was chosen for further analysis.

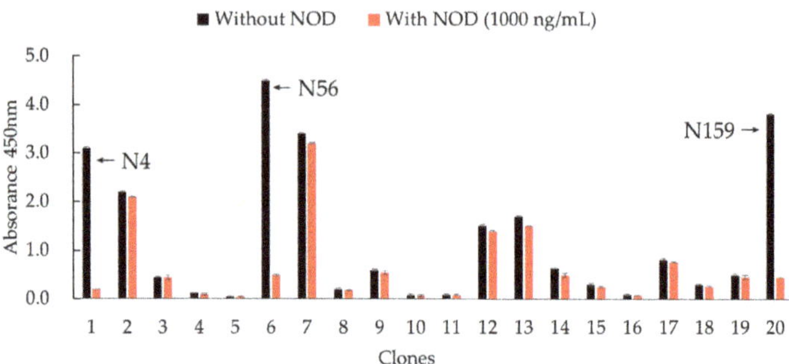

Figure 2. Positive clones identified by ic-ELISA.

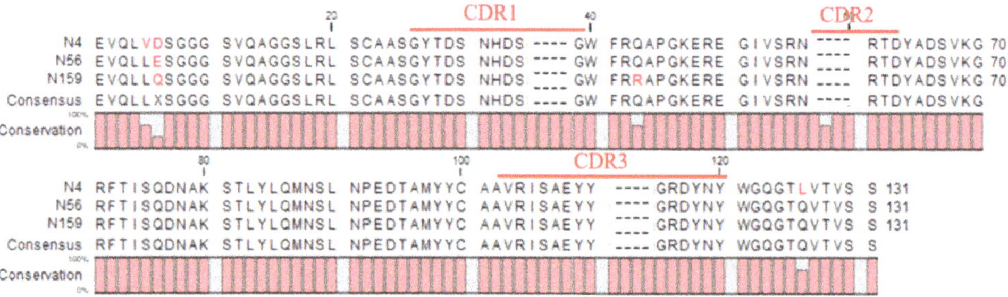

Figure 3. Sequence alignments of the three positive clones.

3.2. Preparation and Characterization Analysis of the Nanobody N56

To better investigate the characteristics of the nanobody N56, it was expressed in 1L of LB medium and purified using the HisPur Ni-NTA resin. The purified N56 were identified by 15% SDS-PAGE (Figure 4a) and Western blotting (Figure 4b). The results showed that the molecular weight of N56 is around 17 kDa, with purity of more than 90%, and the yield was about 4.5 mg/L. It has been observed that Nbs can be highly expressed in *E. coli*, yielding from more than 10 mg to tens of milligrams, thanks to their low molecular weight and good solubility [26]. However, in our work, N56 showed a relatively low yield, not higher than 5 mg/L. We thought it might be related to the ratio of hydrophobic amino acids (22%) in its composition, but after comparison with other Nbs, e.g., Nb-T3-15 (for tetrabromobisphenol A) [27] or Nb-3F9 (for tenuazonic acid) [28], which gave yields up to 30 mg/L and just less then 1 mg/L under the same expression conditions, respectively (Figure S1), this hypothesis was discarded because Nb-T3-15 has a higher hydrophobic amino acids ratio than Nb-3F9. Therefore, maybe the location of the hydrophobic amino acids in the 3D structure of Nbs plays a more important influence on the solubility of nanobodies then their expression level in *E. coli* host cells.

Furthermore, the thermostability and organic solvent tolerance of N56 was evaluated with Mc-MAb as a control. After incubation at different temperatures, from 4 °C to 90 °C, for 10 min, N56 could maintain a binding activity higher than 90%, whereas Mc-MAb lost its activity very quickly. Especially at 25 °C (RT) or 37 °C, two commonly used temperatures during practical detection, N56 exhibited full activity, while Mc-MAb lost nearly 30% of its activity (Figure 5a). Even more, N56 was stable at the high temperature of 90 °C for 1 h, retaining 80% of its activity (Figure 5b). In addition, the tolerance of N56 for the organic solvents methanol (MeOH) and acetonitrile was determined. As shown in Figure 5c,d, with increasing concentration of MeOH or acetonitrile, both N56 and Mc-MAb began to lose their activities; however, N56 as a whole had superior performance than Mc-Mab, in particular, it showed better tolerance to acetonitrile. Generally, the disulfide bonds formed by cysteine residues present in Nbs are thought to contribute to its excellent stability. According to the sequence of N56, four cysteine residues (cys22, cys37, cys100, cys103) might form two disulfide bonds and maintain the stability of N56 structure in harsh environments, at either high temperature or in the presence of organic solvents (Figure 3). This phenomenon was also observed for other reported Nbs for low-molecular-weight chemical contaminants [29].

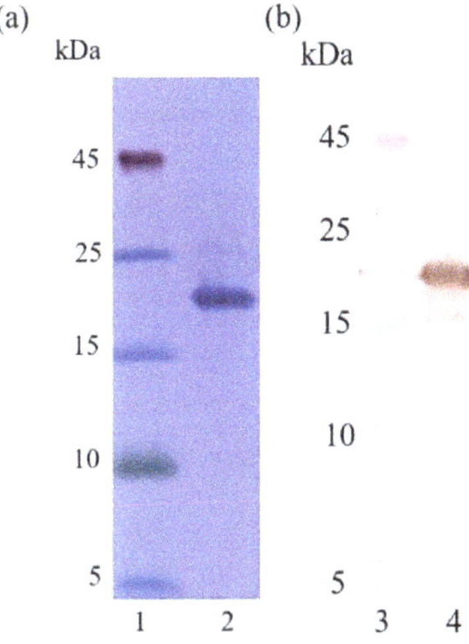

Figure 4. Identification of N56 by SDS-PAGE and Western blotting. (**a**) SDS-PAGE: lane 1, marker; lane 2, purified N56. (**b**) Western blotting: lane 3, marker; lane 4, purified N56.

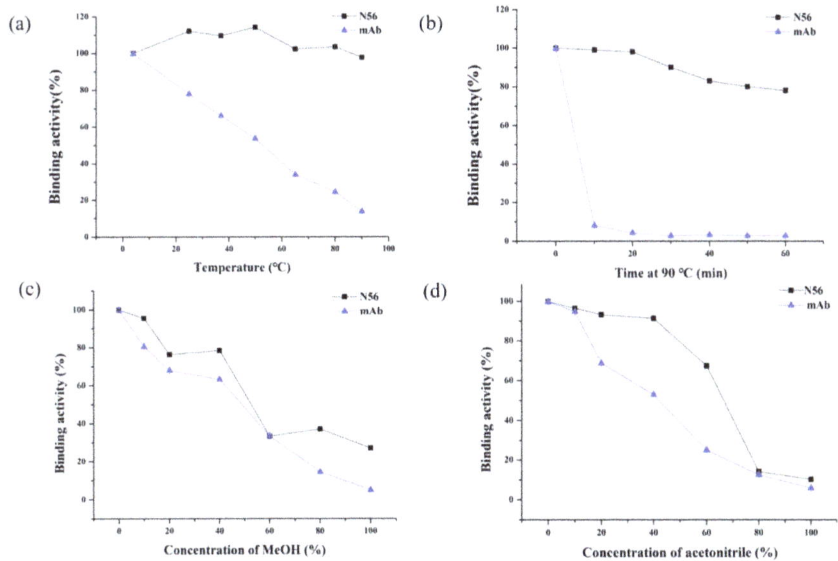

Figure 5. Thermostability and organic solvent tolerance of N56 and Mc-MAb by ic-ELISA: (**a**) N56 and Mc-MAb (1 mg/mL) were incubated at 25, 37, 50, 65, 80, and 90 °C for 10 min; (**b**) N56 and Mc-MAb (1 mg/mL) were incubated at 90 °C for 10, 20, 30, 40, 50, and 60 min; a series concentrations (10%, 20%, 30%, 40%, 50%, 60%, 70% and 80%) of (**c**) methanol (MeOH) and (**d**) acetonitrile were used to dilute (*v/v*) N56 and Mc-Mab (*n* = 3).

3.3. Establishment of an Ic-ELISA Based on the Nanobody N56

In order to develop an ic-ELISA for NOD-R with N56 as the recognition element, various reaction conditions, such as concentrations of nanobody and coating antigen, buffers, pH, and ionic strength, were optimized (Figure 6a–c). After the checkboard titration test, the optimal concentrations of N56 and NOD-R-OVA were determined to be 0.125 μg/mL and 2 μg/mL, respectively. The other conditions were further optimized with the highest Amax/IC_{50} as the selection criterion [28]. Consequently, 10 mM PBS buffer at pH 7.0, at a relative lower ionic strength under a neutral reactive environment, was confirmed as the most suitable buffer for the establishment of ic-ELISA. Finally, an ic-ELISA standard curve based on N56 for NOD-R was established (Figure 7). The IC_{50} value was 9.94 μg/L, and the linear range of detection was 1.74–56.66 μg/L with a limit of detection (LOD) of 0.67 μg/L, which was lower than that of Mc-MAb for NOD-R (1.25 μg/L). This LOD value can also satisfy the detection requirement of 1 μg/L in drinking water.

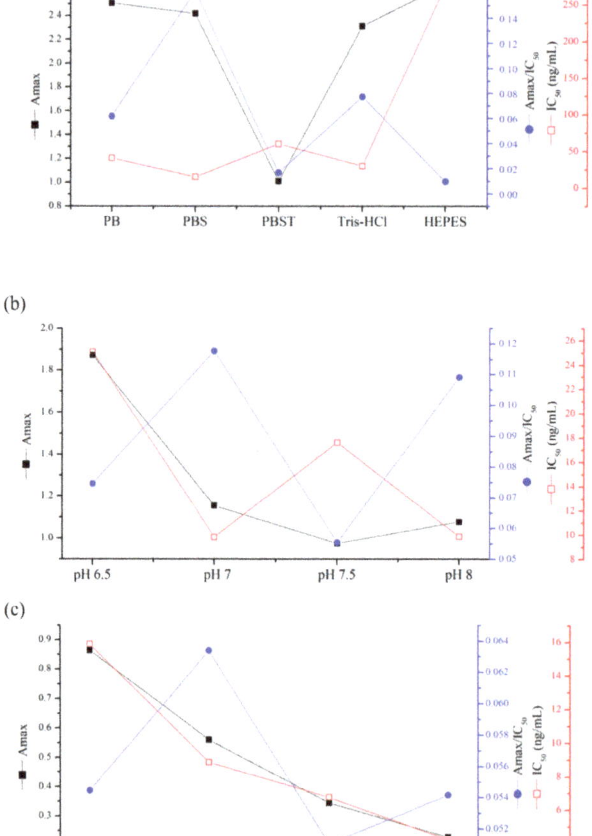

Figure 6. Optimization of assay conditions for the establishment of ic-ELISA: (**a**) types of buffers (PB, PBS, PBST, Tris-HCl, and HEPES), (**b**) pH value (6.5, 7.0, 7.5, and 8.0), and (**c**) ionic strength (5 mM, 10 mM, 20 mM, and 40 mM PBS). The conditions were evaluated by the Amax (maximal absorbance)/IC_{50} ratio.

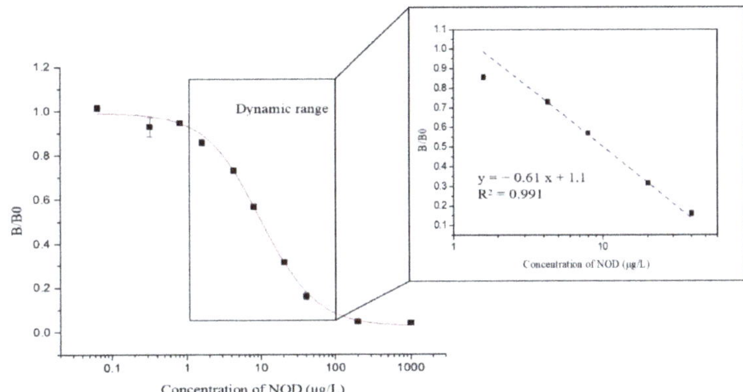

Figure 7. Standard curves of ic-ELISA for NOD-R based on purified N56 under the optimized reaction conditions. The standard curves were normalized by expressing the experimental absorbance values (B) as B/B_0, where B_0 is the absorbance value at zero analyte concentration (*n* = 3).

Besides, the N56-based assay showed relatively significant cross reactivity for MCs. This might result from the common special Adda structural regions ((2S,3S,8S,9S)-3-amino-9-methoxy-2,6,8-trimethyl-10-phenyl deca-4,6-dienoic acid). In this work, NOD-R-OVA was used as the coating antigen, and NOD-R at decreasing concentration as the competitive drug to increase the specificity of the eluted phages; yet, the generation of VHHs against the Adda structure in camel by the immunogen MC-LR-KLH was possible, and these antibodies would consequently recognize NOD-R as well as MCs. In addition, among the MCs capable of binding N56, MC-LR, -YR, -RR, -WR exhibited higher cross reactivity than other MCs (Table 2). This is because these four MCs have an arginine residue as NOD-R besides the Adda region. From another perspective, high cross reactivity might be helpful in the detection of multiple residues of cyanobacteria toxins. It was reported that VHHs for MC-LR prepared by Pírez-Schirmer et al. [30] or Xu et al. [31] could recognize other MCs, but there are no data about their cross reactivity with NOD-R. In this study, we, for the first time, obtained N56 that is able to detect NOD-R and MCs, i.e. two groups of cyanobacteria toxins.

Table 2. CRs of N56 for MCs analogues (*n* = 3).

Algal Toxins	Structure	IC_{50}(μg/L)	CRs
NOD		9.94	100.0%
MC-LR		8.24	120.6%

Table 2. *Cont.*

Algal Toxins	Structure	IC_{50}(μg/L)	CRs
MC-LA	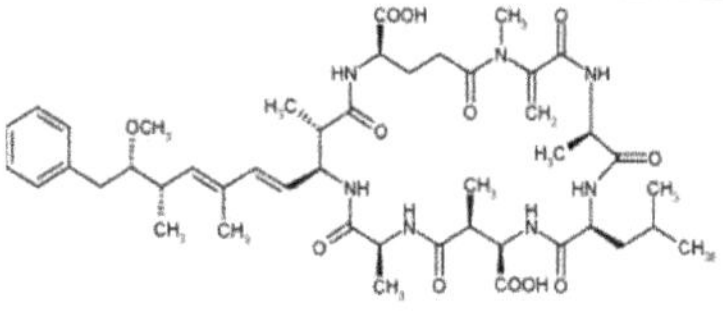	42.41	23.4%
MC-LY	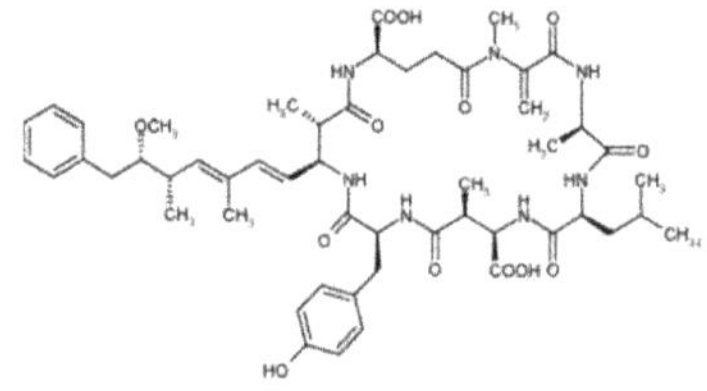	60.78	16.4%
MC-LW	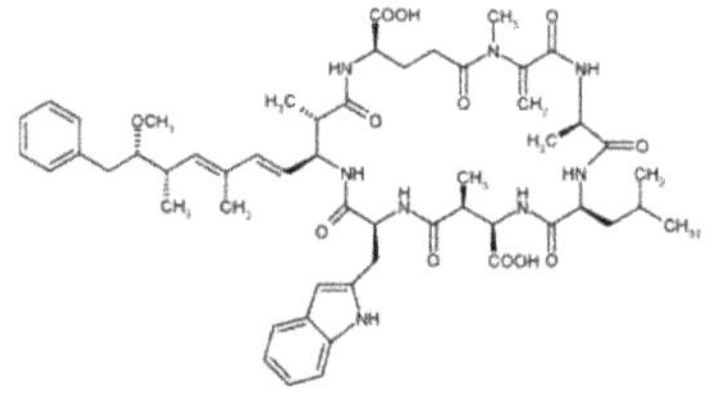	180.81	5.5%
MC-LF	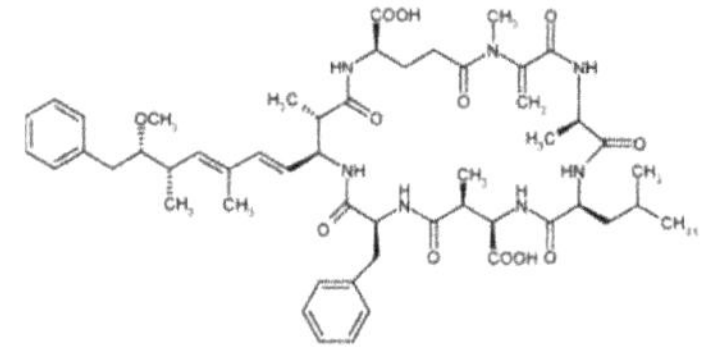	113.79	8.7%
MC-YR	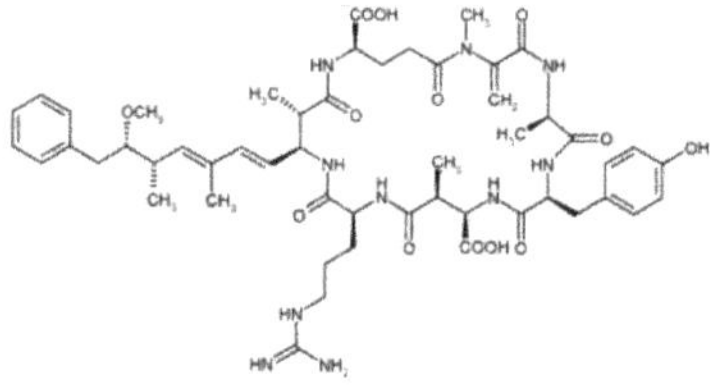	7.85	126.6%
MC-WR	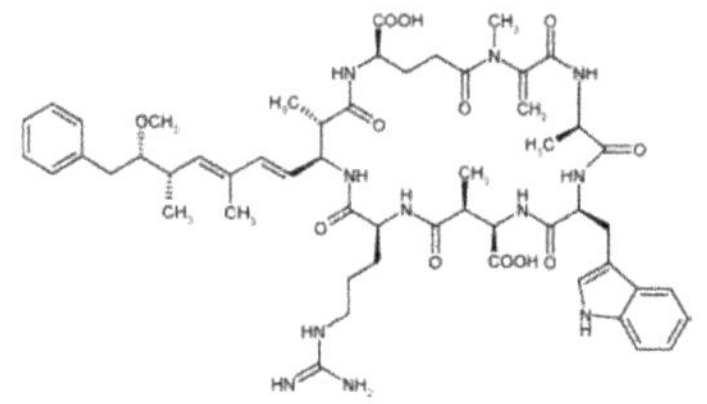	13.50	73.6%
MC-RR	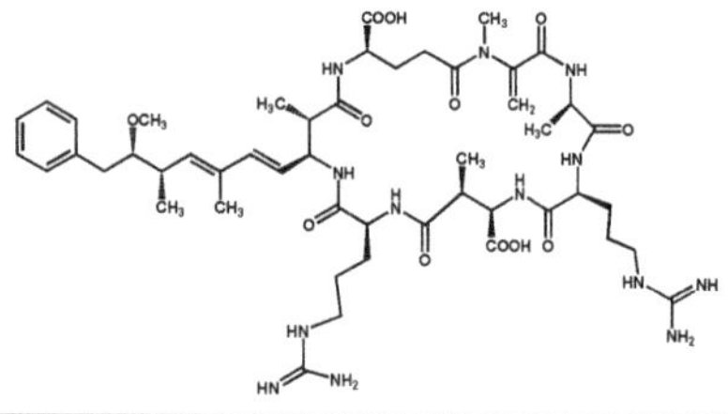	11.87	83.7%

3.4. Water Sample Analysis by ELISA and UPLC–MS/MS

The N56-based ic-ELISA was further employed to analyze water samples spiked with NOD-R at three concentrations (0.5, 1, 2 μg/L). As shown in Table 3, the average recovery ranged from 111.0% to 119.5%, with CVs of 2.0–5.8%. The value of the correlation coefficient (R^2) between ic-ELISA and UPLC–MS/MS was up to 0.999, indicating that the established N56-based ic-ELISA is accurate and reliable for detecting NOD-R in water samples.

Table 3. Reproducibility and accuracy of NOD-R from spiked water samples by ic-ELISA and LC–MS/MS (n = 3).

Spiked Level (μg/L)	Ic-ELISA			UPLC-MS/MS			R^2
	Found ± SD (μg/L)	Recovery (%)	CV (%)	Found ± SD (μg/L)	Recovery (%)	CV (%)	
0	/	/	/	/	/	/	0.999
0.5	0.59 ± 0.02	117.7	3.6	0.53 ± 0.01	106.0	1.9	
1	1.11 ± 0.07	111.0	5.8	1.07 ± 0.02	107.0	1.9	
2	2.39 ± 0.02	119.5	2.0	2.33 ± 0.04	116.5	1.5	

4. Conclusions

In this work, three Nbs for NOD-R were selected from a Bactrian camel phage display library, and N56 showed high stability and good affinity. Then, an ic-ELISA method for NOD-R detection was established, with a LOD lower than 1.0 μg/L, the generally reference value for cyanobacteria toxins. In a thorough comparison with UPLC–MS/MS, the ic-ELISA method exhibited good accuracy and reliability, which suggests its potential for the development of a rapid detection kit for the safety detection of toxins based on N56.

Supplementary Materials: The following are available online at https://www.mdpi.com/article/10.3390/foods10112758/s1, Figure S1: Comparison of the expression levels of three nanobodies (Nb-T3-15, Nb-56, Nb-3F9) by SDS-PAGE; Figure S2: Results of ethical review of animal experiments. (**a**) Original copy. (**b**) English translation; Table S1: Abbreviation list.

Author Contributions: Conceptualization, J.Y. and R.S.; methodology, G.W., R.S. and R.F.; validation, H.W. and R.S.; resources, H.L., F.L., J.Y. and Q.Z.; data curation, Y.W., G.W. and R.S.; writing-original draft preparation, G.W., R.S. and R.F.; writing—review and editing, J.Y., F.W., H.W.; supervision, H.W. and Y.S.; project administration, H.W. and J.Y. All authors have read and agreed to the published version of the manuscript.

Funding: This work was supported by the National Key R&D Program of China (2019YFE0116600), Natural Science Foundation of China (31972157), Key-Area Research and Development Program of Guangdong Province (2019B020211002), Guangdong Province Universities and Colleges Pearl River Scholar Funded Scheme (2017), Lingnan Modern Agricultural Science and Technology Guangdong Laboratory independent scientific research project (NZ2021032).

Institutional Review Board Statement: The study was conducted according to the guidelines of the Declaration of Helsinki, and approved by Experimental Animal Ethics Committee of South China Agricultural University (protocol code 2020F061, 5 August 2020) (Figure S2).

Conflicts of Interest: The authors declare no conflict of interest.

References

1. Wang, C.C.; Petty, E.E.; Smith, C.M. Rapid and Efficient Analysis of Microcystins, Nodularin, Cylindrospermopsin, and Anatoxin-a in Drinking Water by LC Tandem MS. *J. AOAC Int.* **2016**, *99*, 1565–1571. [CrossRef] [PubMed]
2. Chen, G.; Wang, L.; Wang, M.; Hu, T. Comprehensive insights into the occurrence and toxicological issues of nodularins. *Mar. Pollut. Bull.* **2020**, *162*, 111884. [CrossRef]
3. Ohta, T.; Sueoka, E.; Iida, N.; Komori, A.; Suganuma, M.; Nishiwaki, R.; Tatematsu, M.; Kim, S.J.; Carmichael, W.W.; Fujiki, H. Nodularin, a potent inhibitor of protein phosphatases 1 and 2A, is a new environmental carcinogen in male F344 rat liver. *Cancer Res.* **1994**, *54*, 6402–6406.

4. Zhou, Z.; Feng, D.; Liang, W.; Lu, X.; Shi, X.; Zhao, C.; Xu, G. Synthesis of metal-organic framework-5@chitosan material for the analysis of microcystins and nodularin based on ultra-performance liquid chromatography-tandem mass spectrometry. *J. Chromatogr. A* **2020**, *1623*, 461198. [CrossRef] [PubMed]
5. Toruńska-Sitarz, A.; Kotlarska, E.; Mazur-Marzec, H. Biodegradation of nodularin and other nonribosomal peptides by the Baltic bacteria. *Int. Biodeterior. Biodegrad.* **2018**, *134*, 48–57. [CrossRef]
6. Beversdorf, L.J.; Weirich, C.A.; Bartlett, S.L.; Miller, T. Variable Cyanobacterial Toxin and Metabolite Profiles across Six Eutrophic Lakes of Differing Physiochemical Characteristics. *Toxins* **2017**, *9*, 62. [CrossRef] [PubMed]
7. WHO. Cyanobacterial toxins: Microcystin-LR. In *Guidelines for Drinking-Water Quality*; World Health Organization: Geneva, Switzerland, 1998.
8. Fitzgerald, D.J.; Cunliffe, D.A.; Burch, M.D. Development of health alerts for cyanobacteria and related toxins in drinking water in South Australia. Environ Toxicol. *Environ. Toxicol. Int. J.* **1999**, *14*, 203–209. [CrossRef]
9. Jussi, M. Chromatography of microcystins. *Anal. Chim. Acta* **1997**, *352*, 277–298.
10. Marcel, E.; Hans, V.D.; Peter, J. Rapid typing and elucidation of new secondary metabolites of intact cyanobacteria using MALDI-TOF mass spectrometry. *Nat. Biotechnol.* **1997**, *15*, 906–909.
11. Meriluoto, J.; Karlsson, K.; Spoof, L. High-Throughput Screening of Ten Microcystins and Nodularins, Cyanobacterial Peptide Hepatotoxins, by Reversed-Phase Liquid Chromatography-Electrospray Ionisation Mass Spectrometry. *Chromatographia* **2004**, *59*, 291–298.
12. Ouyang, S.; Hu, B.; Zhou, R.; Liu, D.; Peng, D.; Li, Z.; Li, Z.; Jiao, B.; Wang, L. Rapid and sensitive detection of nodularin-R in water by a label-free BLI aptasensor. *Analyst* **2018**, *143*, 4316–4322. [CrossRef] [PubMed]
13. Hou, S.; Ma, J.; Cheng, Y.; Wang, H.; Sun, J.; Yan, Y. Quantum dot nanobead-based fluorescent immunochromatographic assay for simultaneous quantitative detection of fumonisin B1, dexyonivalenol, and zearalenone in grains. *Food Control* **2020**, *117*, 107331. [CrossRef]
14. Samdal, I.A.; Ballot, A.; Løvberg, K.E.; Miles, C.O. Multihapten Approach Leading to a Sensitive ELISA with Broad Cross-Reactivity to Microcystins and Nodularin. *Environ. Sci. Technol.* **2014**, *48*, 8035–8043. [CrossRef]
15. Zhou, Y.; Li, Y.S.; Zhi, B.H.; Lu, S.Y.; Ren, H.L.; Zhang, Y.Y.; Li, Z.H.; Shen, Q.F.; Meng, X.M.; Liu, Z.S.; et al. Detection of nodularin based on a monoclonal antibody in water and aquatic fish samples. *Food Control* **2011**, *22*, 797–800. [CrossRef]
16. Lu, N.; Ling, L.; Guan, T.; Wang, L.; Wang, D.; Zhou, J.; Ruan, T.; Shen, X.; Li, X.; Sun, Y.; et al. Broad-specificity ELISA with a heterogeneous strategy for sensitive detection of microcystins and nodularin. *Toxicon* **2019**, *175*, 44–48. [CrossRef]
17. Hamers-Casterman, C.; Atarhouch, T.; Muyldermans, S.; Robinson, G.; Hammers, C.; Songa, E.B.; Bendahman, N. Naturally occurring antibodies devoid of light chains. *Nat. Cell Biol.* **1993**, *363*, 446–448. [CrossRef]
18. Kim, H.J.; McCoy, M.R.; Majkova, Z.; Dechant, J.E.; Gee, S.J.; Tabares-da Rosa, S.; González-Sapienza, G.G.; Hammock, B.D. Isolation of alpaca anti-hapten heavy chain single domain antibodies for development of sensitive immunoassay. *Anal. Chem.* **2012**, *84*, 1165–1171. [CrossRef]
19. Bever, C.S.; Dong, J.X.; Vasylieva, N.; Barnych, B.; Cui, Y.; Xu, Z.L.; Hammock, B.D.; Gee, S.J. VHH antibodies: Emerging reagents for the analysis of environmental chemicals. *Anal. Bioanal. Chem.* **2016**, *408*, 5985–6002. [CrossRef]
20. Sun, Z.; Duan, Z.; Liu, X.; Deng, X.; Tang, Z. Development of a Nanobody-Based Competitive Dot ELISA for Visual Screening of Ochratoxin A in Cereals. *Food Anal. Methods* **2017**, *10*, 3558–3564. [CrossRef]
21. Zhang, Y.Q.; Xu, Z.L.; Wang, F.; Cai, J.; Dong, J.X.; Zhang, J.R.; Si, R.; Wang, C.L.; Wang, Y.; Shen, Y.D.; et al. Isolation of Bactrian camel Single Domain Antibody for Parathion and Development of One-Step dc-FEIA Method Using VHH-Alkaline Phosphatase Fusion Protein. *Anal. Chem.* **2018**, *90*, 12886–12892. [CrossRef]
22. Ren, W.; Li, Z.; Xu, Y.; Wan, D.; Barnych, B.; Li, Y.; Tu, Z.; He, Q.; Fu, J.; Hammock, B.D. One-Step Ultrasensitive Bioluminescent Enzyme Immunoassay Based on Nanobody/Nanoluciferase Fusion for Detection of Aflatoxin B1 in Cereal. *J. Agric. Food Chem.* **2019**, *67*, 5221–5229. [CrossRef]
23. Wang, K.; Liu, Z.; Ding, G.; Li, J.; Vasylieva, N.; Li, Q.X.; Li, D.; Gee, S.J.; Hammock, B.D.; Xu, T. Development of a one-step immunoassay for triazophos using camel single-domain antibody–alkaline phosphatase fusion protein. *Anal. Bioanal. Chem.* **2019**, *411*, 1287–1295. [CrossRef] [PubMed]
24. Zhang, J.R.; Wang, Y.; Dong, J.X.; Yang, J.Y.; Zhang, Y.Q.; Wang, F.; Si, R.; Xu, Z.L.; Wang, H.; Xiao, Z.L.; et al. Development of a Simple Pretreatment Immunoassay Based on an Organic Solvent-Tolerant Nanobody for the Detection of Carbofuran in Vegetable and Fruit Samples. *Biomolecules* **2019**, *9*, 576. [CrossRef] [PubMed]
25. Yang, Y.Y.; Wang, Y.; Zhang, Y.F.; Wang, F.; Liang, Y.F.; Yang, J.Y.; Xu, Z.L.; Shen, Y.D.; Wang, H. Nanobody-Based Indirect Competitive ELISA for Sensitive Detection of 19-Nortestosterone in Animal Urine. *Biomolecules* **2021**, *11*, 167. [CrossRef]
26. De Meyer, T.; Muyldermans, S.; Depicker, A. Nanobody-based products as research and diagnostic tools. *Trends Biotechnol.* **2014**, *32*, 263–270. [CrossRef] [PubMed]
27. Wang, J.; Bever, C.R.S.; Majkova, Z.; Dechant, J.E.; Yang, J.; Gee, S.J.; Xu, T.; Hammock, B.D. Heterologous Antigen Selection of Camelid Heavy Chain Single Domain Antibodies against Tetrabromobisphenol A. *Anal. Chem.* **2014**, *86*, 8296–8302. [CrossRef]
28. Wang, F.; Li, Z.F.; Yang, Y.Y.; Wan, D.B.; Vasylieva, N.; Zhang, Y.Q.; Cai, J.; Wang, H.; Shen, Y.D.; Xu, Z.L.; et al. Chemiluminescent Enzyme Immunoassay and Bioluminescent Enzyme Immunoassay for Tenuazonic Acid Mycotoxin by Exploitation of Nanobody and Nanobody-Nanoluciferase Fusion. *Anal. Chem.* **2020**, *92*, 11935–11942. [CrossRef]

29. Zabetakis, D.; Olson, M.A.; Anderson, G.; Legler, P.M.; Goldman, E.R. Evaluation of Disulfide Bond Position to Enhance the Thermal Stability of a Highly Stable Single Domain Antibody. *PLoS ONE* **2014**, *9*, e115405. [CrossRef]
30. Pírez-Schirmer, M.; Rossotti, M.; Badagian, N.; Leizagoyen, C.; Brena, B.M.; Gonzalez-Sapienza, G. Comparison of Three Antihapten VHH Selection Strategies for the Development of Highly Sensitive Immunoassays for Microcystins. *Anal. Chem.* **2017**, *89*, 6800–6806. [CrossRef]
31. Xu, C.; Yang, Y.; Liu, L.; Li, J.; Liu, X.; Zhang, X.; Liu, Y.; Zhang, C.; Liu, X. Microcystin-LR nanobody screening from an alpaca phage display nanobody library and its expression and application. *Ecotoxicol. Environ. Saf.* **2018**, *151*, 220–227. [CrossRef]

Article

Encapsulation of Lutein via Microfluidic Technology: Evaluation of Stability and In Vitro Bioaccessibility

Yuanhang Yao [1], Jiaxing Jansen Lin [2], Xin Yi Jolene Chee [1], Mei Hui Liu [1], Saif A. Khan [2] and Jung Eun Kim [1,*]

[1] Department of Food Science and Technology, National University of Singapore, Singapore 117543, Singapore; yuanhang@u.nus.edu (Y.Y.); jolene_chee@sics.a-star.edu.sg (X.Y.J.C.); fstlmh@nus.edu.sg (M.H.L.)

[2] Department of Chemical and Biomolecular Engineering, National University of Singapore, Singapore 117585, Singapore; mdclinj@nus.edu.sg (J.J.L.); saifkhan@nus.edu.sg (S.A.K.)

* Correspondence: fstkje@nus.edu.sg; Tel.: +65-6516-1136

Abstract: Inadequate intake of lutein is relevant to a higher risk of age-related eye diseases. However, lutein has been barely incorporated into foods efficiently because it is prone to degradation and is poorly bioaccessible in the gastrointestinal tract. Microfluidics, a novel food processing technology that can control fluid flows at the microscale, can enable the efficient encapsulation of bioactive compounds by fabricating suitable delivery structures. Hence, the present study aimed to evaluate the stability and the bioaccessibility of lutein that is encapsulated in a new noodle-like product made via microfluidic technology. Two types of oils (safflower oil (SO) and olive oil (OL)) were selected as a delivery vehicle for lutein, and two customized microfluidic devices (co-flow and combination-flow) were used. Lutein encapsulation was created by the following: (i) co-flow + SO, (ii) co-flow + OL, (iii) combination-flow + SO, and (iv) combination-flow + OL. The initial encapsulation of lutein in the noodle-like product was achieved at 86.0 ± 2.7%. Although lutein's stability experienced a decreasing trend, the retention of lutein was maintained above 60% for up to seven days of storage. The two types of device did not result in a difference in lutein bioaccessibility (co-flow: 3.1 ± 0.5%; combination-flow: 3.6 ± 0.6%) and SO and OL also showed no difference in lutein bioaccessibility (SO: 3.4 ± 0.8%; OL: 3.3 ± 0.4%). These results suggest that the types of oil and device do not affect the lutein bioaccessibility. Findings from this study may provide scientific insights into emulsion-based delivery systems that employ microfluidics for the encapsulation of bioactive compounds into foods.

Keywords: lutein; stability; bioaccessibility; encapsulation; microfluidics

Citation: Yao, Y.; Lin, J.J.; Chee, X.Y.J.; Liu, M.H.; Khan, S.A.; Kim, J.E. Encapsulation of Lutein via Microfluidic Technology: Evaluation of Stability and In Vitro Bioaccessibility. *Foods* **2021**, *10*, 2646. https://doi.org/10.3390/foods10112646

Academic Editors: Hong Wu and Hui Zhang

Received: 28 September 2021
Accepted: 29 October 2021
Published: 1 November 2021

1. Introduction

Carotenoids are a family of terpenoid pigments rich in fruits and vegetables and are related to several potential health benefits because of their antioxidant and anti-inflammatory properties [1–5]. Lutein is one of the main carotenoids that can selectively accumulate in the eye, macula and retina in particular, and is known for eye protection effects, especially against photoinduced damage [6–8]. This is mainly because lutein is capable of quenching singlet oxygen and other reactive oxygen species and absorbing blue light [6]. Abundant epidemiological evidence has suggested that lutein intake is positively correlated with a lower risk of age-related macular degeneration and cataracts [9–11]. Nevertheless, lutein can only be obtained from diets, which is often insufficient due to the limited consumption of fruits and vegetables. Additionally, the biological activity of lutein is highly dependent on its gastrointestinal absorption, which may be hindered mainly as a consequence of its physicochemical properties [12,13].

One approach to overcome the challenges of insufficient lutein intake is to develop lutein-enriched staple foods, thus delivering lutein to humans regularly and continuously. However, merely incorporating lutein as an ingredient into foods and beverages hardly exerts the nutritional value of lutein due to its poor solubility and the fact that lutein is prone to oxidative degradation [13,14]. Encapsulation technology has shown to be an attractive strategy to entrap the bioactive compounds within a carrier, which can improve their stability during food processing, storage and gastrointestinal absorption [15–18]. Microfluidics has become a trending topic in innovative food processing in recent years, with nutrients encapsulation being an emerging application of the microfluidic technique [19]. Microfluidics particularly focuses on accurate control over minute volumes of fluids within a system of microchannels [20,21]. This makes it possible to alter the way of working with dispersed food systems, and inherently manipulate structures at a micro-level [19].

Lutein has to be incorporated into mixed micelles for gastrointestinal absorption due to its hydrophobic property. Bioaccessibility describes the fraction of lutein solubilized in the mixed micelles and is usually determined in vitro via a simulated gastrointestinal digestion model [22]. Several lines of studies have indicated that fat-soluble carotenoids such as lutein, when dispersed in dietary oils, obtain greater bioaccessibility than when they are consumed alone [23–25]. This is because oils work as a delivery vehicle for these hydrophobic nutrients: The hydrolysis products of oils—free fatty acids and monoglycerides—together with phospholipids, bile salts and cholesterol, form the mixed micelles in the aqueous digestion fluid [26]; the formation of the mixed micelles facilitates the solubilization of hydrophobic lutein in mixed micelles and makes lutein become accessible during digestion [23,27]. Moreover, several studies have reported that the emulsion-based delivery system shows desired properties such as alleviating the degradation of bioactive compounds, improving the efficiency in micellization and promoting the digestive enzyme activity [28–30].

However, considering microfluidics is a relatively innovative technology, especially in "foods" area, limited studies have investigated its potential in nutrient encapsulation combined with an emulsion-based delivery system. Additionally, the evidence of applying this microfluidic technique to encapsulate carotenoids into foods is lacking. Therefore, this study aimed to encapsulate lutein into a staple food noodle using excipient emulsions via microfluidics-based continuous extrusion technique, and to assess the stability and bioaccessibility of lutein with different microfluidic assemblies and different types of oils.

2. Materials and Methods

2.1. Materials

Food-grade sodium alginate and calcium chloride ($CaCl_2$) were purchased from a local shop (Phoon Huat, Singapore). Soy protein isolate (SPI, New Fujipro SEH; Fuji Oil Co., Ltd., Tokyo, Japan) formulation was generously sponsored by Fuji Oil Holdings Inc. (Izumisano-shi, Japan), and its protein content was about 90%. FloraGlo lutein was sponsored by DSM company (Dutch State Mines, DU, Heerlen, The Netherlands) and was known to be extracted from *Tagetes erecta*. Safflower oil (SO) was purchased from iHerb (Eden Foods, Clinton, Michigan, US) and olive oil (OL) was purchased from the local supermarket (NTUC FairPrice, Singapore). Pure lutein standards, amylase (10 U/mg), pepsin (695 U/mg), pancreatin (P7545, 8 X USP) and bile salt were purchased from Sigma Aldrich (St. Louis, MI, USA). All used chemicals were of analytical-grade and all solvents for lutein extraction were of high-performance liquid chromatography (HPLC)-grade.

2.2. Sample Preparation

Sodium alginate was blended with deionized water to form a 2% (w/v) solution, and was centrifuged at 19,802$\times$ g, 20 °C for 15 min. SPI was blended with deionized water to form a 12% (w/v) solution at room temperature and underwent ultrasonic degassing for 20 min. The pH of the SPI solution was determined to be 7.51. The viscosity of the prepared sodium alginate solution and SPI solution was determined to be 2630 and 10,590 cP, respectively, at 20 rpm speed. $CaCl_2$ solution (3% w/v) was prepared by dissolving $CaCl_2$ in deionized water. Lutein fortified OL (0.5% w/v) or SO (0.5% w/v) was added in the SPI solution to form the emulsion by using the high shear mixer (Silverson L4RT, US) at 8000 rpm for 7 min, and the container was placed under the ice bath to prevent the excessive heat. The ratio of the lutein fortified oil to SPI solution was 10:1 (v/v).

2.3. Assembly of Microfluidic Devices

Assembly of microfluidic devices was performed based on the methods disclosed in a patent (No. 10201906256W). The flow rate was set at 3000 μL/min for sodium alginate solution and 500 μL/min for SPI and lutein-fortified oil emulsion for the co-flow device, while it was set separately at 454.5 μL/min and 45.5 μL/min for the SPI solution and lutein-fortified oil for the combination of co-flow and flow-focusing device (namely combination-flow).

Schematic diagrams of two different microfluidic devices are shown in Figure 1. The co-flow device (Figure 1a) is made up of an inner circular capillary that is fitted in an outer square capillary, with both the inner (SPI and lutein fortified oil emulsion) and outer (sodium alginate) fluids flowing in the same direction. Flow-focusing is similar to the co-flow device, whereby the inner (lutein fortified oil) and outer (SPI) fluids are introduced in opposing directions to each other before exiting through the same capillary. Combination-flow (Figure 1b) consists of the flow-focusing on the left part and co-flow on the right part. The mixture of lutein-fortified oil and SPI is extruded first with flow-focusing and then follows the fate of the co-flow; flowing through the inner capillary with the same direction as sodium alginate, which flows in the outer capillary. The material composition of extrudates are theoretically the same for both two devices. The extruded noodle-like samples were subjected to a $CaCl_2$ bath for 30 min for ionic gelation and were subsequently heated in a deionized water bath (80–85 °C) for 6 min for thermal gelation. Afterwards, all noodle samples were kept in water as dehydration may occur due to the diffusion of saturated water vapor at the surface [31]. All the above procedures were done under dull red light to minimize the photodecomposition of lutein.

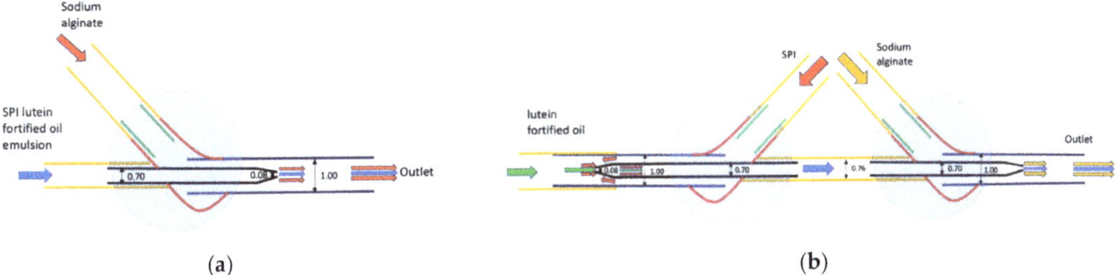

Figure 1. Schematic diagram of microfluidic devices (**a**) Co-flow; (**b**) Combination-flow. SPI: Soy protein isolate. Critical channel diameters are measured in millimeters.

2.4. In Vitro Digestion

The microfluidic noodle was cut into < 2 mm pieces before being subjected to the simulated in vitro digestion. The protocol for digestion was referred from the INFOGEST method with minor modification [32]. All simulated saliva fluid (SSF), simulated gastric fluid (SGF) and simulated intestinal fluid (SIF) solutions were prepared according to the INFOGEST method. Specifically, SSF was comprised of 15.1 mmol/L potassium chloride (KCL), 3.7 mmol/L potassium dihydrogen phosphate (KH_2PO_4), 13.6 mmol/L sodium bicarbonate ($NaHCO_3$), 0.15 mmol/L magnesium chloride hexahydrate ($MgCl_2(H_2O)_6$) and 0.06 mmol/L ammonium carbonate ($(NH_4)_2CO_3$); SGF was comprised of 6.9 mmol/L KCL, 0.9 mmol/L KH_2PO_4, 25 mmol/L $NaHCO_3$, 47.2 mmol/L NaCl, 0.1 mmol/L $MgCl_2(H_2O)_6$ and 0.5 mmol/L $(NH_4)_2CO_3$; SIF was comprised of 6.8 mmol/L KCL, 0.8 mmol/L KH_2PO_4, 85 mmol/L $NaHCO_3$, 38.4 mmol/L NaCl and 0.33 mmol/L $MgCl_2(H_2O)_6$. To start, 5 g of cut noodle samples were mixed with 3.5 mL SSF, 0.5 mL salivary a-amylase (1500 U/mL) solution, 25 uL $CaCl_2$ (0.3 M) and 975 uL deionized water, and were incubated in the water bath at 37 °C, 85 rpm for 2 min. Gastric digestion was started with the oral digesta being acidified, by adding 1 M hydrogen chloride (HCL) (~0.2 mL) to adjust the pH to reach 3.0. It was subsequently mixed with 7.5 mL SGF, 5 uL $CaCl_2$ (0.3 M), 1.6 mL pepsin

(25,000 U/mL) solution and 695 uL deionized water, and was incubated in the water bath at 37 °C, 85 rpm for 1.5 h. Afterwards, to simulate the intestinal digestion, 1 M sodium hydroxide (NaOH) (~0.15 mL) was used to adjust the pH to 7.0, and then was mixed with 11 mL SIF, 5 mL pancreatin solution (800 U/mL), 2.5 mL bile (10 mM), 40 uL $CaCl_2$ (0.3 M) and 1.31 mL deionized water, and was incubated in the water bath at 37 °C, 85 rpm for 2.5 h. Right after the final stage of the simulated gastrointestinal digestion was completed, raw digesta was centrifuged at 19,802× *g*, 20 °C for 45 min. Given that large particles are unable to pass through the mucus layer and be taken up by epithelium cells [33], the aqueous supernatant after filtration (pore size 0.45 µm) was assumed to be comprised of mixed micelles where the lutein solubilized [34]. We considered the lutein solubilized in the micellar fractions as the bioaccessible lutein. During the entire digestion procedure, all the samples were kept in the amber color tubes or the containers were covered with aluminum foil to minimize the photodecomposition of lutein.

2.5. Extraction and Quantification of Lutein

Lutein in digesta, micelle fraction and homogenate were extracted and analyzed as previously reported [35]. Briefly, digesta, micellar fraction or homogenate was extracted with acetone:petroleum ether (1:1, *v*/*v*, second and third times was extracted with petroleum ether alone), vortexed for 2 min and was centrifuged for 10 min at 19,802× *g*, 20 °C. The supernatant layer was collected and the above extraction was repeated three times. All the supernatant layers were combined, and then it was evaporated by nitrogen gas. The final samples were reconstituted in methanol:methyl tert-butyl ether (MTBE) (1:1, *v*/*v*) and were filtered through a 0.45 µm filter. The extraction procedure was fully carried out under dull red light, and 0.1% butylated hydroxytoluene (*w*/*v*) was added in the extraction solvents to minimize lutein degradation.

Lutein was detected by the HPLC (Waters, US) at 4 °C at the wavelength of 450 nm with a YMC carotenoid C30 column, 250 mm × 4.6 mm ID (YMC, Japan), that has been reported previously [35]. The mobile phases were comprised of methanol:MTBE:water (A, 81:15:4, *v*/*v*/*v*) and methanol:MTBE:water (B, 9:87:4, *v*/*v*/*v*). The gradient program was carried out as follows: an initial condition of eluent A:B was 100:0 (%), then there was a linear increase till A:B was 81:19 (%) at 3 min, followed by an A:B of 47:53 (%) at 25 min, and then a rapid increase till A:B was 0:100 (%) at 27 min, held for 10 min and finally back to the initial condition in 3 min, allowing for a 10 min hold as re-equilibration. The flow rate was set as 1 mL/min and the injection volume was 80 µL.

2.6. Optical Microscopy

Images of microfluidic noodle with two types of devices (co-flow and combination-flow) were obtained using a microscope digital camera DP74 mounted on an Olympus BX51 light microscope. The images were viewed under 4× magnification.

2.7. Storage Stability

The stability of lutein was represented by the retention of lutein in the microfluidic noodle at each storage day 1, 2, 3, 4, 5, 6 and 7 under 4 °C as compared to the initial added lutein content. The storage stability was calculated as follows:

$$Stability(\%) = \frac{100 \times C_{sample}}{C_{intial}} \quad (1)$$

where C_{sample} is the remaining lutein content in the microfluidic noodle samples at each storage day, and C_{intial} corresponds to the initial added lutein content.

2.8. Bioaccessibility, Release and Micellarization of Lutein

The fraction of lutein solubilized in the mixed micelles phase after passing through the simulated in vitro digestion was taken to be bioaccessibility and was calculated as follows:

$$Bioaccessibility(\%) = \frac{100 \times C_{micelles}}{C_{intial}} \tag{2}$$

The release rate was determined as the lutein content in the digesta released from the initial food matrix and was calculated as follows:

$$Release(\%) = \frac{100 \times C_{digesta}}{C_{intial}} \tag{3}$$

The micellarization rate was determined as transfer of lutein from the digesta to the mixed micelles and was calculated as follows:

$$Micellarization(\%) = \frac{100 \times C_{micelles}}{C_{digesta}} \tag{4}$$

where $C_{micelles}$ is lutein content in the micellar fraction, $C_{digesta}$ is lutein content in the digesta, and C_{intial} is the initial added lutein content in the microfluidic noodle. All the determinations of lutein bioaccessibility, release and micellarization were conducted on day 1.

A schematic representation of the stability, bioaccessibility, release and micellarization of lutein are shown in Figure 2.

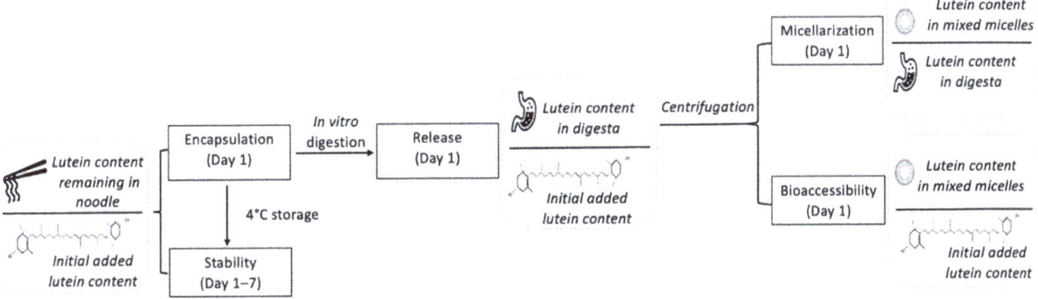

Figure 2. A schematic representation of the stability, bioaccessibility, release and micellarization of lutein.

2.9. Statistical Analysis

All determinations were performed in triplicates and all data were expressed as mean ± SE. Analysis of the variance followed by Tukey test was performed using SPSS software (SPSS Inc., US), and $p < 0.05$ was considered as statistically significant.

3. Results and Discussion

3.1. Structure Characteristics of the Microfluidic Noodle

The noodle-like structures were created with the co-flow device and the combination-flow device and are shown in Figure 3, and their microscope images viewed under 4× magnification are shown in Figure 4. For the co-flow device, two distinct layers were observed. The outer layer is calcium alginate and inner layer is the SPI and lutein fortified oil emulsion. For the combination-flow, two distinct layers of SPI and calcium alginate were observed and oil droplets were seen within the SPI layer.

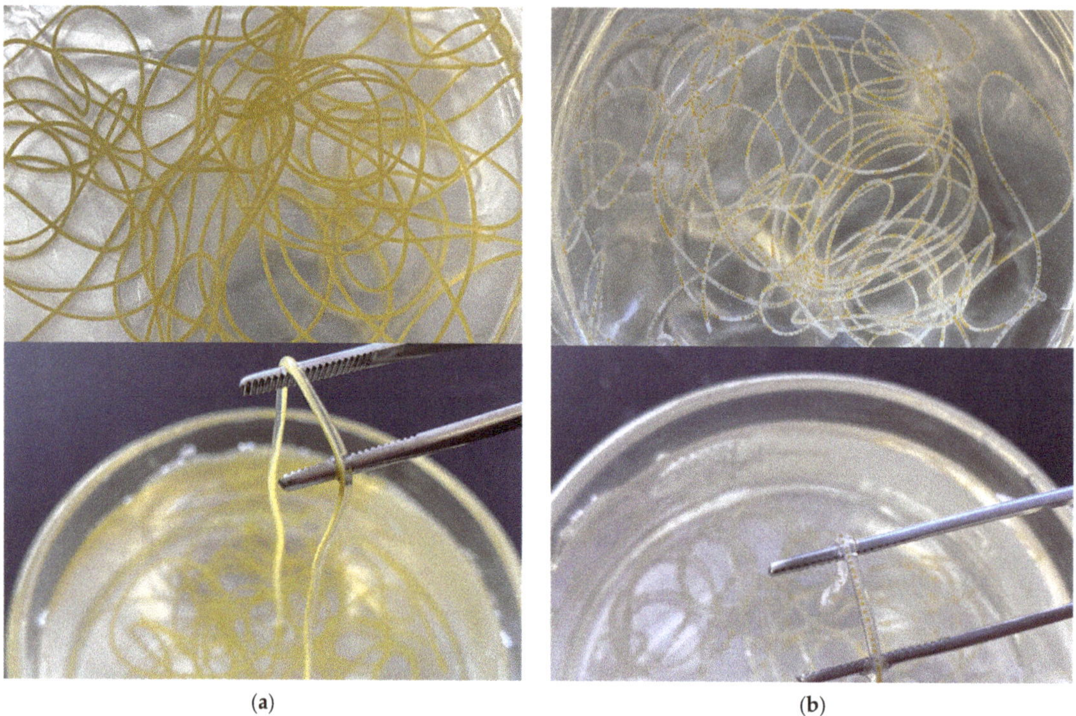

(**a**) (**b**)

Figure 3. Microfluidic noodle (**a**) created via co-flow device; (**b**) created via combination-flow.

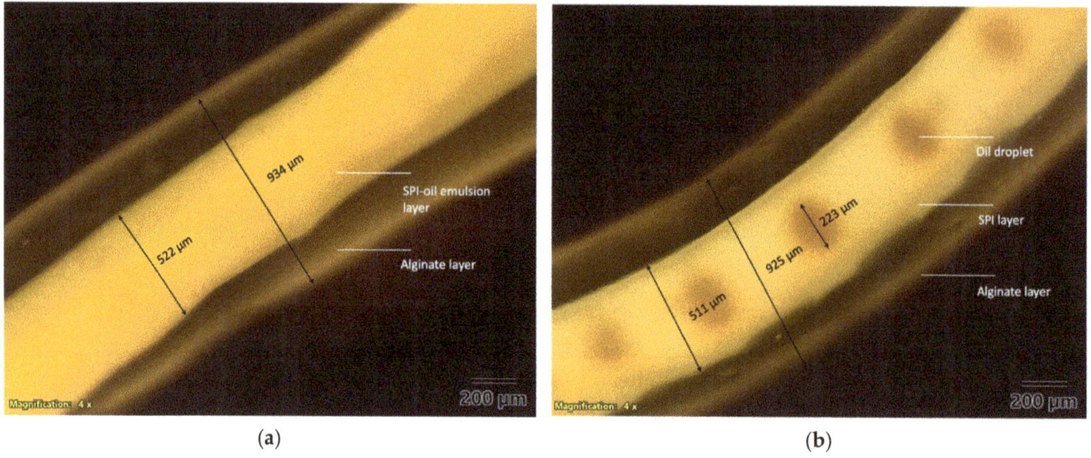

(**a**) (**b**)

Figure 4. Microscope image of microfluidic noodle (**a**) created via co-flow device; (**b**) created via combination-flow.

The gel-like outer layer was formed by ionic cross-linking of the sodium alginate layer being extruded into a hardening bath of $CaCl_2$ that promotes gelation. The core-shell structured noodle comprised a gelled alginate shell and an SPI core. The alginate shell was measured to be about 0.9 mm in diameter and the SPI core was around 0.5 mm in diameter. The oil droplet that fortified with lutein was approximately 0.2 mm in diameter and was dispersed in the SPI core for the combination-flow. Therefore, the formation of an alginate shell may act as a physical barrier and help inhibit the degradation of its encapsulated compounds [36].

3.2. Stability of Lutein

The percentage of encapsulated lutein in the microfluidic noodle with four different treatments are presented in Table 1. The highest level was with combination-flow + SO, being 92.1 ± 1.6%, followed by co-flow + SO and co-flow + OL, being 88.7 ± 3.6% and 83.3 ± 1.5% respectively. Among which, combination-flow + OL showed the lowest encapsulated lutein level, being 80.0 ± 0.6%. Compared to OL, SO resulted in a higher lutein encapsulated content in the microfluidic noodle. It is worth noting that here the encapsulated lutein percentages were the same data present as the stability of lutein on day 1. Given that the microfluidic noodle was subjected to thermal processing (80–85 °C) for six minutes on day 1, the differences in the encapsulated content were assumed mainly to be attributed to the different degradation rate of lutein under high temperatures. Therefore, our findings suggest that SO may be beneficial in protecting lutein from degradation during heating. Theoretically, the encapsulated lutein content is approximately 17.4 mg in the 100 g microfluidic noodle, and 15.0 ± 0.5 mg of lutein was detected in the extruded noodle (86.0 ± 2.7%), calculated by the average level with four different treatments. After up to seven days of storage, although the stability of lutein experienced a decreasing trend, the retention of lutein in all four types of microfluidic noodle still maintained above 60% (Figure 5). This result indicates that microfluidic encapsulation may be a good strategy to ensure the stability of bioactive compounds.

Table 1. Encapsulated lutein in noodle-like product via microfluidic techniques.

Co-Flow + OL (%)	Co-Flow + SO (%)	Combination-Flow + OL (%)	Combination-Flow + SO (%)
83.3 ± 1.5 ^{ab}	88.7 ± 3.6 ^{ab}	80.0 ± 0.6 ^{b}	92.1 ± 1.6 ^{a}

Notes: Tukey tests were carried out and significant differences ($p < 0.05$) exist among those with different letters (a, b). OL: olive oil; SO: safflower oil.

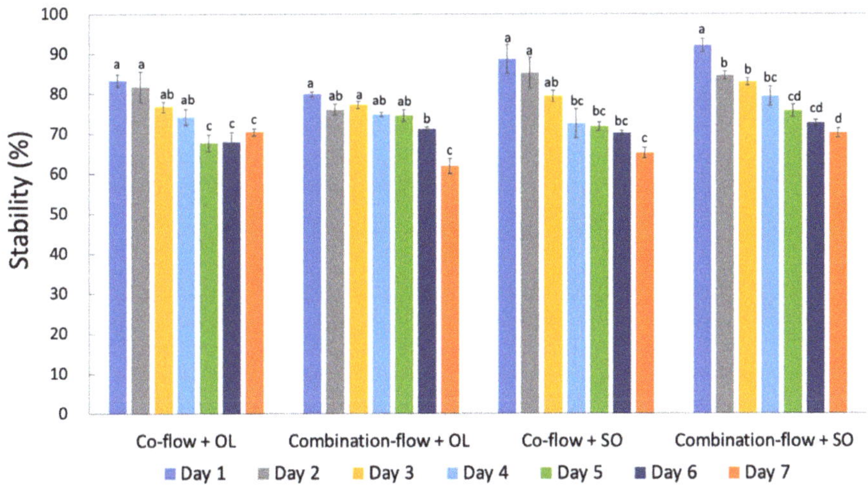

Figure 5. Storage stability of lutein in the microfluidic noodle. Notes: Tukey tests were carried out and bars with different letters (a, b, c, d) indicate significant differences among each storage day ($p < 0.05$). OL: olive oil; SO: safflower oil.

In this study, SPI was used to create oil-in-water emulsion and served as a vehicle for encapsulation of lutein. However, the interfacial layer formed by SPI is electrically charged and is particularly sensitive to pH and ionic strength [37], leading to an unstable emulsion system. For example, oil–water phase separation was observed during the storage [38], and significant droplet flocculation occurred following the simulated gastric digestion [39]. Additionally, the stability of the encapsulated compounds is hardly ensured with these protein-stabilized emulsions. This is because only a thin layer of protein molecules coat the lipid droplets. Therefore, any encapsulated compounds located close to the droplet surfaces are prone to chemical degradation, which could be promoted by compounds such as acids, transition metals, or enzyme components present in the aqueous phase [40].

To overcome this shortage, this study fabricated a core-shell structured food material, comprising a gelled alginate shell to protect the SPI core from any adverse environmental conditions and improve its stability during the long-term storage. A previous study reported that trapping β-carotene-loaded oil droplets in protein-alginate microparticles was observed with less oil-water phase separation and had a relatively constant particle size, as compared to the emulsion without the alginate coating [38]. Similarly, another study described that the microbeads maintained good physical integrity when they were coated with alginate [31]. These suggested a potential advantage of alginate in stability improvement. During seven days of storage, no free oil fortified with lutein was observed in the water medium where the noodle samples were stored. Consistent with our observation, Liu et al. reported that nearly all of the β-carotene-loaded oil droplets were fully trapped inside the alginate microparticles [38]. This indicated that alginate coating ensured a good encapsulation efficiency. Ivana et al. also demonstrated that the free oil enriched with carotenoids was only about 10%, and this value was measured as the free oil left on the surface [41]. However, it is worth noting that in our study, the amount of free lutein left on the surface was minimized by using the microfluidic technique, given that lutein was directly extruded into the core of the noodle-like sample. Meanwhile, studies showed that applying an interfacial layer of alginate outside the protein-stabilized emulsions was more effective at inhibiting carotenoids degradation [37,38]. In our study, the retention of lutein was still above 60% in all four types of microfluidic noodle after seven days of storage. Consistent with our result, a study reported that the content of β-carotene was able to maintain around 55% with an alginate coating after storage for twelve days. However, it fell down to only 0.2% with the emulsion alone [40]. Moreover, emerging evidence demonstrated the improvement in lutein stability with multilayered emulsions by covalently attaching polysaccharides to proteins [42]. A previous study fabricated the whey protein isolate-flaxseed gum-chitosan stabilized lutein emulsions by using layer-by-layer electrostatic deposition, and they observed that the retention of lutein was as high as 69% after seven days of storage at a higher temperature 70 °C [43]. This is possibly attributed to the multilayer biopolymer, which provides a physical barrier to the diffusion of oxygen, pro-oxidant, and free radicals [36], and thus inhibits the oxidation of carotenoids.

3.3. Bioaccessibility, Release and Micellarization of Lutein

The co-flow and combination-flow devices did not result in a difference in lutein bioaccessibility (co-flow: 3.1 ± 0.5%; combination-flow: 3.6 ± 0.6%). SO and OL also showed no differences in lutein bioaccessibility (SO: 3.4 ± 0.8%; OL: 3.3 ± 0.4%). These results suggest that both types of oil and device do not influence on the bioaccessibility of lutein. However, during the gastrointestinal digestion, the co-flow device showed higher lutein release (co-flow: 64.3 ± 4.5%; combination-flow: 44.3 ± 1.6%), while lower micellarization (co-flow: 4.8 ± 0.2%; combination-flow: 8.1 ± 0.7%) as compared with the combination-flow device. Moreover, compared to OL, SO resulted in less lutein released from the noodle matrix (SO: 48.7 ± 3.0%; OL: 59.9 ± 6.3%) but greater lutein formed into micelles (SO: 7.2 ± 1.0%; OL: 5.7 ± 0.5%). Specific data of the bioaccessibility, release and micellarization of the encapsulated lutein are presented in Table 2.

Table 2. Bioaccessibility, release and micellarization of lutein in microfluidic noodle following the in vitro digestion.

Lutein	Lutein in Micelles (μg)	Lutein in Digesta (μg)	Bioaccessibility (%)	Release (%)	Micellarization (%)
Co-flow + OL	29.8 ± 2.2	640.8 ± 21.3	3.4 ± 0.3 ab	73.7 ± 2.5 a	4.6 ± 0.3 c
Combination-flow + OL	27.1 ± 1.6	401.8 ± 12.4	3.1 ± 0.2 ab	46.2 ± 1.4 bc	6.8 ± 0.4 b
Co-flow + SO	23.7 ± 1.8	477.6 ± 24.1	2.7 ± 0.2 b	54.9 ± 2.8 b	5.0 ± 0.4 bc
Combination-flow + SO	34.8 ± 1.7	369.1 ± 22.2	4.0 ± 0.2 a	42.4 ± 2.6 c	9.4 ± 0.5 a
Device Type	-	-	$p = 0.051$	$p < 0.05$	$p < 0.05$
Oil Type	-	-	$p > 0.05$	$p < 0.05$	$p < 0.05$

Notes: Theoretically, 870 μg lutein was initially added in every 5 g of microfluidic noodle. Lutein content in micelles and digesta were calculated based on every 5 g of the noodle sample. The bioaccessibility, release and micellarization of lutein were all determined on day 1. Lutein bioaccessibility was determined as the fraction of lutein solubilized in the mixed micelles after passing through the simulated in vitro digestion. Lutein release was determined as the lutein content in the digesta from the initial food matrix. Lutein micellarization was determined as transfer of lutein from the digesta to the mixed micelles. Tukey tests were carried out in each column and significant differences ($p < 0.05$) exist among those with different letters (a, b, c). OL: olive oil; SO: safflower oil.

Compared to the co-flow, the combination-flow device had an approximately 31% lower lutein release rate. This is possibly because the droplet of lutein-fortified oil is tightly trapped within the SPI layer and further surrounded by an alginate layer when the noodle is created with the combination-flow device. As described above, the lutein-fortified oil droplet was shown to be encapsulated in a core-shell structure with SPI being the core and gelled alginate being the shell. In particular, the droplet of the lutein-fortified oil (0.2 mm in diameter) was tightly trapped at the center of the SPI core (0.5 mm in diameter), and a relatively thick alginate shell (0.9 mm in diameter) fully covered the SPI core. Thus, it may make lutein difficult to be released from this complex multilayer structure. However, lutein is relatively easier to be released when the noodle is created with the co-flow device. Since lutein-fortified oil was pre-mixed with SPI solution and formed an emulsion, to be released from the noodle matrix, lutein only has to penetrate the outer alginate layer. On the contrary, the co-flow device resulted in about 41% lower lutein micellarization as compared with the combination-flow device. This is mainly explained by the SPI-stabilized emulsion as an inner layer of the noodle made using the co-flow device. When subjected to in vitro digestion, especially during the gastric phase, due to the weakening of electrostatic repulsion and protein hydrolysis, protein stabilized emulsions are prone to flocculate [44,45]. This flocculation phenomenon may further restrict the lipase access to lipid droplet surfaces [38], resulting in less generation of lipid-digestion products, which are the components of micelles. Consequently, the formed mixed micelles could be insufficient to solubilize the lutein molecules, and thereby result in lower lutein micellarization with the co-flow device.

Compared to OL, SO showed an approximately 19% lower lutein release rate. Linoleic acid is a polyunsaturated omega-6 fatty acid and accounts for 55–77% fatty acids in the SO [46]. Dietary oils rich in polyunsaturated fatty acids (PUFAs) were reported to make carotenoids more susceptible to oxidation. PUFAs are prone to oxidation and the generated radicals are likely to attack carotenoids, thereby leading to a variety of carotenoid decomposition products [47,48]. Since SO is rich in PUFAs, greater lutein degradation may happen when lutein was co-digested with SO than OL. This may also explain the lower lutein content determined in the digesta which was considered as the released fraction. On the contrary, SO resulted in about 26% higher lutein micellarization as compared with OL. A previous study reported that PUFAs-rich oils promoted the formation of smaller droplet size in emulsions because of the presence of more unsaturated linkages increasing the interface flexibility [49]. Emulsions with a smaller droplet size tend to have a higher extent of lipid digestion, and consequently, a greater mixed micelles formation [50,51]. Although the detail underlying the mechanism explaining the effect of oil types on lutein micellarization cannot be elucidated from this present study, further research may be required to investigate the differences (if any) of lipid digestion extent and the formed micelles size between OL and SO.

Lutein generally shows a low bioaccessibility, which ranges from 9 to 59% determined from raw fruits and vegetables [52]; in this study, the bioaccessibility of lutein from the lutein-fortified noodle was about 3 to 4%. A previous study formulated a carotenoids-rich milk beverage and the bioaccessibility of lutein was also reported to be lower or equal to 7% [27]. Several reasons possibly attribute to this relatively low lutein bioaccessibility. Firstly, lutein degradation may happen during the noodle thermal processing, considering carotenoids are sensitive to high temperatures [53,54]. Therefore, lutein loss is more inclined to happen with the processed food rather than in raw fruits and vegetables. Secondly, lutein can remain in the lipid droplets and be trapped inside the alginate layer. Thirdly, trapped and undigested lipids can result in insufficient fatty acids and monoacylglycerols to form mixed micelles [55], and in turn hinder the micellarization of lutein. Lastly, the increased viscosity due to the presence of alginate may induce lutein aggregation and precipitation, thereby making lutein hard to solubilize in the mixed micelles [56].

As mentioned earlier, an advantage of this fabricated multilayer structure lies in the release rate of its encapsulated compounds could be engineered to fit different purposes. A previous study evaluated the stability of alginate-coated microbeads with different pH to mimic the gastrointestinal tract, and reported that the alginate-coated microbeads maintained physical integrity in acidic conditions but started to disintegrate when pH reaches 7 [31]. This indicates that the multilayer structure of the noodle may be able to delay the release of lutein in the gastric phase at a pH of about 3. This may in turn help protect lutein from degradation. When following the intestinal phase near pH 7, the structure may disintegrate and lutein will be released for absorption. Furthermore, abundant evidence has been reported that consuming 10–20 mg lutein supplementation resulted in a beneficial effect on visual health [57–59]. A meta-analysis reported that every 0.3 mg increment of lutein/zeaxanthin intake reduced the risk of nuclear cataracts, cortical cataracts and posterior subcapsular cataracts by 3%, 1% and 3%, respectively [60], and lutein intake was reported to be safe up to 20 mg per day [61]. According to USDA, a serving size of dry pasta is about two ounces (57 g); after considering the morphology and compositions of this microfluidic noodle, we recommend one cup as a portion, which is about 128 g for consumption. This present study managed to encapsulate about 15 mg lutein per 100 g of the new noodle-like food products, which indicates that this lutein-fortified noodle may be a possible alternative food item to compensate for the insufficient lutein intake in humans. However, it is also worth noting that only around 0.5 mg lutein may be successfully incorporated into the mixed micelles upon consuming 100 g of this noodle. Therefore, further study on the application of microfluidic technology in encapsulation is necessary, especially to improve the bioaccessibility of encapsulated compounds.

4. Conclusions

In conclusion, lutein is successfully encapsulated in new noodle-like food product using excipient emulsions via a novel microfluidic technology, and is relatively stable for storage. *In vitro* results suggest that types of oil and device do not affect the lutein bioaccessibility. Findings from this study may provide insights into an emulsion-based delivery system via microfluidics that is a potential for lutein encapsulation in commercial applications, such as functional foods.

Author Contributions: Conceptualization, S.A.K. and J.E.K.; methodology, Y.Y., X.Y.J.C. and J.J.L.; formal analysis, Y.Y.; investigation, Y.Y. and J.J.L.; writing—original draft preparation, Y.Y.; writing—review and editing, J.E.K.; supervision, M.H.L., S.A.K. and J.E.K.; project administration, M.H.L., S.A.K. and J.E.K.; funding acquisition, S.A.K. and J.E.K.; All authors have read and agreed to the published version of the manuscript.

Funding: We acknowledge Fuji Oil Holdings Inc. for the sponsorship of soy protein isolate powders and DSM company for the sponsorship of FloraGLO Lutein. This research was funded by the Cross-Faculty Research Grant for Interdisciplinary Research, National University of Singapore and the Industry Alignment Fund (Pre-Positioning): Programme Food Structure Engineering for Nutrition. This research was also partially supported by the Health Innovation Program: Future Foods for Health, National University Health System and National University of Singapore.

Conflicts of Interest: Authors Y.Y., X.Y.J.C. and J.E.K. declare no conflict of interest; Authors J.J.L., M.H.L. and S.A.K. are inventors of the patent (No. 10201906256W) mentioned in this manuscript used to make the microfluidic devices. The funders have no role in the design of the study; in the collection, analyses, or interpretation of data; in the writing of the manuscript, or in the decision to publish the results.

References

1. Young, A.J.; Lowe, G.L. Carotenoids—Antioxidant properties. *Antioxidants* **2018**, *7*, 28. [CrossRef]
2. Johnson, E.J. The role of carotenoids in human health. *Nutr. Clin. Care* **2002**, *5*, 56–65. [CrossRef]
3. Perera, C.O.; Yen, G.M. Functional properties of carotenoids in human health. *Int. J. Food Prop.* **2007**, *10*, 201–230. [CrossRef]
4. Di Pietro, N.; Di Tomo, P.; Pandolfi, A. Carotenoids in cardiovascular disease prevention. *JSM Atheroscler.* **2016**, *1*, 1002.
5. Ciccone, M.M.; Cortese, F.; Gesualdo, M.; Carbonara, S.; Zito, A.; Ricci, G.; De Pascalis, F.; Scicchitano, P.; Riccioni, G. Dietary intake of carotenoids and their antioxidant and anti-inflammatory effects in cardiovascular care. *Mediat. Inflamm.* **2013**, *2013*, 782137. [CrossRef] [PubMed]
6. Roberts, J.E.; Dennison, J. The photobiology of lutein and zeaxanthin in the eye. *J. Ophthalmol.* **2015**, *2015*, 687173. [CrossRef]
7. Koushan, K.; Rusovici, R.; Li, W.; Ferguson, L.R.; Chalam, K.V. The role of lutein in eye-related disease. *Nutrients* **2013**, *5*, 1823–1839. [CrossRef]
8. Junghans, A.; Sies, H.; Stahl, W. Biophysics. Macular pigments lutein and zeaxanthin as blue light filters studied in liposomes. *Arch. Biochem. Biophys.* **2001**, *391*, 160–164. [CrossRef]
9. Wu, J.; Cho, E.; Willett, W.C.; Sastry, S.M.; Schaumberg, D.A. Intakes of lutein, zeaxanthin, and other carotenoids and age-related macular degeneration during 2 decades of prospective follow-up. *JAMA Ophthalmol.* **2015**, *133*, 1415–1424. [CrossRef]
10. Brown, L.; Rimm, E.B.; Seddon, J.M.; Giovannucci, E.L.; Chasan-Taber, L.; Spiegelman, D.; Willett, W.C.; Hankinson, S.E. A prospective study of carotenoid intake and risk of cataract extraction in US men. *Am. J. Clin. Nutr.* **1999**, *70*, 517–524. [CrossRef]
11. Moeller, S.M.; Voland, R.; Tinker, L.; Blodi, B.A.; Klein, M.L.; Gehrs, K.M.; Johnson, E.J.; Snodderly, D.M.; Wallace, R.B.; Chappell, R. Associations between age-related nuclear cataract and lutein and zeaxanthin in the diet and serum in the Carotenoids in the Age-Related Eye Disease Study (CAREDS), an ancillary study of the women's health initiative. *Arch. Ophthalmol.* **2008**, *126*, 354–364. [CrossRef]
12. Sy, C.; Gleize, B.; Dangles, O.; Landrier, J.F.; Veyrat, C.C.; Borel, P. Effects of physicochemical properties of carotenoids on their bioaccessibility, intestinal cell uptake, and blood and tissue concentrations. *Mol. Nutr. Food Res.* **2012**, *56*, 1385–1397. [CrossRef] [PubMed]
13. Becerra, M.O.; Contreras, L.M.; Lo, M.H.; Díaz, J.M.; Herrera, G.C. Lutein as a functional food ingredient: Stability and bioavailability. *J. Funct. Foods* **2020**, *66*, 103771. [CrossRef]
14. Yao, K.; McClements, D.J.; Xiang, J.; Zhang, Z.; Cao, Y.; Xiao, H.; Liu, X. Improvement of carotenoid bioaccessibility from spinach by co-ingesting with excipient nanoemulsions: Impact of the oil phase composition. *Food Funct.* **2019**, *10*, 5302–5311. [CrossRef]
15. Soukoulis, C.; Bohn, T. A comprehensive overview on the micro-and nano-technological encapsulation advances for enhancing the chemical stability and bioavailability of carotenoids. *Crit. Rev. Food Sci. Nutr.* **2018**, *58*, 1–36. [CrossRef]
16. Zhao, C.; Wei, L.; Yin, B.; Liu, F.; Li, J.; Liu, X.; Wang, J.; Wang, Y. Encapsulation of lycopene within oil-in-water nanoemulsions using lactoferrin: Impact of carrier oils on physicochemical stability and bioaccessibility. *Int. J. Biol. Macromol.* **2020**, *153*, 912–920. [CrossRef]
17. Champagne, C.P.; Fustier, P. Microencapsulation for the improved delivery of bioactive compounds into foods. *Curr. Opin. Biotechnol.* **2007**, *18*, 184–190. [CrossRef]
18. Dias, M.I.; Ferreira, I.C.; Barreiro, M.F. Microencapsulation of bioactives for food applications. *Food Funct.* **2015**, *6*, 1035–1052. [CrossRef]
19. He, S.; Joseph, N.; Feng, S.; Jellicoe, M.; Raston, C.L. Application of microfluidic technology in food processing. *Food Funct.* **2020**, *11*, 5726–5737. [CrossRef]
20. Cheng, S.; Deng, J.; Zheng, W.; Jiang, X. Microfluidics for biomedical applications. In *Encyclopedia of Biomedical Engineering*; Narayan, R., Ed.; Elsevier: Oxford, UK, 2019; pp. 368–383.
21. Olanrewaju, A.; Beaugrand, M.; Yafia, M.; Juncker, D. Capillary microfluidics in microchannels: From microfluidic networks to capillaric circuits. *Lab Chip* **2018**, *18*, 2323–2347. [CrossRef]
22. McClements, D.J.; Zou, L.; Zhang, R.; Salvia-Trujillo, L.; Kumosani, T.; Xiao, H. Enhancing nutraceutical performance using excipient foods: Designing food structures and compositions to increase bioavailability. *Compr. Rev. Food Sci. Food Saf.* **2015**, *14*, 824–847. [CrossRef]
23. Mashurabad, P.C.; Palika, R.; Jyrwa, Y.W.; Bhaskarachary, K.; Pullakhandam, R. Dietary fat composition, food matrix and relative polarity modulate the micellarization and intestinal uptake of carotenoids from vegetables and fruits. *J. Food Sci. Technol.* **2017**, *54*, 333–341. [CrossRef]
24. Pullakhandam, R.; Failla, M.L. Micellarization and intestinal cell uptake of β-carotene and lutein from drumstick (*Moringa oleifera*) leaves. *J. Med. Food.* **2007**, *10*, 252–257. [CrossRef] [PubMed]
25. Schweiggert, R.M.; Mezger, D.; Schimpf, F.; Steingass, C.B.; Carle, R. Influence of chromoplast morphology on carotenoid bioaccessibility of carrot, mango, papaya, and tomato. *Food Chem.* **2012**, *135*, 2736–2742. [CrossRef]

26. Kotake-Nara, E.; Nagao, A. Effects of mixed micellar lipids on carotenoid uptake by human intestinal Caco-2 cells. *Biosci. Biotechnol. Biochem.* **2012**, *76*, 875–882. [CrossRef] [PubMed]
27. González-Casado, S.; Martín-Belloso, O.; Elez-Martínez, P.; Soliva-Fortuny, R. In vitro bioaccessibility of colored carotenoids in tomato derivatives as affected by ripeness stage and the addition of different types of oil. *J. Food Sci.* **2018**, *83*, 1404–1411. [CrossRef]
28. Teixé-Roig, J.; Oms-Oliu, G.; Ballesté-Muñoz, S.; Odriozola-Serrano, I.; Martín-Belloso, O. Improving the in vitro bioaccessibility of β-carotene using pectin added nanoemulsions. *Foods.* **2020**, *9*, 447. [CrossRef] [PubMed]
29. McClements, D.J. Enhanced delivery of lipophilic bioactives using emulsions: A review of major factors affecting vitamin, nutraceutical, and lipid bioaccessibility. *Food Funct.* **2018**, *9*, 22–41. [CrossRef]
30. McClements, D.J.; Li, Y. Structured emulsion-based delivery systems: Controlling the digestion and release of lipophilic food components. *Adv. Colloid Interface Sci.* **2010**, *159*, 213–228. [CrossRef]
31. Trif, M.; Vodnar, D.C.; Mitrea, L.; Rusu, A.V.; Socol, C.T. Design and development of oleoresins rich in carotenoids coated microbeads. *Coatings* **2019**, *9*, 235. [CrossRef]
32. Minekus, M.; Alminger, M.; Alvito, P.; Ballance, S.; Bohn, T.; Bourlieu, C.; Carriere, F.; Boutrou, R.; Corredig, M.; Dupont, D. A standardised static in vitro digestion method suitable for food—An international consensus. *Food Funct.* **2014**, *5*, 1113–1124. [CrossRef]
33. Yuan, X.; Liu, X.; McClements, D.J.; Cao, Y.; Xiao, H. Enhancement of phytochemical bioaccessibility from plant-based foods using excipient emulsions: Impact of lipid type on carotenoid solubilization from spinach. *Food Funct.* **2018**, *9*, 4352–4365. [CrossRef]
34. Granado-Lorencio, F.; López-López, I.; Herrero-Barbudo, C.; Blanco-Navarro, I.; Cofrades, S.; Pérez-Sacristán, B.; Delgado-Pando, G.; Jiménez-Colmenero, F. Lutein-enriched frankfurter-type products: Physicochemical characteristics and lutein in vitro bioaccessibility. *Food Chem.* **2010**, *120*, 741–748. [CrossRef]
35. Toh, D.W.K.; Loh, W.W.; Sutanto, C.N.; Yao, Y.; Kim, J.E. Skin carotenoids status and plasma carotenoids: Biomarkers of dietary carotenoids, fruits and vegetables for middle-aged and older Singaporean adults. *Br. J. Nutr.* **2021**, *126*, 1398–1407. [CrossRef]
36. McClements, D.J.; Decker, E.A. Lipid oxidation in oil-in-water emulsions: Impact of molecular environment on chemical reactions in heterogeneous food systems. *J. Food Sci.* **2000**, *65*, 1270–1282. [CrossRef]
37. Zhang, C.; Xu, W.; Jin, W.; Shah, B.R.; Li, Y.; Li, B. Influence of anionic alginate and cationic chitosan on physicochemical stability and carotenoids bioaccessibility of soy protein isolate-stabilized emulsions. *Food Res. Int.* **2015**, *77*, 419–425. [CrossRef]
38. Liu, W.; Wang, J.; McClements, D.J.; Zou, L. Encapsulation of β-carotene-loaded oil droplets in caseinate/alginate microparticles: Enhancement of carotenoid stability and bioaccessibility. *J. Funct. Foods* **2018**, *40*, 527–535. [CrossRef]
39. Matalanis, A.; McClements, D.J. Impact of encapsulation within hydrogel microspheres on lipid digestion: An in vitro study. *Food Biophys.* **2012**, *7*, 145–154. [CrossRef]
40. Zhang, Z.; Zhang, R.; McClements, D.J. Encapsulation of β-carotene in alginate-based hydrogel beads: Impact on physicochemical stability and bioaccessibility. *Food Hydrocoll.* **2016**, *61*, 1–10. [CrossRef]
41. Savic Gajic, I.M.; Savic, I.M.; Gajic, D.G.; Dosic, A. Ultrasound-assisted extraction of carotenoids from orange peel using olive oil and its encapsulation in ca-alginate beads. *Biomolecules* **2021**, *11*, 225. [CrossRef]
42. Steiner, B.M.; McClements, D.J.; Davidov-Pardo, G. Encapsulation systems for lutein: A review. *Trends Food Sci. Technol.* **2018**, *82*, 71–81. [CrossRef]
43. Xu, D.; Aihemaiti, Z.; Cao, Y.; Teng, C.; Li, X. Physicochemical stability, microrheological properties and microstructure of lutein emulsions stabilized by multilayer membranes consisting of whey protein isolate, flaxseed gum and chitosan. *Food Chem.* **2016**, *202*, 156–164. [CrossRef]
44. Zhang, R.; Zhang, Z.; Zhang, H.; Decker, E.A.; McClements, D.J. Influence of emulsifier type on gastrointestinal fate of oil-in-water emulsions containing anionic dietary fiber (pectin). *Food Hydrocoll.* **2015**, *45*, 175–185. [CrossRef]
45. Kenmogne-Domguia, H.B.; Meynier, A.; Viau, M.; Llamas, G.; Genot, C. Gastric conditions control both the evolution of the organization of protein-stabilized emulsions and the kinetic of lipolysis during in vitro digestion. *Food Funct.* **2012**, *3*, 1302–1309. [CrossRef]
46. Matthaus, B.; Özcan, M.; Al Juhaimi, F. Fatty acid composition and tocopherol profiles of safflower (*Carthamus tinctorius* L.) seed oils. *Nat. Prod. Res.* **2015**, *29*, 193–196. [CrossRef] [PubMed]
47. Nagao, A.; Kotake-Nara, E.; Hase, M. Effects of fats and oils on the bioaccessibility of carotenoids and vitamin E in vegetables. *Biosci. Biotechnol. Biochem.* **2013**, *77*, 1055–1060. [CrossRef] [PubMed]
48. Clark, R.M.; Yao, L.; She, L.; Furr, H.C. A comparison of lycopene and astaxanthin absorption from corn oil and olive oil emulsions. *Lipids* **2000**, *35*, 803–806. [CrossRef]
49. Verkempinck, S.; Salvia-Trujillo, L.; Moens, L.; Carrillo, C.; Van Loey, A.; Hendrickx, M.; Grauwet, T. Kinetic approach to study the relation between in vitro lipid digestion and carotenoid bioaccessibility in emulsions with different oil unsaturation degree. *J. Funct. Foods* **2018**, *41*, 135–147. [CrossRef]
50. Salvia-Trujillo, L.; Qian, C.; Martín-Belloso, O.; McClements, D.J. Influence of particle size on lipid digestion and β-carotene bioaccessibility in emulsions and nanoemulsions. *Food Chem.* **2013**, *141*, 1472–1480. [CrossRef]
51. Zhang, R.; Zhang, Z.; Zou, L.; Xiao, H.; Zhang, G.; Decker, E.A.; McClements, D.J. Enhancement of carotenoid bioaccessibility from carrots using excipient emulsions: Influence of particle size of digestible lipid droplets. *Food Funct.* **2016**, *7*, 93–103. [CrossRef]

52. Jeffery, J.L.; Turner, N.D.; King, S.R. Carotenoid bioaccessibility from nine raw carotenoid-storing fruits and vegetables using an in vitro model. *J. Sci. Food Agric.* **2012**, *92*, 2603–2610. [CrossRef]
53. Achir, N.; Randrianatoandro, V.A.; Bohuon, P.; Laffargue, A.; Avallone, S. Kinetic study of β-carotene and lutein degradation in oils during heat treatment. *Eur. J. Lipid Sci. Technol.* **2010**, *112*, 349–361. [CrossRef]
54. Ahmad, F.T.; Asenstorfer, R.E.; Soriano, I.R.; Mares, D.J. Effect of temperature on lutein esterification and lutein stability in wheat grain. *J. Cereal Sci.* **2013**, *58*, 408–413. [CrossRef]
55. Li, Q.; Li, T.; Liu, C.; Dai, T.; Zhang, R.; Zhang, Z.; McClemnets, D.J. Enhancement of carotenoid bioaccessibility from tomatoes using excipient emulsions: Influence of particle size. *Food Biophys.* **2017**, *12*, 172–185. [CrossRef]
56. Yonekura, L.; Nagao, A. Soluble fibers inhibit carotenoid micellization in vitro and uptake by Caco-2 cells. *Biosci. Biotechnol. Biochem.* **2009**, *73*, 196–199. [CrossRef] [PubMed]
57. Weigert, G.; Kaya, S.; Pemp, B.; Sacu, S.; Lasta, M.; Werkmeister, R.M.; Dragostinoff, N.; Simader, C.; Garhöfer, G.; Schmidt-Erfurth, U. Effects of lutein supplementation on macular pigment optical density and visual acuity in patients with age-related macular degeneration. *Investig. Ophthalmol. Vis. Sci.* **2011**, *52*, 8174–8178. [CrossRef]
58. Murray, I.J.; Makridaki, M.; van der Veen, R.L.; Carden, D.; Parry, N.R.; Berendschot, T.T. Lutein supplementation over a one-year period in early AMD might have a mild beneficial effect on visual acuity: The CLEAR study. *Investig. Ophthalmol. Vis. Sci.* **2013**, *54*, 1781–1788. [CrossRef] [PubMed]
59. Zhang, P.C.; Wu, C.R.; Wang, Z.L.; Wang, L.Y.; Han, Y.; Sun, S.L.; Li, Q.S.; Ma, L. Effect of lutein supplementation on visual function in nonproliferative diabetic retinopathy. *Asia Pac. J. Clin. Nutr.* **2017**, *26*, 406–411.
60. Ma, L.; Hao, Z.-X.; Liu, R.-R.; Yu, R.-B.; Shi, Q.; Pan, J.-P. A dose–response meta-analysis of dietary lutein and zeaxanthin intake in relation to risk of age-related cataract. *Graefe Arch. Clin. Exp. Ophthalmol.* **2014**, *252*, 63–70. [CrossRef]
61. Shao, A.; Hathcock, J.N. Risk assessment for the carotenoids lutein and lycopene. *Regul. Toxicol. Pharmacol.* **2006**, *45*, 289–298. [CrossRef]

Article

Development, Characterization, Stability and Bioaccessibility Improvement of 7,8-Dihydroxyflavone Loaded Zein/Sophorolipid/Polysaccharide Ternary Nanoparticles: Comparison of Sodium Alginate and Sodium Carboxymethyl Cellulose

Yufeng Chen [1], Jingchong Peng [1], Yueqi Wang [2,*], Daniel Wadhawan [3], Lijun Wu [1], Xiaojing Gao [1], Yi Sun [1] and Guobin Xia [4,*]

[1] College of Food Science and Technology, Zhejiang University of Technology, Hangzhou 310014, China; wuyycyf@163.com (Y.C.); pjchs6g@163.com (J.P.); wulj416@163.com (L.W.); gaoxiaojing66@163.com (X.G.); sunyi971219@163.com (Y.S.)

[2] Key Lab of Aquatic Product Processing, Ministry of Agriculture and Rural Affairs of the People's Republic of China, South China Sea Fisheries Research Institute, Chinese Academy of Fishery Sciences, Guangzhou 510300, China

[3] Department of Natural Science and Mathematics, University of Houston, Houston, TX 77002, USA; dkwadhawan2@uh.edu

[4] Department of Pediatrics Section of Neonatology, Texas Children's Hospital, Houston, TX 77030, USA

* Correspondence: wangyueqi@scsfri.ac.cn (Y.W.); gxxia@texaschildrens.org (G.X.)

Citation: Chen, Y.; Peng, J.; Wang, Y.; Wadhawan, D.; Wu, L.; Gao, X.; Sun, Y.; Xia, G. Development, Characterization, Stability and Bioaccessibility Improvement of 7,8-Dihydroxyflavone Loaded Zein/Sophorolipid/Polysaccharide Ternary Nanoparticles: Comparison of Sodium Alginate and Sodium Carboxymethyl Cellulose. *Foods* **2021**, *10*, 2629. https://doi.org/10.3390/foods10112629

Academic Editors: Hong Wu and Hui Zhang

Received: 25 September 2021
Accepted: 25 October 2021
Published: 29 October 2021

Abstract: In this study, two polysaccharides [sodium alginate (ALG) and sodium carboxymethyl cellulose (CMC)] were selected to establish zein/sophorolipid/ALG (ALG/S/Z) and zein/sophorolipid/ALG (CMC/S/Z) nanoparticles to encapsulate 7,8-dihydroxyflavone (7,8-DHF), respectively. The results showed that polysaccharide types significantly affected performance of ternary nanoparticles, including CMC/S/Z possessed lower polydispersity index, particle size and turbidity, but higher zeta potential, encapsulation efficiency and loading capacity compared to ALG/S/Z. Compared to zein/sophorolipid nanoparticles (S/Z), both ALG/S/Z and CMC/S/Z had better stability against low pH (pH 3~4) and high ionic strengths (150~200 mM NaCl). Hydrophobic effects, electrostatic interactions and hydrogen bonding were confirmed in ternary nanoparticles fabrication via Fourier-transform infrared spectroscopy. Circular dichroism revealed that CMC and ALG had no evident impact on secondary structure of zein in S/Z, but changed surface morphology of S/Z as observed by scanning electron microscope. Encapsulated 7,8-DHF exhibited an amorphous state in ternary nanoparticles as detected by X-ray diffraction and differential scanning calorimetry. Furthermore, compared to S/Z, ALG/S/Z, and CMC/S/Z remarkably improved the storage stability and bioaccessibility of 7,8-DHF. CMC/S/Z possessed a greater storage stability for 7,8-DHF, however, ALG/S/Z exhibited a better in vitro bioaccessibility of 7,8-DHF. This research provides a theoretical reference for zein-based delivery system application.

Keywords: Zein; 7,8-dihydroxyflavone; sophorolipid; polysaccharide; ternary nanoparticles

1. Introduction

7,8-dihydroxyflavone (7,8-DHF) (Figure S1), a naturally occurring infrequent flavone monomeric compound from *Tridax procumbens* and *Godmania aesculifolia*,, was authenticated as a high-affinity tropomyosin receptor kinase B (TrkB) agonist [1,2]. It could mimic the physiological roles of brain-derived neurotrophic factor (BDNF) and its associated downstream signaling pathway [3]. Current literature confirms that 7,8-DHF's role in attenuating many BDNF- relevant human illnesses, including Parkinson's disease, Alzheimer's disease, obesity, and depression [4–8]. However, glucuronidation, sulfation, and methylation of

7,8-DHF in the intestinal tract or liver caused its extremely low oral bioavailability [9]. In our previous study, we testified that 7,8-DHF's permeability coefficient was lower than 3×10^{-6} cm/s and had an active efflux mediated by multidrug resistance-related proteins (MRPs, especially MRP 2 outflow) and P-glycoprotein (P-gp) [10]. Thus, the low chemical instability, poor water solubility, and intestinal efflux of 7,8-DHF limit its application as a nutraceutical ingredient in functional food and beverage products.

To resolve the previously mentioned limitations and challenges, constructing an effective delivery vehicle using food-grade polymer is a novel strategy for boosting the utilization of water-insoluble functional components, such as liposomes, microemulsions, nano-emulsions, nanosuspensions, and nanoparticles [11–14]. Among these carriers, protein-based polymers, such as ovalbumin [15], casein [16], gliadin [17], and zein [18], have been utilized to create colloidal delivery systems for phenolic compounds to improve their physicochemical stability, water insolubility, and bioaccessibility. Specifically, these polymers possess several benefits such as high levels of biocompatibility, biodegradability, and label-friendliness.

As a primary storage protein found in corn, zein occupies for approximately 45~50% of total protein content, and has a relatively high percentage of hydrophobic groups. Zein's high level of hydrophobic groups means that it is soluble in 60~95% ethanol-water solutions, but insoluble in water [19]. The varying solubility of zein indicates that zein colloidal delivery systems can be established by antisolvent precipitation (ASP) [20]. Nevertheless, when exposed to a certain temperature, ionic strength, and pH range, zein nanoparticles are highly susceptible to aggregation because of the strong hydrophobic attractions among them. Previous studies have shown that food-grade biopolymers and surfactants can improve the aggregation stability of zein nanoparticles by establishing a protective layer around them. Among these applications, zein-based ternary nanoparticles exhibited more advantages compared to binary nanoparticles. Dai et al. and Wei et al. report that zein/polysaccharide/surfactant ternary nanoparticles functioned as delivery systems for coenzyme Q_{10} and curcumin, enhancing their structural stability and in vitro bioaccessibility [21,22]. Zhang et al. reported a one-step assembly method for building zein/caseinate/alginate nanoparticles, as propolis were successfully encapsulated in these carriers, their bioaccessibility was improved [23]. In addition, zein/caseinate/sodium alginate nanoparticles were also established to improve the controlled release and physicochemical properties of curcumin [24].

In our previous study, we utilized sophorolipid, a surfactant molecule, to function as fabrication stabilizer for zein binary nanoparticles. These nanoparticles displayed stability at varying salt concentrations (0~100 mM NaCl) and a wide pH range (5~9) [25]. However, at low pH conditions, their stability was low. Therefore, we aim to use biomacromolecules like polysaccharide in this study to construct zein ternary nanoparticles and solve the aforementioned defects. Sodium alginate (ALG) is a natural anionic polysaccharide with a structure of $(1 \rightarrow 4)$-b-d-mannan and $(1 \rightarrow 4)$-a-l-guluronopyranoys homopolymer sequence isolated from *Phaeophyceae*. ALG can be used as a carrier for biological food components [26]. In addition, sodium carboxymethyl cellulose (CMC) is also a suitable anionic polymer for the delivery of bioactive ingredients [27]. Due to their widespread availability and functional properties, CMC and ALG are favorable candidates for the protection of coated zein/sophorolipid binary nanoparticles when encapsulating 7,8-DHF.

The first objective of our research was to explore the effect of polysaccharides type (CMC and ALG) on the formation and performance of zein ternary nanoparticles. Subsequently, to improve in vitro bioaccessibility and storage stability of 7,8-DHF, zein ternary nanoparticles encapsulation were manufactured. Moreover, the microstructure and chemical structure of these ternary nanoparticles were analyzed by a series of characterization techniques. For precision diet intervention or functional food field, this research provides valuable insights to develop more effective delivery systems for 7,8-DHF, which will allow more 7,8-DHF to be absorbed by the human intestines, exerting greater biological effects for human heath. For functional beverage industry, stable zein-surfactant-polysaccharide delivery systems allow hydrophobic active components, such as 7,8-DHF, to be better dispersed in the water-phase beverage system to develop more functional beverage products.

2. Materials and Methods

2.1. Materials and Chemicals

Pancreatin (4 × USP specification) and zein (≥95%) and were bought from Sigma-Aldrich (Missouri, USA). 7,8-dihydroxyflavone (≥98%) was purchased from TCI Co., Ltd. (Tokyo, Japan). Sophorolipid was purchased from the Boliante Chemical Company (Xian, China). Sodium alginate (>90%) was obtained from Macklin (Shanghai, China). Sodium carboxymethyl cellulose was purchased from Aladdin (Shanghai, China). Bile salts and pepsin (activity 3000~3500 U mg^{-1}) were obtained from Sangon Biotech Co., Ltd. (Shanghai, China). Other utilized reagents and chemicals were of analytical grade.

2.2. Zein/Sophorolipid/Polysaccharide Ternary Nanoparticles Preparation

Zein/sophorolipid/polysaccharide ternary nanoparticles and zein/sophorolipid binary nanoparticles (S/Z) were fabricated based on our previous study [25]. In detail, zein and sophorolipid (mass ratio 1:1, *w*/*w*) were both dissolved in 80% ethanol/water solution to prepare the stock solution (1% sophorolipid +1% zein, *w*/*v*). For preparation of zein/sophorolipid/polysaccharide ternary nanoparticles, the stock solution was quickly added into the polysaccharide (ALG or CMC) aqueous solution (antisolvent) in a volume ratio of 1:3 (*v*/*v*) under continuous stirring at 800 rpm for 30 min. Subsequently, ethanol was removed by a rotary evaporator at appropriate temperature. The mass ratio (*w*/*w*) of zein to polysaccharide was set to 30:1, 20:1, 10:1, 5:1, 3:1, 2:1, and 1:1, respectively. The prepared CMC-based ternary nanoparticles were expressed as CMC/S/Z 30:1, CMC/S/Z 20:1, CMC/S/Z 10:1, CMC/S/Z 5:1, CMC/S/Z 3:1, CMC/S/Z 2:1, and CMC/S/Z 1:1, respectively. And the ALG-based ternary nanoparticles were denominated as ALG/S/Z 30:1, ALG/S/Z 20:1, ALG/S/Z 10:1, ALG/S/Z 5:1, ALG/S/Z 3:1, ALG/S/Z 2:1, ALG/S/Z 1:1, respectively. In addition, the S/Z was also fabricated by ASP methods; the difference was that antisolvent was deionized water without ALG or CMC. The final concentration of zein in every nanoparticle was 2.5 mg/mL. Then, the pH value of each nanoparticle dispersion was adjusted to 4.0 for dispersions stability investigation. The CMC/S/Z 5:1 and ALG/S/Z 5:1 were used for the following structure characterization. The freshly prepared samples were stored at 4 °C, and a portion of each sample was freeze-dried into lyophilized powder for the following testing and analysis.

2.3. Polydispersity Index (PDI), Particle Size, Zeta Potential and Turbidity

PDI, particle size, and zeta potential of fresh dispersions were characterized using a dynamic light scattering (DLS) instrument (Nano-ZS90 analyzer, Malvern, UK) at 25 °C. PDI and particle size was detected via light intensity at a fixed scattering angle of 90°, and the refractive index of water was set at 1.45. The zeta potential was calculated by Smoluchowski model. The turbidity of complex particles was tested at 600 nm using an ultraviolet-visible spectrophotometer at 25 °C.

2.4. Physical Stability of Ternary Nanoparticles

2.4.1. pH Influence

The influence of pH on physical stability of each nanoparticle dispersion was evaluated within a pH range of 3 to 9 adjusting by either 2 M HCl or 2 M NaOH.

2.4.2. Ionic Strength Influence

Each nanoparticle dispersion was blended with NaCl to obtain samples with 0, 25, 50, 100, 150 and 200 mM NaCl concentrations and stored for 24 h for observing physical stability.

Size changes within nanoparticle dispersions were recorded using DLS at 25 °C.

2.5. Encapsulation of 7,8-DHF into Ternary Nanoparticles

7,8-DHF encapsulation was conducted according to the methods described in Section 2.2. 7,8-DHF, sophorolipid, and zein were dissolved at a mass ratio of 1:10:10 and 1:5:5 in 80% ethanol/water solution, respectively. The mass ratio of zein to polysaccharide was 5:1 in the final reaction system. 7,8-DHF encapsulation in binary and ternary nanoparticles were denoted as DHF-S/Z, DHF-CMC/S/Z, and DHF-ALG/S/Z. Loaded complex particles were reserved at 4 °C, with other samples freeze-dried for 48 h to conduct further analysis.

2.6. Encapsulation Efficiency (EE) and Loading Capacity (LC)

The EE and LC of encapsulated 7,8-DHF were assessed by UPLC based on our previously described method [25]. Specifically, the nanoparticles were centrifuged at 10,000× g for 10 min at 4 °C using a centrifugal machine (3K15, Sigma, Osterode, Germany). The supernatant (containing loaded 7,8-DHF) was removed and diluted 5-fold with methanol. And an equal volume of the initial suspension was diluted in 5-fold methanol to obtain initial 7,8-DHF. Then, EE and LC were calculated according to the following equation:

$$\text{EE (\%)} = \text{loaded 7,8-DHF/initial 7,8-DHF} \times 100 \quad (1)$$

$$\text{LC (\%)} = \text{loaded 7,8-DHF/weight of vehicle} \times 100 \quad (2)$$

2.7. 7,8-DHF Loaded Ternary Nanoparticles Characterization

2.7.1. Fourier-Transform Infrared (FTIR) Spectroscopy

7,8-DHF and lyophilized sample under analysis were prepared by adding 99% KBr disc and scanned on an FTIR spectrometer (Avatar 370, Nicolet, Madison, WI, USA). The spectral scanning range was 500~4000 cm^{-1} at a resolution of 4 cm^{-1}. The analytical results were carried out by OMNIC software version 8.0.

2.7.2. Circular Dichroism (CD) Spectrum

Secondary structural characteristics of nanoparticle dispersions under analysis were measured using a CD spectrometer (J-1500, JASCO, Tokyo, Japan). The concentration of complex dispersions was 0.2 mg/mL corresponding to zein. The buffer in CD experiments was deionized water (pH = 7). The secondary structure scanning region was 190~260 nm with a 0.1 cm path length. The bandwidth was 1.0 nm and scanning speed was 50 nm/min. The data were evaluated by Spectra Manager™ II Software equipped with CD spectrometer.

2.7.3. Differential Scanning Calorimetry (DSC)

Thermal behavior of 7,8-DHF and freeze-dried samples were studied via DSC (Mettler Toledo, Zurich, Switzerland). 2~10 mg of samples were accurately weighed and hermetically sealed in aluminum pans. An empty crucible under the same condition was used as a reference. Scanning calorimetry was performed from a range of 25 to 200 °C in N2 atmosphere at a heating rate of 10 °C /min under 30 mL/min flow.

2.7.4. X-ray Diffraction (XRD)

The crystalline characteristic of lyophilized nanoparticles and selected 7,8-DHF were recorded on an X-ray diffractometer (Bruker D8, Karlsruhe, Germany). This diffractometer was carried at 40 kV accelerating voltage and 40 mA tube current to produce copper Kα radiation. Soller slit and divergence slit were set at 2.5° and 0.5°, respectively, the 2θ angle was ranged from 5° to 90°.

2.7.5. Transmission Electron Microscopy (TEM)

10-fold diluted fresh nanoparticle dispersions were deposited on a copper grid with formvar-carbon coating. Then, the samples were air-dried for 5 min and stained with

2% uranyl acetate. TEM (JEM-1200 EX, Tokyo, Japan) was performed for microscopic observation at 120 kV accelerating voltage.

2.7.6. Field Emission Scanning Electron Microscope (FE-SEM)

The surface morphology of polysaccharide samples and freeze-dried nanoparticles was captured by FE-SEM (GeminiSEM 300, ZEISS, Germany). Before analysis, 3–6 nm of a thick gold layer was covered on the sample surfaces. The electron microscope acceleration voltage was 15.0 kV.

2.8. Storage Stability of 7,8-DHF

7,8-DHF, DHF-S/Z, DHF-ALG/S/Z, and DHF-CMC/S/Z were performed at 5 °C for 72 h under dark and 25 °C for 15 days under light, respectively. At an appropriate point in time, the samples were acquired with 7,8-DHF presence being measured by UPLC. The storage stability was calculated based on the following equation:

$$\text{Retention rate } (\%) = \text{retained 7,8-DHF concentration}/\text{initial 7,8-DHF concentration} \times 100 \quad (3)$$

2.9. In Vitro Simulated Gastrointestinal Digestion

Based on Yuan et al.'s study via some amendments [28], briefly, 10 mL of simulated gastric fluid (SGF, 3.2 mg/mL pepsin and 2 mg/mL NaCl, pH = 2.5) and 10 mL of 7,8-DHF, DHF-S/Z, DHF-ALG/S/Z, and DHF-CMC/S/Z were mixed for incubating in a 37 °C water bath shaker for 60 min at 100 rpm. After SGF digestion, 10 mL of above-mentioned simulated gastric digestive fluids were rapidly adjusted to pH 7.4 using 2 M NaOH. Whereafter, 10 mL of simulated intestinal fluid (SIF, 4 mg/mL pancreatin, 5 mg/mL bile salts, 6.8 mg/mL K2HPO4 and 8.8 mg/mL NaCl, pH = 7.4) was added into above-mentioned simulated gastric digestive fluids and incubated for 120 min at same temperature and revolving speed. Finally, the final digestive solution was centrifuged by 20,000× *g* centrifugal force for 1 h, and the supernatant (the mixed micelle phase containing 7,8-DHF) was collected. The bioaccessibility (%) was calculated based on the following equation:

$$\text{Bioaccessibility } (\%) = \text{7,8-DHF concentration in the micelles phases}/\text{7,8-DHF concentration in the formulation} \times 100 \quad (4)$$

2.10. Statistical Analysis

Mean ± SD was presented via at least three times for all data. One-way ANOVA followed by Tukey's honestly significant difference *post hoc* tests were utilized to assess potential significant differences among groups. A $p < 0.05$ indicated significant differences between groups. Data analysis was carried out by Origin 2021 (Origin Lab Co., Northampton, MA, USA) and GraphPad Prism version 8.0 (GraphPad Sofware, San Diego, CA, USA).

3. Results and Discussion

3.1. Effect of Polysaccharides on Ternary Nanoparticles Fabrication

The effects of varying concentrations and types of polysaccharide on particle size of S/Z were illustrated in Figure 1. S/Z was approximate 82.81 nm particle size. The particle size of the ternary nanoparticles was differentially altered from that of S/Z after introduction with CMC and ALG into the binary system. After supplementation with CMC, the particle size of CMC/S/Z stepwise grew as the concentration of CMC increased (Figure 1A). Moreover, the incorporation of ALG resulted in a significant increase in particle size ($p < 0.05$). And in comparison to CMC/S/Z, ALG/S/Z showed a greater particle size at the identical polysaccharide level (Figure 1B). The above phenomenon might be attributed to the distinctive molecular structure of two studied polysaccharides and interaction force among molecules. The changes in particle size of zein/sophorolipid/polysaccharide ternary nanoparticles are possibly a result of sufficient anionic CMC and ALG being absorbed on the surface of S/Z through the electrostatic attractions. The changes in PDI value of CMC/S/Z and ALG/S/Z was analogous to that of particle size. As the concentration of CMC and ALG increased, the PDI value of CMC/S/Z and ALG/S/Z gradually increased as well. However, ALG/S/Z showed a higher PDI value at the same polysaccharide level.

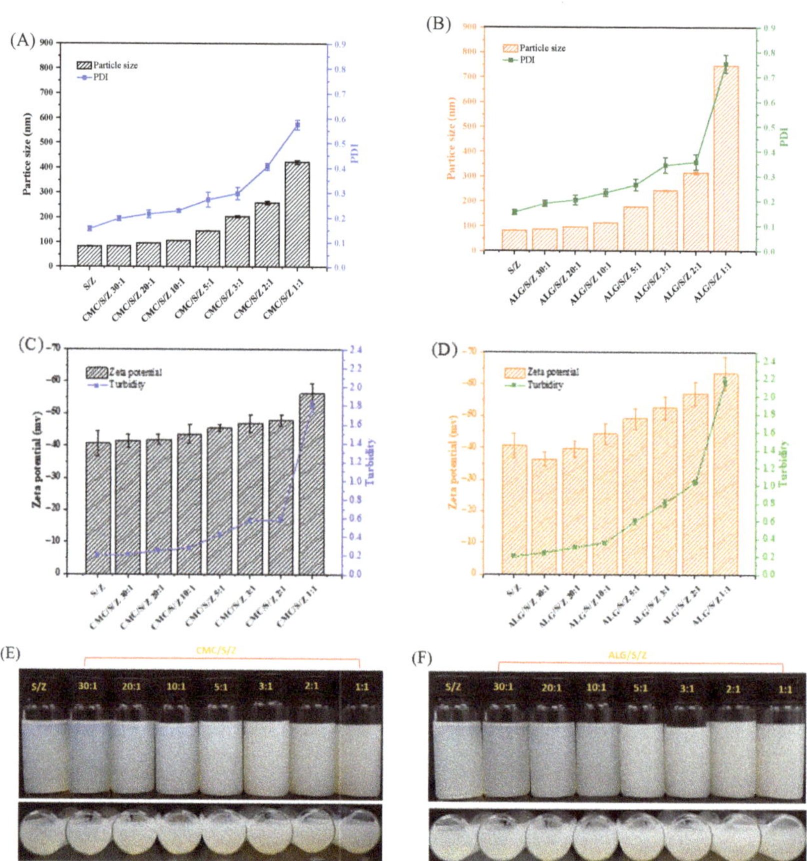

Figure 1. Particle size, PDI, zeta potential and turbidity of CMC/S/Z and ALG/S/Z with different zein to polysaccharide mass ratios. Particle size and PDI (**A**,**B**). Zeta-potential and turbidity (**C**,**D**). The photograph of each group appearance (**E**,**F**).

The zeta-potential of S/Z was −40.63 mV. When CMC and S/Z were combined, their charge values became more negative as the fraction of present CMC increased (Figure 1C). On the other hand, the negative charges of ALG/S/Z initially declined and grew whereafter when the level of ALG was increased (Figure 1D). When the mass ratio of polysaccharide to zein was above 1:10, the negative charges of ALG/S/Z were higher than that of CMC/S/Z. This difference is potentially caused by the different molecular structures of the two polysaccharides. The zeta-potential of zein/sophorolipid/polysaccharide ternary nanoparticles had been dominated by negatively charged CMC and ALG. It is reported that electrostatic attraction played a critical role in the formation of protein-surfactant-polysaccharide complexes, for instance, zein-rhamnolipid-propylene glycol alginate [29], and zein-lecithin-propylene glycol alginate [30]. Meanwhile, the uniform trend of turbidity in ternary nanoparticles was consistent with that of PDI value and particle size, as the increasing trend of turbidity was primarily caused by micro-aggregation of complexes [31].

3.2. CMC and ALG Prevents S/Z Ternary Nanoparticles Precipitation at pH = 4

Our previous work confirms that S/Z exhibits weak stability in low pH conditions. To overcome this deficiency, we explore the influence of polysaccharide type (CMC and ALG) and concentration on S/Z stability at pH = 4. As shown in (Figure 2A), S/Z ternary nanoparticles were extremely unstable and aggregated together as expected. This phenomenon could be elucidated by the lack of appropriate electrostatic repulsion, which is required to conquer attractive interactions (e.g., van der Waals) among nanoparticles. Furthermore, at low pH values, the pKa of hydrophilic sugar residues present on sophorolipid molecules were reported to promote this instability [32]. After CMC was added, the PDI value, particle size, and turbidity of CMC/S/Z initially declined and subsequently jumped until a mass of zein to CMC was 5:1, the CMC/S/Z were possessed the lowest values for particle size, PDF, and turbidity (particle size ≈ 341.50 nm, PDI ≈ 0.335 and turbidity ≈ 2.257) (Figure 2A,C). Moreover, the incorporation of ALG similarly decreased and then increased PDI value, particle size and turbidity of ALG/S/Z. The particle size for ALG/S/Z was smallest at the 10:1 (353.62 nm), while the PDI value (0.335) and turbidity (2.509) were lowest at 5:1 mass ratio of zein to ALG (Figure 2B,D). As seen from photograph (Figure 2E,F), no floccules were observed at the bottom of the container at varying zein to polysaccharide mass ratios (10:1, 5:1, 3:1, and 2:1). A reason for lacking floccules at these mass ratios might be a result of appropriate CMC and ALG participation, which potentially increased the steric or electrostatic repulsions for S/Z against aggregation. Nevertheless, at a zein to polysaccharide mass of 1:1, there were sediments in both CMC/S/Z and ALG/S/Z. The underlying mechanism for these observations is bridging flocculation, i.e., the ability of a single CMC or ALG molecule to adsorb to the surfaces of two or more S/Z. Therefore, CMC or ALG adsorption results in the formation of clusters [33]. Furthermore, with the CMC and ALG concentrations raised, the charge values of CMC/S/Z and ALG/S/Z became increasingly negative. The intensification of these charges revalidates the idea that CMC/S/Z and ALG/S/Z experience sufficient electrostatic repulsion to avoid precipitation at varying mass ratios of 10:1~2:1. Based on these results, we subsequently selected CMC/S/Z and ALG/S/Z at a mass ratio of zein to polysaccharide at 5:1 as the optimum ratio to study in the following research.

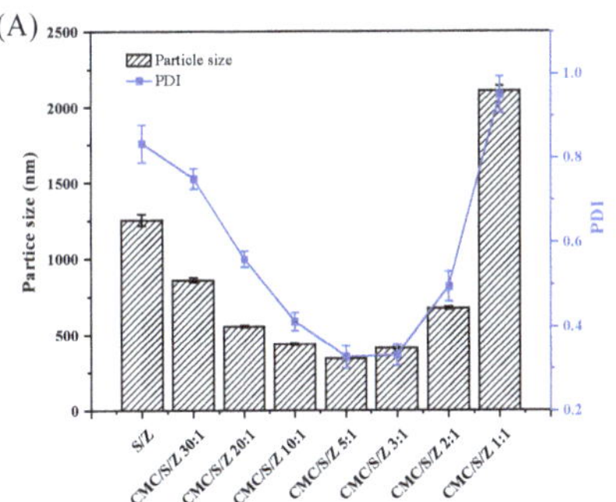

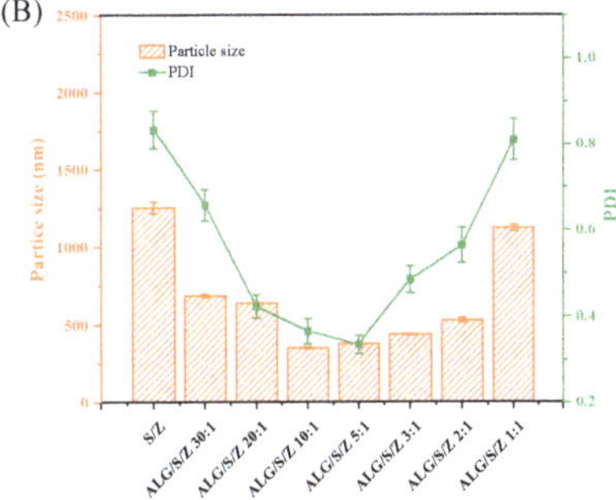

Figure 2. *Cont.*

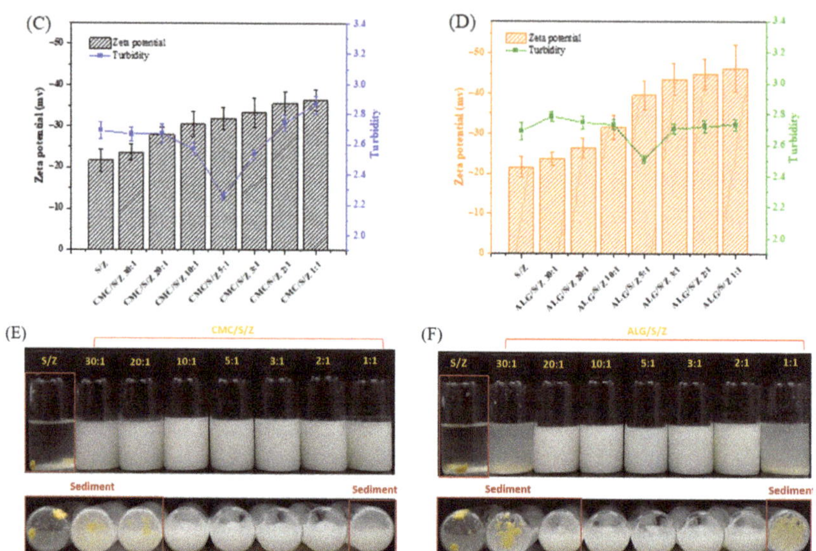

Figure 2. Particle size, PDI, zeta potential and turbidity of CMC/S/Z and ALG/S/Z with different zein to polysaccharide mass ratios at pH = 4. Particle size and PDI (**A**,**B**). Zeta-potential and turbidity (**C**,**D**). The photograph of each group appearance (**E**,**F**).

3.3. Physical-Chemical Stability Study on Ternary Nanoparticles

The optimum delivery system for food nutraceuticals must accommodate flexible pH and ionic environments during beverages processing, storage, and gut passage. Therefore, pH and salt stability testing of delivery systems is necessary to evaluate colloidal particles' functionality.

3.3.1. Effect of pH

As seen from Figure 3A, S/Z exhibited extensive aggregation at pH 3.0 and 4.0, exhibiting an increased particle size. This increase in size could be due to the fact that weakened electrostatic repulsion was weakened among binary nanoparticles. However, after adding CMC and ALG, compared to S/Z, the stability of both nanoparticles was enhanced noteworthily within pH 3.0 to 4.0. Most notably, from pH 3 to 9, both CMC/S/Z and ALG/S/Z had excellent stability. It is reported that certain zein-surfactant or zein-polysaccharide binary nanoparticles, such as zein-rhamnolipid, zein-alginate, and zein-chondroitin sulfate, experienced a significant size increase at low pH levels [34–36]. Unexpectedly, zein-surfactant-polysaccharide ternary nanoparticles exhibited superior pH stability. There are some yellow floccules in the bottom of the container of S/Z at pH 3.0 and 4.0 but no floccules in CMC/S/Z and ALG/S/Z (Figure S4).

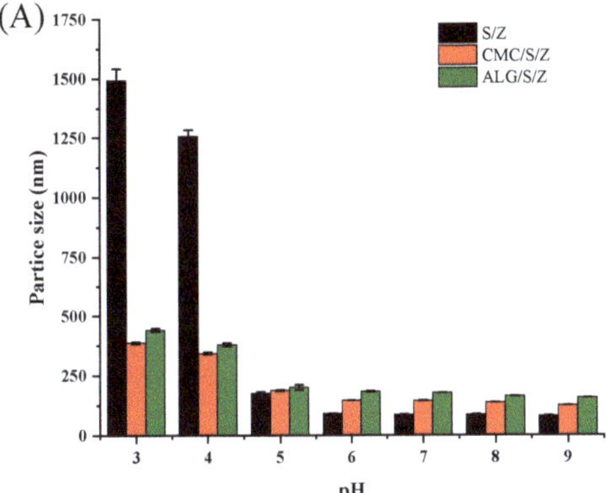

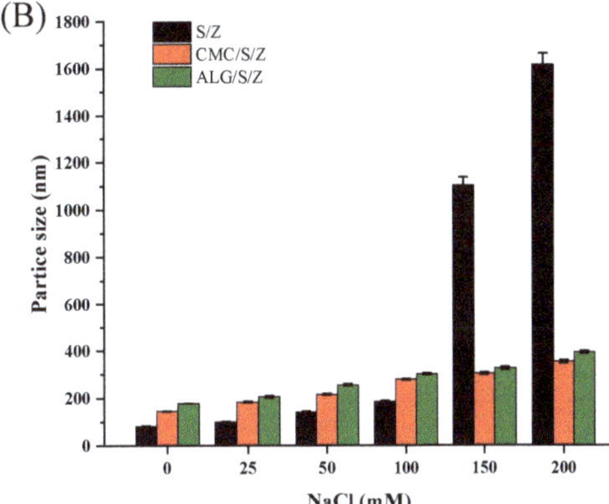

Figure 3. Particle size of colloidal particles at different pH conditions (**A**) and different NaCl concentrations (**B**).

3.3.2. Effect of Ionic Strengths

Particle size of S/Z was relatively low by 0 to 100 mM NaCl concentrations (Figure 3B). However, when increased to higher ionic strengths at 150~200 mM NaCl concentrations, precipitates were produced along with increasing particle size. This forming aggregation patterns for nanoparticles at high ionic strengths could be explained by increasing counterions (such as Cl^- and Na^+), depressed electrostatic repulsion via electrostatic screening and nanoparticle charge neutralization [21]. However, in the existence of CMC and ALG, the particle size of CMC/S/Z and ALG/S/Z was increased with the NaCl concentration increased. And adding CMC and ALG reduced the sensitivity of S/Z to aggregation under high ionic strength condition. There are two possible reasons for the improved physical stability of CMC/S/Z and ALG/S/Z. Firstly, CMC and ALG prevented counterion neutralization and strong electrostatic repulsions, which could potentially restrain complex particles from coalescence via enhancing repulsive forces among particles. Secondly, an increase of the steric repulsion among particles could have occurred as polysaccharide molecules were attached to S/Z. Some floccules were also observed at the bottom of the tube for S/Z at 150 and 200 mM NaCl (Figure S5). However, there were no floccules at the bottom of the tube for CMC/S/Z and ALG/S/Z, in which stable colloidal systems were both developed.

3.4. Encapsulation of 7,8-DHF

The influence of polysaccharide types (CMC and ALG) and 7,8-DHF concentration on PDI value, particle size, EE and LC value in different delivery systems were summarized in Table 1. When the mass ratio of 7,8-DHF to zein was 1:5, the EE and LC of DHF-S/Z were 82.42 and 7.49%, respectively. However, compared to DHF-S/Z, DHF-CMC/S/Z and DHF-ALG/S/Z exhibited better EE value, especially DHF-CMC/S/Z nanoparticles (88.63%) ($p < 0.05$). These findings suggested that polysaccharide addition improved the embedding ability of S/Z via non-covalent interactions. There are two hydroxyl groups in the 7,8-DHF molecule that can interact with the hydroxyl groups of polysaccharides and the tyrosyl of zein. These interactions can form hydrogen bonds, possibly resulting in an increase of EE. These phenomena indicated that polysaccharides and S/Z showed a synergistic effect on the EE of 7,8-DHF. When mass ratio of 7,8-DHF to zein was 1:10, three colloidal delivery systems possessed high EE for 7,8-DHF (above 98.21%). In contrast to unloaded colloidal delivery systems, the particle size of loaded colloidal delivery systems was growing when increasing the mass ratio of 7,8-DHF to zein. In general, 7,8-DHF, as a hydrophobic compound, is expected to be located within hydrophobic region of zein, ultimately possessing an impact on the size and stability of particles. 7,8-DHF encapsulation might impair interactions among hydrophobic groups of zein and increase particle size.

Additionally, the encapsulation of 7,8-DHF caused an increase of PDI value in contrast to unloaded complexes. A uniform result was reported by Sun et al. [37]. The formation of free 7,8-DHF aggregates might help to explain the above observed phenomenon.

Table 1. EE, LC, particle size and PDI of 7,8-DHF in different colloidal systems.

Zein: 7,8-DHF (*w*/*w*)	Colloidal Systems	EE (%)	LC (%)	Particle Size (nm)	PDI
Without 7,8-DHF	S/Z	-		82.81 ± 0.61 [a]	0.155 ± 0.010 [a]
	CMC/S/Z	-		145.6 ± 0.75 [c]	0.228 ± 0.006 [b]
	ALG/S/Z	-		178.2 ± 0.35 [d]	0.266 ± 0.024 [c]
5:1	S/Z	82.42 ± 3.72 [a]	7.49 ± 0.22 [a]	114.7 ± 3.01 [b]	0.271 ± 0.016 [c]
	CMC/S/Z	88.63 ± 3.01 [b]	7.38 ± 0.13 [a]	177.4 ± 3.04 [a]	0.363 ± 0.023 [de]
	ALG/S/Z	84.15 ± 2.63 [a]	7.01 ± 0.16 [a]	214.3 ± 3.21 [e]	0.394 ± 0.022 [e]
10:1	S/Z	98.21 ± 1.31 [c]	4.68 ± 0.40 [b]	106.9 ± 1.11 [b]	0.201 ± 0.021 [b]
	CMC/S/Z	99.51 ± 0.24 [c]	4.33 ± 0.12 [b]	168.4 ± 3.62 [dc]	0.334 ± 0.014 [d]
	ALG/S/Z	98.71 ± 1.12 [c]	4.29 ± 0.35 [b]	200.1 ± 2.01 [e]	0.352 ± 0.017 [de]

Values are the means ± SD (n = 3). [a–e]: Different letters in the same column indicate significant differences ($p < 0.05$) based on one-way ANOVA analysis followed by Tukey's honest significant difference post hoc tests.

3.5. Characterization of Loaded Nanoparticles

3.5.1. FTIR

FTIR is a versatile tool for monitoring changes within the functional groups of biopolymers and evaluating intermolecular interactions between components and particles. As shown in Figure 4A, the spectrum of CMC and ALG included diversiform representative carbohydrate peaks. The broad 3446 and 3441 cm^{-1} peaks represented hydroxyl groups (O-H) stretching, and the sharp 1604 and 1615 cm^{-1} peaks are associated with carboxyl (-COO-) symmetrical stretching vibration [38]. The featured peaks of CMC and ALG spectra appear at 1418 and 1417 cm^{-1}, corresponding to rhamnogalacturonan moiety [39]. Multiple simultaneous vibration peaks in CMC and ALG at 900~1350 cm^{-1} were accredited to characteristic peaks of polysaccharides [24]. In our previous study, we confirmed that there are electrostatic attractions, hydrophobic interactions, and hydrogen bonding in S/Z. The FTIR spectrum of S/Z nanoparticles showed typical characteristic peaks at 3368, 1657, and 1545 cm^{-1}, respectively. When CMC and ALG were added, the peaks of O-H stretching vibration (3100~3500 cm^{-1} peaks) shifted from 3368 (S/Z) to 3382 (CMC/S/Z) and 3402 cm^{-1} (ALG/S/Z) [40]. This characteristic peak migration implied that there was a strong hydrogen bond between -OH groups in polysaccharides and amide group of glutamine in zein [41]. According to Liu et al.'s report [42], the 1657 cm^{-1} peak of zein at was the C=O stretching (amide I). 1545 cm^{-1} peak was primarily associated with bending of N-H coupled with the stretching of C-N (amide II). With CMC and ALG incorporation, the amide I and amide II characteristic peaks of ternary nanoparticles were switched to (1650 and 1651 cm^{-1}) and (1544 and 1544 cm^{-1}), respectively. These results revealed electrostatic attractions were related to the establishment process of CMC/S/Z and ALG/S/Z. Based on these results, we confirmed that ALG/S/Z possessed stronger hydrogen bonding and electrostatic attraction than CMC/S/Z. As seen from Figure 4B, the peaks at 3114, 1626, 1575, 1405, 1195, and 1071 cm^{-1} were the typical peaks of 7,8-DHF, which have been confirmed in our previous study [25,43]. Expectedly, these characteristic peaks of 7,8-DHF were vanished in both binary and ternary nanoparticles, indicating that DHF-S/Z, DHF-CMC/S/Z, and DHF-ALG/S/Z samples successfully encapsulated for 7,8-DHF.

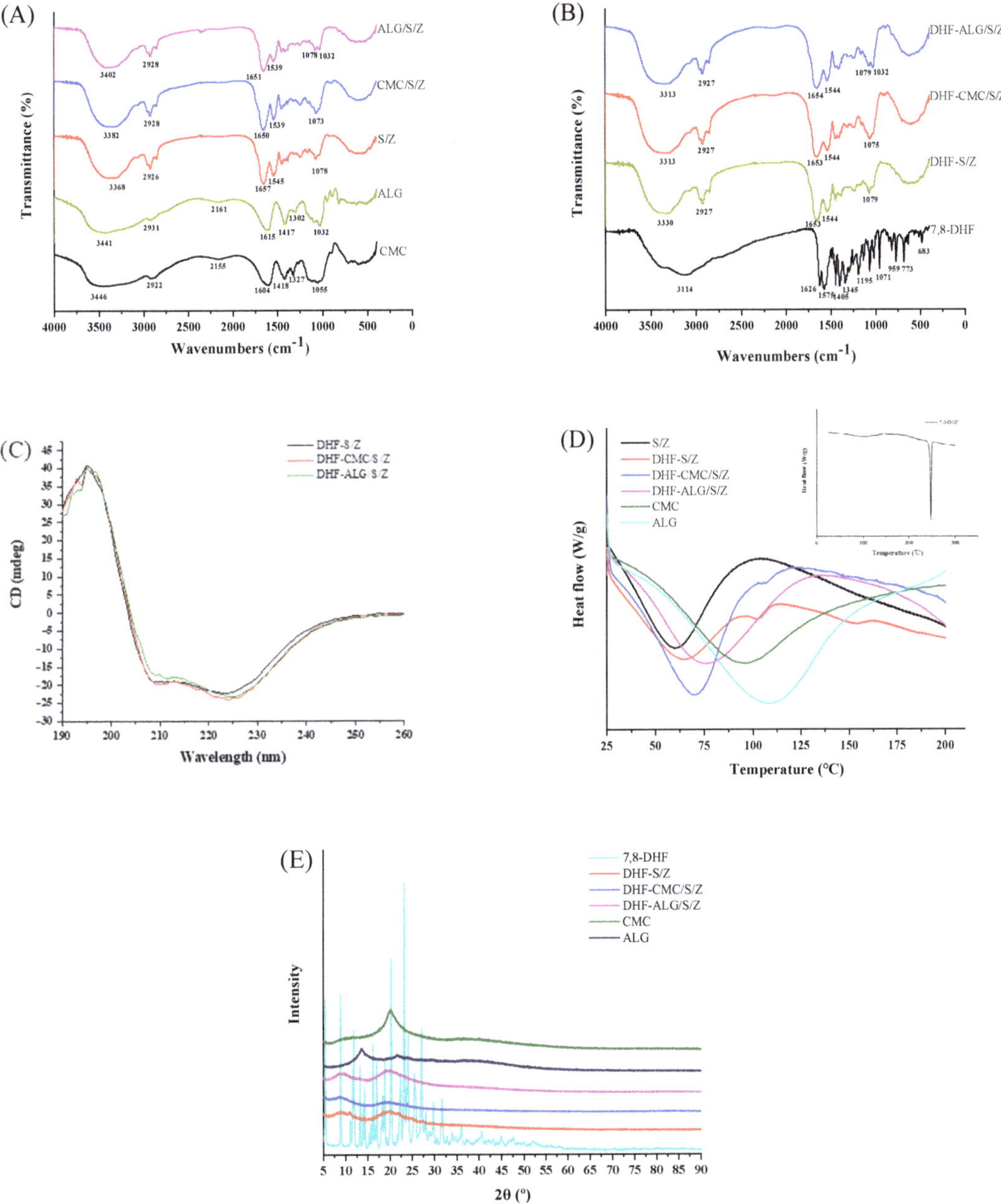

Figure 4. FTIR spectra of CMC, ALG and bare complex nanoparticles (**A**), free 7,8-DHF and 7,8-DHF loaded in each complex nanoparticles (**B**), CD spectra of zein in DHF-Z, DHF-Z/S-CMC, and DHF-ALG/S/Z (**C**), DSC analysis of free 7,8-DHF, CMC, ALG, S/Z, and 7,8-DHF loaded each complex nanoparticles (**D**), XRD spectra of free 7,8-DHF, polysaccharides and loaded each complex nanoparticles (**E**).

3.5.2. CD Spectrum

In this study, CD spectrum analysis was applied to measure conformational changes (secondary structure) of zein in complexation (260 nm~190 nm). As seen from (Figure 4C), two peaks at 209 and 223 nm with a zero-crossing were around 203 nm (typical secondary structure) in the zein spectrum [44]. As shown in (Table 2), the α-helix, β-sheet and β-turn content of DHF-S/Z changed from 25.3 to 24.7 and 24.5%, 25.8 to 26.1 and 26.6% and 28.3 to 29.1 and 27.9%, respectively after adding CMC and ALG. These results suggested that the addition of CMC and ALG had no apparent impact on zein's secondary structure in DHF-S/Z. The behavior of CMC and ALG could possibly be the result of sophorolipid adsorption on the zein surface. This adsorption could have possibly resulted in insufficient contact between the polysaccharides and zein, which was difficult to further change the secondary structure of zein. In general, the amide I band shifts in FTIR spectra could also reflect secondary structure changes of the protein, including the band of α-helix, β-sheet, and β-turn [45]. In the FTIR spectra of our study (Figure 4B), the band of amide I did not show significant change among DHF-S/Z, DHF-CMC/S/Z and DHF-ALG/S/Z. This lack of significant change further confirmed that CMC and ALG did not impact the secondary structure of zein.

Table 2. Secondary structure of zein in DHF-S/Z, DHF-CMC/S/Z and DHF-ALG/S/Z.

Sample	Content (%)			
	α-Helix	β-Sheet	β-Turns	Unordered
DHF-S/Z	25.3 ± 0.23	25.8 ± 0.21	20.6 ± 0.19	28.3 ± 0.26
DHF-CMC/S/Z	24.7 ± 0.18	26.1 ± 0.17	20.1 ± 0.21	29.1 ± 0.20
DHF-ALG/S/Z	24.5 ± 0.16	26.6 ± 0.23	21.0 ± 0.26	27.9 ± 0.24

3.5.3. DSC

Thermal properties of individual components and composite nanoparticles were studied via DSC. As shown in (Figure 4D), the embedded thermograms of 7,8-DHF displayed a narrow and sharp peak at 246.24 °C. This temperature peak was probably caused by the melting of 7,8-DHF crystals [46]. Furthermore, the representative endothermal peak of CMC and ALG was at approximately 96.33 and 113.33°C, respectively. These findings confirmed that ALG had a higher thermostability than CMC because of its specific carbohydrate structure. However, the endothermic peak of S/Z was at approximately 60.33 °C, showing a low thermostability. After 7,8-DHF was encapsulated, no endothermic peaks of 7,8-DHF were found in DHF-S/Z, DHF-CMC/S/Z, and DHF-ALG/S/Z. The lack of endothermic peaks verified that 7,8-DHF was defined as an amorphous form rather than a crystalline form. Similar literature has reported in recent study on curcumin [47] and hyperoside [48]. In addition, the endothermic peak of DHF-S/Z nanoparticles was increased from 60.33 °C to 64.66 °C compared to S/Z. This increase in melting temperatures could be attributed to intermolecular interactions among 7,8-DHF, zein and sophorolipids [29]. Most importantly, after adding CMC and ALG, the endothermic peak of DHF-CMC/S/Z and DHF-ALG/S/Z rose to 70.01 and 76.33 °C in comparison to DHF-S/Z, respectively. The higher endothermic peak of DHF-CMC/S/Z and DHF-ALG/S/Z manifested that they possessed better thermal stability than DHF-S/Z. This new peak might be the result of CMC and ALG interaction enhancements for hydrophobic, electrostatic, or hydrogen bond interactions among different components in nanoparticles, which ultimately leads to a higher endothermic peak temperature.

3.5.4. XRD

X-ray diffraction ranging from 5° to 90° at 2θ values was applied to examine 7,8-DHF's physical state in different nanoparticles. As presented in (Figure 4E), 7,8-DHF was highly crystalline with multiple sharp diffraction peaks at the range of 5~40°. On the one hand, when the diffraction angles of ALG in the 2θ range were 13.2° and 21.8°, two flat peaks without sharp diffraction maximum appeared in the XRD spectrum. On the other hand, CMC only had a flat peak at a diffraction angle of 20.2°, while no sharp diffraction peaks were emerged. This XRD hump indicated that both pure polysaccharides were totally in an amorphous form [21,32]. However, no obvious sharp diffraction peaks about crystalline form for 7,8-DHF were found in both binary and ternary nanoparticles. Such behavior indicated 7,8-DHF was completely loaded into nanoparticles with an amorphous state. Interestingly, in comparison to DHF-S/Z, the hump for DHF-ALG/S/Z was distinctly increased, while the hump of DHF-CMC/S/Z was only slightly increased as the diffraction angle was at 19.6°. Thus, the polysaccharide types applied to construct nanocomposites can significantly affect XRD patterns. Moreover, no characteristic peaks for these two polysaccharides were observed. Collectively, due to changes in the interactions (hydrophobic, hydrogen bonding, and electrostatic) among zein, polysaccharide, and surfactant molecule, distinguishable behaviors of 7,8-DHF in DHF-S/Z, DHF-CMC/S/Z, and DHF-ALG/S/Z were observed.

3.6. Micromorphology

The microstructural features of loaded composite nanoparticles were analyzed via TEM. As observed in (Figure 5A), the diameter of DHF-S/Z was roughly 100 nm, this finding was in agreement with the results of dynamic light scattering. After adding CMC and ALG, DHF-CMC/S/Z and DHF-ALG/S/Z showed a similar spherical structure that was comparable to the DHF-S/Z. However, the diameter of DHF-CMC/S/Z and DHF-ALG/S/Z was larger than that of DHF-S/Z, exhibiting a particle size of 100~200 nm (Figure 5B,C). These findings demonstrate that CMC and ALG were absorbed on the surface of DHF-S/Z. Additionally, the diameter of DHF-CMC/S/Z was larger than that of DHF-ALG/S/Z, which was consistent with the results of DLS.

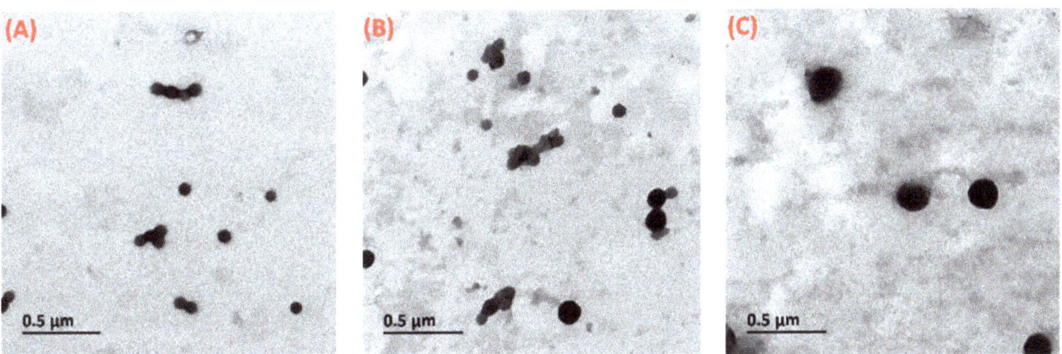

Figure 5. TEM images of DHF-S/Z (**A**), DHF-CMC/S/Z (**B**) and DHF-ALG/S/Z (**C**), 30,000× magnification times.

Furthermore, FE-SEM was applied to further observe the differences in surface microscopic morphology among individual composite nanoparticles (Figure 6). As seen from (Figure 6A), the surface morphology of DHF-S/Z expressed an irregular and rough shape consisting of many interlinked independent complexes. This morphological change was possibly attributed to sophorolipid adsorption on the surface of zein particles. The microphotograph of CMC and ALG depicted a shape similar to that of a silk ribbon (Figure 6D,E). When CMC and ALG were incorporated into DHF-S/Z, the micromorphology of DHF-CMC/S/Z and DHF-ALG/S/Z were spherical with uniform size and smooth surface (Figure 6B,C). The particle sizes of DHF-CMC/S/Z and DHF-ALG/S/Z were larger than that of DHF-S/Z. The polysaccharide coating on the outer layer of DHF-S/Z nanoparticles via electrostatic attraction possibly caused this size difference [49]. The above results were consistent with DLS measurements.

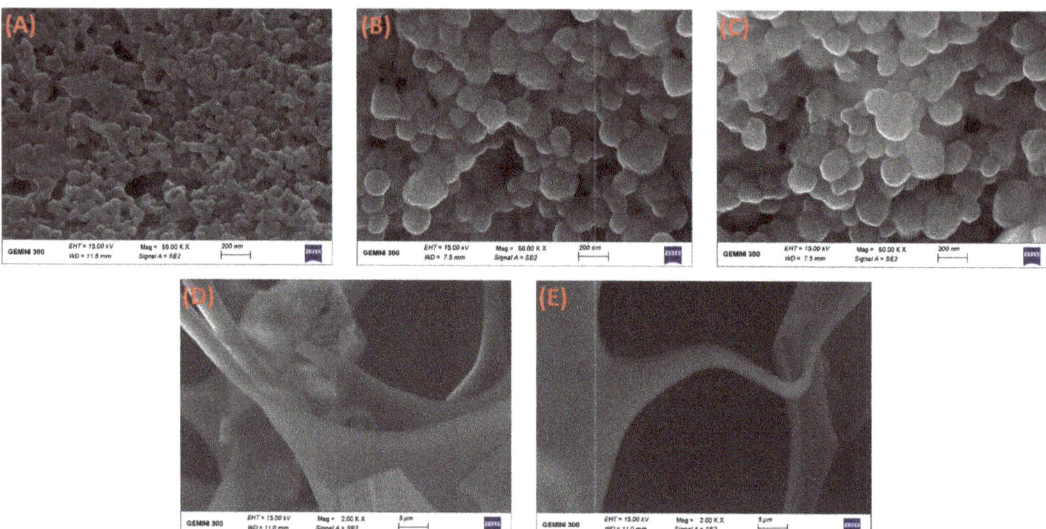

Figure 6. FE-SEM images of DHF-S/Z (**A**), DHF-CMC/S/Z (**B**), DHF-ALG/S/Z (**C**) (50,000× magnification times), individual CMC (**D**) and ALG (**E**) (2000× magnification times).

3.7. A Graphic Illustration for the Formation and Stability Mechanism of Nanoparticles

Diverse technologies and measurements including EE, PDI, particle size, zeta potential, turbidity, CD, TEM, FE-SEM, DSC, XRD, and FTIR were used to make clear the formation and stability mechanism of zein/sophorolipid/polysaccharide ternary delivery system (Figure 7). After zein and sophorolipid were rapidly added into polysaccharide (polysaccharide: zein mass ratio $\leq$ 1:2), sufficient CMC or ALG acted as a shielding effect (electrostatic repulsion) and steric hindrance stabilizer was coated onto the surface of S/Z particles to prevent their sedimentation at low pH 3~4 range condition, showing a homogeneous PDI, turbidity and size based on DLS and UV. Certain internal drives (electrostatic interaction, hydrogen bonding and hydrophobic effect) participated in the formation of DHF-CMC/S/Z and DHF-ALG/S/Z, and CMC and ALG had no significant impact on secondary structure of zein in S/Z according to CD, FTIR, and FE-SEM. Upon employing EE, XRD, DSC, TEM analysis, 7,8-DHF was shown to be successfully encapsulated in CMC/S/Z and ALG/S/Z with relatively uniform sphericity, displaying a good entrapment efficiency.

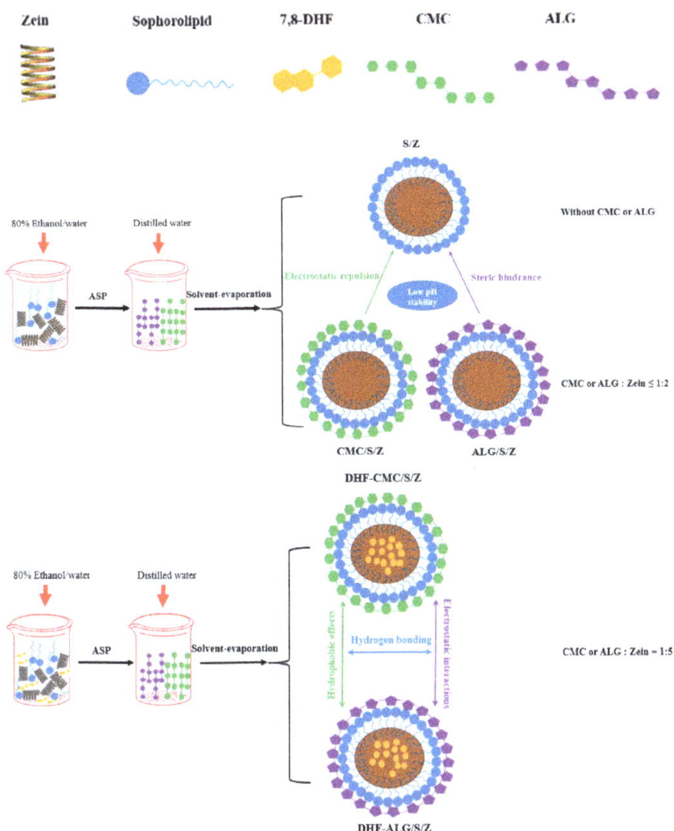

Figure 7. An illustration of the formation and stability mechanism of DHF-CMC/S/Z and DHF-ALG/S/Z.

3.8. Storage Stability of 7,8-DHF

During storage, preventing food neutraceuticals from heat or light exposure is challenging, but critically necessary for mitigating degradation. To meet this application end, short and long term storage were investigated under varying environments for 7,8-DHF, loaded binary and ternary nanoparticles. As shown in (Figure 8A), free 7,8-DHF was mostly degraded at 25 °C with light exposure for 15 days post-storage. Encapsulation of the 7,8-DHF in S/Z nanoparticles strengthened storage stability of 7,8-DHF (26.26 %). The addition of CMC and ALG further enhanced the stability of encapsulated 7,8-DHF, in particular DHF-CMC/S/Z (58.75%). At 50 °C under dark conditions (Figure 8B), a similar effect was observed. The active groups of 7,8-DHF were possibly protected within the hydrophobic lumen of DHF-S/Z, DHF-CMC/S/Z, and DHF-ALG/S/Z nanoparticles as a mechanism [50], besides, due to the different chemical structures of CMC and ALG, DHF-CMC/S/Z have higher EE, which caused more 7,8-DHF to be protected in the hydrophobic part. These results are in agreement with previous studies that introduced curcumin being embedded in zein and quaternized chitosan complexes [51], along with work showing that quercetagetin was loaded using the zein-hyaluronic acid binary complexes [52].

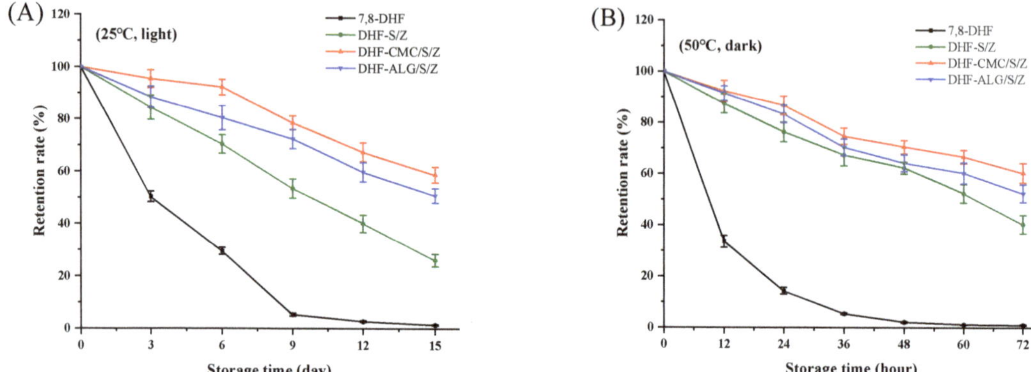

Figure 8. Storage stability of free 7,8-DHF, DHF-S/Z, DHF-CMC/S/Z and DHF-ALG/S/Z at 25 °C under light (**A**), at 50 °C under dark (**B**).

3.9. *In Vitro Simulated Gastrointestinal Digestion*

A gastrointestinal tract (GIT) model was applied to study the digestive fate and bioaccessibility of 7,8-DHF in different formulations. Particle size changes were monitored at a specific digestion time (30, 60, 120, and 180 min), and the results were presented in (Figure 9A). The mean particle size of DHF-S/Z was significantly increased after 60 min SGF digestion ($p < 0.05$). This finding was possibly due to the fact that the S/Z were exposed to ionic strength along with low pH and partially digestion via pepsase. The low pH and ionic strength exposure likely weakened electrostatic repulsion forces among the nanoparticles [52]. Particle size reduction in DHF-S/Z post-SIF-exposure was attributed to the fact that SIF contains bile salt with strong emulsifying ability. Bile salt can bind many biopolymer molecules and induce bridging flocculation [21]. The particle size of DHF-CMC/S/Z was increased after exposure to the stomach phase but remained relatively constant during incubation in SIF. This behavior suggested that the existence of CMC strengthened intestinal stability of DHF-CMC/S/Z. However, the particle size of DHF-ALG/S/Z remained fairly steady throughout simulated GIT, only showing a large increase at 30 min during the SGF incubation. The different influence of polysaccharide type (CMC or ALG) on gastrointestinal fate of colloidal carriers can be attributed to different molecular characteristics. FE-SEM microscopic observation further confirmed that exposure to simulated gastrointestinal conditions had a significant effect on the morphology of the 7,8-DHF-loaded nano-complexes (Figure 9C). For DHF-S/Z, irregular shapes have developed after they were added into the SGF and SIF, similar to an anomalous sheet structure. This shape acquisition was due to irregular aggregates appearing after SGF digestion. Furthermore, DHF-CMC/S/Z and DHF-ALG/S/Z exhibited a spherical shape. After gastrointestinal digestion, the nanoparticles possessed a relatively spherical morphology, resembling the cross-linked structure of large nanoparticles. Overall, CMC and ALG effectively protected the stability of DHF-S/Z nanoparticles through the GIT. Particularly, ALG performed extremely well.

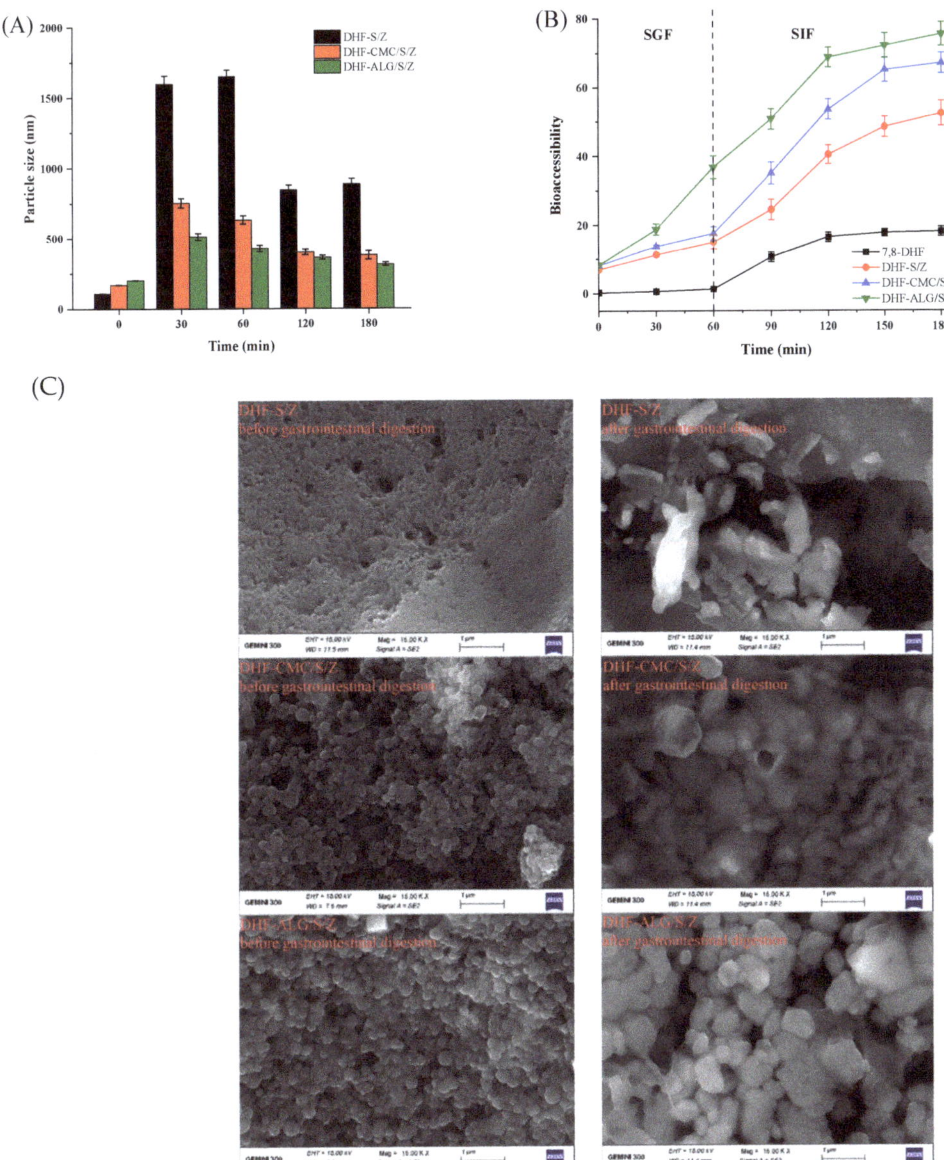

Figure 9. Influence of in vitro digestion time on the particle size (**A**) and bio-accessibility of DHF-S/Z, DHF-CMC/S/Z, and DHF-ALG/S/Z (**B**), FE-SEM images of DHF-S/Z, DHF-CMC/S/Z, and DHF-ALG/S/Z, 15,000× magnification times (**C**).

After being exposed to simulated gastrointestinal conditions, 7,8-DHF bioaccessibility was measured after centrifugation and collection of micelle phases. As shown in Figure 9C, the S/Z was broken down when exposed to SGF digestion, and the core of 7,8-DHF was released, resulting in a low bioaccessibility. The bioaccessibility of DHF-S/Z gradually increased in SIF digestion, primarily due to the emulsibility of bile salts and sophorolipids. Another reason for this increase was the binding of hydrolyzed peptides of zein protein [33]. In the presence of CMC, low levels of bioaccessibility were exhibited in SGF digestion. This behavior was introduced by the vulnerability of CMC to acid, pepsin enzymes, and the alkali ions of SGF juices [53]. However, the bioaccessibility of DHF-CMC/S/Z was higher than that of DHF-S/Z, indicating a synergistic effect existed among CMC, sophorolipid and zein on controlling the release of 7,8-DHF. Furthermore, the presence of ALG further increased bioaccessibility to 75.46% (Figure 9B), showing the highest bioaccessibility of 7,8-DHF after GIT digestion compared to DHF-CMC/S/Z and DHF-S/Z. These results indicated that ALG increased the solubilization capacity of mixed micelles in small intestine fluids. Collectively, our results demonstrated that encapsulating 7,8-DHF in polysaccharide coated S/Z nanoparticles can promote an appreciable increase in its in vitro bioaccessibility.

4. Conclusions

In the present study, we compared the performance of ternary nanoparticles in the existence of two selected polysaccharides (CMC and ALG) for encapsulation of bioactive 7,8-DHF but with low bioavailability. CMC/S/Z exhibited lower PDI, particle size and turbidity, but higher zeta potential and loading capacity compared to ALG/S/Z. Furthermore, both polysaccharides supplementations promoted the EE value of 7,8-DHF in S/Z, especially CMC. Invitation of polysaccharides displayed positive effects on the formation and physical stability (pH and ionic strength stability) of ternary complexes. The formation of ternary complexes mainly occurred via hydrophobic effects, hydrogen bonding and electrostatic interactions. More significantly, compared to S/Z, ALG/S/Z, and CMC/S/Z obviously enhanced the storage stability and in vitro bioaccessibility of 7,8-DHF. CMC/S/Z possessed a higher storage stability for 7,8-DHF. In contrast, ALG/S/Z had a better in vitro bioaccessibility of 7,8-DHF. Collectively, the results of this study indicate that selected polysaccharides containing composite nanoparticles are efficient at encapsulating, retaining, and delivering 7,8-DHF, and might therefore be utilized in dietary supplements and functional foods. Future work will focus on the applicability of DHF-ALG/S/Z and DHF-CMC/S/Z in complex water-phase beverage systems. Besides, the transepithelial transport mechanism of DHF-ALG/S/Z in an Caco-2 cell model, and in vivo pharmacokinetic studies in rat will be also studies.

Supplementary Materials: The following are available online at https://www.mdpi.com/article/10.3390/foods10112629/s1, Figure S1: The chemical structure of 7,8-DHF. Figure S2: Size distributions of nanoparticles with different zein to polysaccharide mass ratios, CMC/S/Z (A), ALG/S/Z (B). Figure S3: Size distributions of nanoparticles with different zein to polysaccharide mass ratios at pH = 4, CMC/S/Z (A), ALG/S/Z (B). Figure S4: The photograph of each colloidal particle at different pH conditions. Figure S5: The photograph of each colloidal particle at different NaCl concentrations.

Author Contributions: Conceptualization, methodology, resources, funding acquisition and writing—original draft preparation, Y.C.; software and formal analysis, J.P. and X.G.; investigation and data curation, L.W. and Y.S.; writing—review and editing, D.W.; project administration, Y.W.; validation and supervision, G.X. All authors have read and agreed to the published version of the manuscript.

Funding: This research was funded by the Zhejiang Province Public Welfare Technology Application Research Project -National Cooperation Project (grant number LGJ21C20001), Fund of Key Laboratory of Aquatic Product Processing, Ministry of Agriculture and Rural Affairs, China (grant number NYJG202104), Zhejiang Province Xinmiao Talents Program (grant number 2021R403035) and College students' Innovation and Entrepreneurship Training program of China (grant number 2021052 and 2021011).

Institutional Review Board Statement: Not applicable.

Informed Consent Statement: Not applicable.

Data Availability Statement: Not applicable.

Acknowledgments: All authors are thankful to their representative universities/institutes for the support and services used in this study.

Conflicts of Interest: The authors declare no conflict of interest.

References

1. Andero, R.; Ressler, K.J. Fear extinction and bdnf: Translating animal models of ptsd to the clinic. *Genes Brain Behav.* **2012**, *11*, 503–512. [CrossRef] [PubMed]
2. Colombo, P.S.; Flamini, G.; Christodoulou, M.S.; Rodondi, G.; Vitalini, S.; Passarella, D.; Fico, G. Farinose alpine primula species: Phytochemical and morphological investigations. *Phytochemistry* **2014**, *98*, 151–159. [CrossRef] [PubMed]
3. Jang, S.; Liu, X.; Yepes, M.; Shepherd, K.R.; Miller, G.W.; Liu, Y.; Wilson, W.D.; Xiao, G.; Blanchi, B.; Sun, Y.E. A selective trkb agonist with potent neurotrophic activities by 7,8-dihydroxyflavone. *Proc. Natl. Acad. Sci. USA* **2010**, *107*, 2687–2692. [CrossRef] [PubMed]
4. Li, X.H.; Dai, C.F.; Chen, L.; Zhou, W.T.; Han, H.L.; Dong, Z.F. 7,8-dihydroxyflavone ameliorates motor deficits via suppressing α-synuclein expression and oxidative stress in the mptp-induced mouse model of parkinson's disease. *CNS Neurosci. Ther.* **2016**, *22*, 617–624. [CrossRef] [PubMed]
5. Aytan, N.; Choi, J.; Carreras, I.; Crabtree, L.; Nguyen, B.; Lehar, M.; Blusztajn, J.K.; Jenkins, B.G.; Dedeoglu, A. Protective effects of 7,8-dihydroxyflavone on neuropathological and neurochemical changes in a mouse model of alzheimer's disease. *Eur. J. Pharmacol.* **2018**, *828*, 9–17. [CrossRef]
6. Chan, C.B.; Tse, M.C.L.; Liu, X.; Zhang, S.; Schmidt, R.; Otten, R.; Liu, L.; Ye, K. Activation of muscular trkb by its small molecular agonist 7,8-dihydroxyflavone sex-dependently regulates energy metabolism in diet-induced obese mice. *Chem. Biol.* **2015**, *22*, 355–368. [CrossRef]
7. Chen, C.; Wang, Z.; Zhang, Z.; Liu, X.; Kang, S.S.; Zhang, Y.; Ye, K. The prodrug of 7,8-dihydroxyflavone development and therapeutic efficacy for treating alzheimer's disease. *Proc. Natl. Acad. Sci. USA* **2018**, *115*, 578–583. [CrossRef]
8. Zhang, J.; Yao, W.; Dong, C.; Yang, C.; Ren, Q.; Ma, M.; Han, M.; Hashimoto, K. Comparison of ketamine, 7,8-dihydroxyflavone, and ana-12 antidepressant effects in the social defeat stress model of depression. *Psychopharmacology* **2015**, *232*, 4325–4335. [CrossRef]
9. Liu, C.; Chan, C.B.; Ye, K. 7,8-dihydroxyflavone, a small molecular trkb agonist, is useful for treating various bdnf-implicated human disorders. *Transl. Neurodegener.* **2016**, *5*, 1–9. [CrossRef]
10. Chen, Y.; Xue, F.; Xia, G.; Zhao, Z.; Chen, C.; Li, Y.; Zhang, Y. Transepithelial transport mechanisms of 7,8-dihydroxyflavone, a small molecular trkb receptor agonist, in human intestinal caco-2 cells. *Food Funct.* **2019**, *10*, 5215–5227. [CrossRef]
11. Chen, Y.; Xia, G.; Zhao, Z.; Xue, F.; Gu, Y.; Chen, C.; Zhang, Y. 7,8-dihydroxyflavone nano-liposomes decorated by crosslinked and glycosylated lactoferrin: Storage stability, antioxidant activity, in vitro release, gastrointestinal digestion and transport in caco-2 cell monolayers. *J. Funct. Foods* **2020**, *65*, 103742. [CrossRef]
12. Behjati, J.; Yazdanpanah, S. Nanoemulsion and emulsion vitamin d3 fortified edible film based on quince seed gum. *Carbohyd. Polym.* **2021**, *262*, 117948. [CrossRef]
13. Niu, F.; Hu, D.; Gu, F.; Du, Y.; Zhang, B.; Ma, S.; Pan, W. Preparation of ultra-long stable ovalbumin/sodium carboxymethylcellulose nanoparticle and loading properties of curcumin. *Carbohyd. Polym.* **2021**, *271*, 118451. [CrossRef]
14. Zhou, W.; Zhang, Y.; Li, R.; Peng, S.; Ruan, R.; Li, J.; Liu, W. Fabrication of caseinate stabilized thymol nanosuspensions via the ph-driven method: Enhancement in water solubility of thymol. *Foods* **2021**, *10*, 1074. [CrossRef] [PubMed]
15. Gao, J.; Liu, C.; Shi, J.; Ni, F.; Shen, Q.; Xie, H.; Wang, K.; Lei, Q.; Fang, W.; Ren, G. The regulation of sodium alginate on the stability of ovalbumin-pectin complexes for vd3 encapsulation and in vitro simulated gastrointestinal digestion study. *Food Res. Int.* **2021**, *140*, 110011. [CrossRef] [PubMed]
16. Zhang, J.; Liu, Y.; Liu, X.; Li, Y.; Yin, X.; Subirade, M.; Zhou, P.; Liang, L. The folic acid/β-casein complex: Characteristics and physicochemical implications. *Food Res. Int.* **2014**, *57*, 162–167. [CrossRef]
17. Chen, S.; Ma, Y.; Dai, L.; Liao, W.; Zhang, L.; Liu, J.; Gao, Y. Fabrication, characterization, stability and re-dispersibility of curcumin-loaded gliadin-rhamnolipid composite nanoparticles using ph-driven method. *Food Hydrocolloid.* **2021**, *118*, 106758. [CrossRef]
18. Xiao, Y.; Ho, C.; Chen, Y.; Wang, Y.; Wei, Z.; Dong, M.; Huang, Q. Synthesis, characterization, and evaluation of genistein-loaded zein/carboxymethyl chitosan nanoparticles with improved water dispersibility, enhanced antioxidant activity, and controlled release property. *Foods* **2020**, *9*, 1604. [CrossRef] [PubMed]
19. Li, Y.; Xu, G.; Li, W.; Lv, L.; Zhang, Q. The role of ultrasound in the preparation of zein nanoparticles/flaxseed gum complexes for the stabilization of pickering emulsion. *Foods* **2021**, *10*, 1990. [CrossRef] [PubMed]
20. Wei, Y.; Wang, C.; Liu, X.; Mackie, A.; Zhang, L.; Liu, J.; Mao, L.; Yuan, F.; Gao, Y. Impact of microfluidization and thermal treatment on the structure, stability and in vitro digestion of curcumin loaded zein-propylene glycol alginate complex nanoparticles. *Food Res. Int.* **2020**, *138*, 109817. [CrossRef]
21. Dai, L.; Wei, Y.; Sun, C.; Mao, L.; Mcclements, D.J.; Gao, Y. Development of protein-polysaccharide-surfactant ternary complex particles as delivery vehicles for curcumin. *Food Hydrocolloid.* **2018**, *85*, 75–85. [CrossRef]
22. Wei, Y.; Zhang, L.; Yu, Z.; Lin, K.; Yang, S.; Dai, L.; Liu, J.; Mao, L.; Yuan, F.; Gao, Y. Enhanced stability, structural characterization and simulated gastrointestinal digestion of coenzyme Q_{10} loaded ternary nanoparticles. *Food Hydrocolloid.* **2019**, *94*, 333–344. [CrossRef]
23. Zhang, H.; Fu, Y.; Xu, Y.; Niu, F.; Li, Z.; Ba, C.; Jin, B.; Chen, G.; Li, X. One-step assembly of zein/caseinate/alginate nanoparticles for encapsulation and improved bioaccessibility of propolis. *Food Funct.* **2019**, *10*, 635–645. [CrossRef] [PubMed]

24. Liu, Q.; Jing, Y.; Han, C.; Zhang, H.; Tian, Y. Encapsulation of curcumin in zein/caseinate/sodium alginate nanoparticles with improved physicochemical and controlled release properties. *Food Hydrocolloid.* **2019**, *93*, 432–442. [CrossRef]
25. Chen, Y.; Xia, G.; Zhao, Z.; Xue, F.; Chen, C.; Zhang, Y. Formation, structural characterization, stability and in vitro bioaccessibility of 7,8-dihydroxyflavone loaded zein-/sophorolipid composite nanoparticles: Effect of sophorolipid under two blending sequences. *Food Funct.* **2020**, *11*, 1810–1825. [CrossRef] [PubMed]
26. Afshar, M.; Dini, G.; Vaezifar, S.; Mehdikhani, M.; Movahedi, B. Preparation and characterization of sodium alginate/polyvinyl alcohol hydrogel containing drug-loaded chitosan nanoparticles as a drug delivery system. *J. Drug Deliv. Sci. Technol.* **2020**, *56*, 101530. [CrossRef]
27. Sheng, Y.; Gao, J.; Yin, Z.; Kang, J.; Kong, Y. Dual-drug delivery system based on the hydrogels of alginate and sodium carboxymethyl cellulose for colorectal cancer treatment. *Carbohyd. Polym.* **2021**, *269*, 118325. [CrossRef]
28. Yuan, Y.; Li, H.; Liu, C.; Zhang, S.; Xu, Y.; Wang, D. Fabrication and characterization of lutein-loaded nanoparticles based on zein and sophorolipid: Enhancement of water solubility, stability, and bioaccessibility. *J. Agric. Food Chem.* **2019**, *67*, 11977–11985. [CrossRef]
29. Wei, Y.; Yu, Z.; Lin, K.; Sun, C.; Dai, L.; Yang, S.; Mao, L.; Yuan, F.; Gao, Y. Fabrication and characterization of resveratrol loaded zein-propylene glycol alginate-rhamnolipid composite nanoparticles: Physicochemical stability, formation mechanism and in vitro digestion. *Food Hydrocolloid.* **2019**, *95*, 336–348. [CrossRef]
30. Dai, L.; Sun, C.; Wei, Y.; Zhan, X.; Mao, L.; Gao, Y. Formation and characterization of zein-propylene glycol alginate-surfactant ternary complexes: Effect of surfactant type. *Food Chem.* **2018**, *258*, 321–330. [CrossRef]
31. Patel, A.R.; Ten-Hoorn, J.S.; Hazekamp, J.; Blijdenstein, T.B.; Velikov, K.P. Colloidal complexation of a macromolecule with a small molecular weight natural polyphenol: Implications in modulating polymer functionalities. *Soft Matter* **2013**, *9*, 1428–1436. [CrossRef]
32. Peng, S.; Li, Z.; Zou, L.; Liu, W.; Liu, C.; Mcclements, D.J. Enhancement of curcumin bioavailability by encapsulation in sophorolipid-coated nanoparticles: An in vitro and in vivo study. *J. Agric. Food Chem.* **2018**, *66*, 1488–1497. [CrossRef]
33. Yao, K.; Chen, W.; Song, F.; Mcclements, D.J.; Hu, K. Tailoring zein nanoparticle functionality using biopolymer coatings: Impact on curcumin bioaccessibility and antioxidant capacity under simulated gastrointestinal conditions. *Food Hydrocolloid.* **2018**, *79*, 262–272. [CrossRef]
34. Hu, K.; Mcclements, D.J. Fabrication of biopolymer nanoparticles by antisolvent precipitation and electrostatic deposition: Zein-alginate core/shell nanoparticles. *Food Hydrocolloid.* **2015**, *44*, 101–108. [CrossRef]
35. Dai, L.; Li, R.; Wei, Y.; Sun, C.; Mao, L.; Gao, Y. Fabrication of zein and rhamnolipid complex nanoparticles to enhance the stability and in vitro release of curcumin. *Food Hydrocolloid.* **2018**, *77*, 617–628. [CrossRef]
36. Yuan, Y.; Li, H.; Liu, C.; Zhu, J.; Xu, Y.; Zhang, S.; Fan, M.; Zhang, D.; Zhang, Y.; Zhang, Z. Fabrication of stable zein nanoparticles by chondroitin sulfate deposition based on antisolvent precipitation method. *Int. J. Biol. Macromol.* **2019**, *139*, 30–39. [CrossRef]
37. Sun, C.; Dai, L.; Gao, Y. Binary complex based on zein and propylene glycol alginate for delivery of quercetagetin. *Biomacromolecules* **2016**, *17*, 3973–3985. [CrossRef] [PubMed]
38. Zhao, W.; Yuan, P.; She, X.; Xia, Y.; Komarneni, S.; Xi, K.; Che, Y.; Yao, X.; Yang, D. Sustainable seaweed-based one-dimensional (1d) nanofibers as high-performance electrocatalysts for fuel cells. *J. Mater. Chem. A* **2015**, *3*, 14188–14194. [CrossRef]
39. Xu, Y.; Liu, L. Structural and functional properties of soy protein isolates modified by soy soluble polysaccharides. *J. Agric. Food Chem.* **2016**, *64*, 7275–7284. [CrossRef] [PubMed]
40. Cerqueira, M.A.; Souza, B.W.; Teixeira, J.A.; Vicente, A.A. Effect of glycerol and corn oil on physicochemical properties of polysaccharide films—A comparative study. *Food Hydrocolloid.* **2012**, *27*, 175–184. [CrossRef]
41. Luo, Y.; Zhang, B.; Whent, M.; Yu, L.L.; Wang, Q. Preparation and characterization of zein/chitosan complex for encapsulation of α-tocopherol, and its in vitro controlled release study. *Colloids Surf. B Biointerfaces* **2011**, *85*, 145–152. [CrossRef] [PubMed]
42. Liu, F.; Ma, C.; Mcclements, D.J.; Gao, Y. A comparative study of covalent and non-covalent interactions between zein and polyphenols in ethanol-water solution. *Food Hydrocolloid.* **2017**, *63*, 625–634. [CrossRef]
43. Chen, Y.; Zhao, Z.; Xia, G.; Xue, F.; Chen, C.; Zhang, Y. Fabrication and characterization of zein/lactoferrin composite nanoparticles for encapsulating 7,8-dihydroxyflavone: Enhancement of stability, water solubility and bioaccessibility. *Int. J. Biol. Macromol.* **2020**, *146*, 179–192. [CrossRef]
44. Liu, E.; Su, Z.; Yang, C.; Ji, Y.; Liu, B.; Meng, X. Fabrication, characterization and properties of dha-loaded nanoparticles based on zein and plga. *Food Chem.* **2021**, *360*, 129957. [CrossRef]
45. Lancelot, E.; Fontaine, J.; Grua-Priol, J.; Assaf, A.; Thouand, G.; Le-Bail, A. Study of structural changes of gluten proteins during bread dough mixing by raman spectroscopy. *Food Chem.* **2021**, *358*, 129916. [CrossRef]
46. Narwal, S.; Kumar, A.; Chaudhary, M.; Budhwar, V. Formulation of eutectic mixture of curcumin with salicylic acid for improving its dissolution profile. *Res. J. Pharm. Technol.* **2021**, *14*, 1875–1879.
47. Jiang, F.; Yang, L.; Wang, S.; Ying, X.; Ling, J.; Ouyang, X.K. Fabrication and characterization of zein-alginate oligosaccharide complex nanoparticles as delivery vehicles of curcumin. *J. Mol. Liq.* **2021**, *342*, 116937. [CrossRef]
48. Wang, X.; Li, M.; Liu, F.; Peng, F.; Li, F.; Lou, X.; Jin, Y.; Wang, J.; Xu, H. Fabrication and characterization of zein-tea polyphenols-pectin ternary complex nanoparticles as an effective hyperoside delivery system: Formation mechanism, physicochemical stability, and in vitro release property. *Food Chem.* **2021**, *364*, 130335. [CrossRef] [PubMed]

49. Sun, C.; Dai, L.; Gao, Y. Interaction and formation mechanism of binary complex between zein and propylene glycol alginate. *Carbohyd. Polym.* **2017**, *157*, 1638–1649. [CrossRef]
50. Wang, W.; Liu, F.; Gao, Y. Quercetagetin loaded in soy protein isolate–κ-carrageenan complex: Fabrication mechanism and protective effect. *Food Res. Int.* **2016**, *83*, 31–40. [CrossRef]
51. Liang, H.; Zhou, B.; He, L.; An, Y.; Lin, L.; Li, Y.; Liu, S.; Chen, Y.; Li, B. Fabrication of zein/quaternized chitosan nanoparticles for the encapsulation and protection of curcumin. *RSC Adv.* **2015**, *5*, 13891–13900. [CrossRef]
52. Chen, S.; Han, Y.; Huang, J.; Dai, L.; Du, J.; Mcclements, D.J.; Mao, L.; Liu, J.; Gao, Y. Fabrication and characterization of layer-by-layer composite nanoparticles based on zein and hyaluronic acid for codelivery of curcumin and quercetagetin. *ACS Appl. Mater. Inter.* **2019**, *11*, 16922–16933. [CrossRef] [PubMed]
53. Parris, N.; Cooke, P.H.; Hicks, K.B. Encapsulation of essential oils in zein nanospherical particles. *J. Agric. Food Chem.* **2005**, *53*, 4788–4792. [CrossRef] [PubMed]

MDPI

Article

Application of Solution Blow Spinning for Rapid Fabrication of Gelatin/Nylon 66 Nanofibrous Film

Zhichao Yang [1,2], Chaoyi Shen [1,2], Yucheng Zou [3], Di Wu [1,2,4,*], Hui Zhang [3] and Kunsong Chen [1,2]

1 Zhejiang Provincial Key Laboratory of Horticultural Plant Integrative Biology, College of Agriculture & Biotechnology, Zhejiang University, Hangzhou 310058, China; 22016147@zju.edu.cn (Z.Y.); scyzju@zju.edu.cn (C.S.); akun@zju.edu.cn (K.C.)
2 The State Agriculture Ministry Laboratory of Horticultural Plant Growth, Development and Quality Improvement, College of Agriculture & Biotechnology, Zhejiang University, Hangzhou 310058, China
3 College of Biosystems Engineering and Food Science, Zhejiang University, Hangzhou 310058, China; yuczou@zju.edu.cn (Y.Z.); hubert0513@zju.edu.cn (H.Z.)
4 Zhongyuan Institute, Zhejiang University, Zhengzhou 450000, China
* Correspondence: di_wu@zju.edu.cn

Abstract: Gelatin (GA) is a natural protein widely used in food packaging, but its fabricated fibrous film has the defects of a high tendency to swell and inferior mechanical properties. In this work, a novel spinning technique, solution blow spinning (SBS), was used for the rapid fabrication of nanofiber materials; meanwhile, nylon 66 (PA66) was used to improve the mechanical properties and the ability to resist dissolution of gelatin films. Morphology observations show that GA/PA66 composite films had nano-diameter from 172.3 to 322.1 nm. Fourier transform infrared spectroscopy and X-ray indicate that GA and PA66 had strong interaction by hydrogen bonding. Mechanical tests show the elongation at break of the composite film increased substantially from 7.98% to 30.36%, and the tensile strength of the composite film increased from 0.03 MPa up to 1.42 MPa, which indicate that the composite films had the highest mechanical strength. Water vapor permeability analysis shows lower water vapor permeability of 9.93 g mm/m^2 h kPa, indicates that GA/PA66 film's water vapor barrier performance was improved. Solvent resistance analysis indicates that PA66 could effectively improve the ability of GA to resist dissolution. This work indicates that SBS has great promise for rapid preparation of nanofibrous film for food packaging, and PA66 can be applied to the modification of gelatin film.

Keywords: solution blow spinning; gelatin; nylon 66; modification; rapid fabrication

Citation: Yang, Z.; Shen, C.; Zou, Y.; Wu, D.; Zhang, H.; Chen, K. Application of Solution Blow Spinning for Rapid Fabrication of Gelatin/Nylon 66 Nanofibrous Film. *Foods* **2021**, *10*, 2339. https://doi.org/10.3390/foods10102339

Academic Editor: Christos Ritzoulis

Received: 14 September 2021
Accepted: 28 September 2021
Published: 30 September 2021

1. Introduction

Solution blow spinning (SBS) is a rising preparation technique to produce polymer films with ultrafine fibers and high specific surface area [1]. A regular SBS apparatus consists of a high velocity air source, a syringe pump, a concentric nozzle system, and a collector [2]. The polymer solution in the inner nozzle is elongated into fine fiber by the surrounding high velocity airflow in the outer nozzle. Moreover, because SBS uses high velocity airflow to form fibers, there is no requirement for the conductivity of the solution compared to electrospinning. As the solvent rapidly evaporates in the process of moving towards the collector, ultrafine fibers deposited and formed a nanofibrous film on the collector [3]. The SBS has served a wide range of purposes that include biosensors, the aerospace industry, impurity removers, and wearable electronics [4]. The development of nanofiber films for food packaging using SBS, however, has undergone little research. For the present, the utilization of nanofiber film in food packaging is primarily manufactured by the electrospinning method [5–9]. Nonetheless, previous works show that the fiber production efficiency of SBS could be much higher than electrospinning. Tandon et al. [10] found that SBS was three times more efficient in fiber production than electrospinning; and in the study of Sett et al. [11], the fiber productivity of SBS was even up to 30 times higher than that of electrospinning. Shen et al. [12] also found that the SBS process could reach a rapid feed rate of 3 mL/h, while the feed rate of the electrospinning is generally between 0.1 mL/h and 0.6 mL/h. In this regard, the SBS technology has demonstrated its promising potential for the fast production of large area nanofiber film for food packaging.

Gelatin (GA) is a single-stranded protein obtained by hydrolysis of collagen, with favorable biocompatibility, biodegradability and nontoxicity, and has been utilized extensively in the food, pharmaceutical and photographic industries [13–15]. However, due to the solubility in aqueous solution and poor mechanical properties, gelatin nanofibers usually require modification by chemical cross-linking agents [16,17]. Bigi et al. [18] investigated the mechanical and swelling properties of glutaraldehyde (GTA) crosslinked gelatin films, and found that the use of GTA allowed to regulate the physical-chemical properties of gelatin films. But the residue and cytotoxicity of chemical agents have limited its application in the food field [19].

Various studies have shown that many synthetic polymers such as poly(caprolactone) (PCL), poly(glycolic acid) (PLA), poly(lactic-co-glycolic acid) (PLGA) can be used with natural polymers to produce polymer composites with improved performance [20–24]. Among these synthetic polymers, nylon 66 (polyamide 66, PA66) is a multifunctional synthetic thermoplastic polymer broadly used in textiles, parachutes, biomedical fields, and functional materials [25]. It was reported that PA66 has been successfully used to modify epoxy resins to achieve tensile toughness enhancement [26], however there has been limited work related to the use of PA66 for the modification of natural polymers, especially gelatin. Owing to its anti-abrasion resistance, high mechanical strength, light weight, relatively low cost and, very importantly, biological safety [27–31], PA66 can be used as a potential modification material for gelatin.

In this work, SBS was used to rapidly fabricate GA/PA66 composite nanofiber films to address the need for food packaging materials to be prepared on a large scale. The feasibility of PA66 as the modifier to enhance comprehensive performance of gelatin in mechanical properties and the ability to resist dissolution was evaluated. The morphology, diameter distribution, molecular interaction, crystal structures, thermal stability, mechanical strength, water contact angle (WCA) and water vapor permeability (WVP) of the GA, GA/PA66 and PA66 films were characterized. In addition, the prospects of SBS for rapid preparation of food packaging are discussed.

2. Materials and Methods

2.1. Chemicals

Gelatin (GA, Bloom 250 g) was obtained from Aladdin, Inc. (Shanghai, China). Nylon 66 (PA66, pellets) was obtained from Sigma Aldrich (St. Louis, MO, USA). 88% (v/v) Formic acid and glacial acetic acid (analytical reagent grade) were acquired from Sinopharm Chemical Reagent Co., Ltd. (Shanghai, China).

2.2. Spinning Solution

GA (12% w/v) solutions were obtained through dissolving 1.2 g GA in 10 mL glacial acetic acid after sufficient stirring overnight. PA66 (12% w/v) solution was obtained through adding 1.2 g PA66 pellets into 10 mL formic acid with stirring overnight. Blend solutions at a final concentration of 12% (w/v) with gelatin/PA66 weight ratios of 1/2, 1/1 and 2/1 were obtained through dissolving gelatin and PA66 in glacial acetic acid and formic acid after stirring overnight. The solution was stirred for 6 h and ultrasonic treated for 30 min before the SBS process.

2.3. Solution Blow Spinning (SBS) Process

The JNS-SBS-01 SBS apparatus (Janus New-Materials Co., LTD, Nanjing, China) was employed to fabricate the nanofibrous films. Solutions were added into a syringe, and the feed rate of syringe pump was 3 mL/h. The distance between nozzle and reticular collector was 20 cm. An air pressure of 0.06 MPa was used to provide a stable shear force. The SBS process was performed at 30 °C with a relative humidity of approximately 50%. Obtained films were GA film, PA66 film and gelatin/PA66 (2:1, 1:1, 1:2) films, respectively.

2.4. Fiber Morphology

The nanofiber morphology of the films was observed by FE-SEM (GeminiSEM 300, ZEISS, Oberkochen, Germany). The average diameters of nanofiber and their distributions were obtained through randomly measuring 100 fibers with Nano Measurer 1.2 (Fudan University, Shanghai, China).

2.5. Fourier Transform Infrared (FTIR) Analysis

FTIR analysis was carried out by method of KBr pellet to determine interactions among the components. The FTIR spectrum was obtained by using a Nicolet 170-SX instrument (Thermo Nicolet Ltd., Waltham, MA, USA). The wavenumber range was 4000–400 cm^{-1} and an average of 32 scans at 4 cm^{-1} were performed for each measurement.

2.6. X-ray Diffraction (XRD) Analysis

The crystal structure of the nanofiber films can be examined through XRD. XRD analysis was performed through an X'Pert Pro diffractometer (PA Analytical B.V., Eindhoven, The Netherlands) with the conditions at 40 kV, 35 mA. The diffraction range was 5–90° (2θ) and the rate of scanning was 2° min^{-1}.

2.7. Thermal Analysis

DSC was performed through a TA Q200 instrument (TA Instruments, Newcastle, DE, USA) with the heating range of 20 °C to 300 °C. TGA was performed through a TA Q500 instruments (TA Instruments, Newcastle, DE, USA) with the heating range of 50 to 600 °C. The rate of heating for DSC and TGA was 10 °C min^{-1}.

2.8. Mechanical Strength

The mechanical strength of the nanofibrous films was measured through a mechanical testing instrument (Instron 5944, Norwood, MA, USA) quipped with a loadcell of 10 N used at ambient temperature. The force rate was 1 mm min^{-1}. The films were tailored into strips with a thickness of 0.1 mm.

2.9. Water Contact Angle (WCA)

A OCA20 device (Data Physics Co., Ltd., Bad Vilbel, Germany) was employed to investigate the WCA of the nanofibrous films through the sessile drop method [32]. The nanofibrous films were fixed on the object slides, and then a droplet of distilled water (3.5 μL) was deposited on the nanofibrous films. The water drops at 0 and 3 s were both recorded and averaged to determine the values of the WCA. The WCA value was computed from three positions on the films [33].

2.10. Water Vapor Permeability (WVP)

The ASTM E96 gravimetric method was used to calculate the WVP values. Fibrous film was fixed on top of a permeation cup filled with distilled water and kept in a desiccator loaded with dry silica gel. Following the attainment of steady state (approximately 1 h), the permeation cups were weighed every 12 h during a 5-day period. The WVP was obtained according to the following Equation [34]:

$$\text{WVP} = \frac{\Delta M \times d}{\Delta t \times A \times \Delta p} \quad (1)$$

where $\Delta M/\Delta t$ is the weight of water loss per unit time (g h^{-1}), d is the film thickness (mm), A is the area of the film exposed to moisture (m^2), and Δp is the water vapor pressure difference crossing the film (3.1671 kPa at 25 °C), respectively.

2.11. Solvent Resistance

The GA, PA66, and GA/PA66 films were tailored into round sheets and immersed in 1 × PBS for 24 h at room temperature, respectively. After drying at room temperature, the morphology of the samples was observed using field-emission scanning microscopy.

2.12. Statistical Analysis

All data were prepared in triplicate and expressed as the mean values ± standard deviation (SD). One-way ANOVA was performed using SPSS 19.0 statistical software (IBM Corp., Armonk, NY, USA) followed by Duncan's multiple comparison test; $p < 0.05$ had statistical significance.

3. Results and Discussion

3.1. Morphologies of Nanofibrous Films

Figure 1 shows the pictures of each film and the morphologies and diameter distributions of the solution blowing spun GA, PA66 films and GA/PA66 films with various weight ratios (2/1, 1/1 and 1/2 w/w). The average diameter of pure GA fibers was over 1000.0 nm, and the average diameter of pure PA66 fibers was 223.2 nm; with the weight ratio of PA66 increased from 1:2 to 2:1, the diameters increased from 172.3 nm to 322.1 nm. It can be observed that the addition of PA66 resulted in a significant reduction in fiber diameter, thus creating a denser network structure, which could improve the water vapor barrier performance. This was probably due to the fact that the addition of PA66 reduced the viscosity of the solution and promoted the elongation of the nanofibers, resulting in a thinner diameter [12]. In a similar study of modification of gelatin by Meng et al. [22], they found the average diameter of GA/poly (lactic-co-glycolic acid) (PLGA) nanofibers also decreased with the addition of PLGA. Moreover, the pure GA fibers were straight, while after the addition of PA66, the GA/PA66 composite fibers showed a curly appearance. In addition, the usage of pure gelatin solution for spinning resulted in a large number of droplets, so the film had hard lumps, leading to low solution utilization; while after the introduction of PA66, the droplets were significantly reduced and the phenomenon of hard lump formation disappeared, making the solution utilization improve. Therefore, this indicates that the addition of PA66 led to the improvement of GA spinnability during the SBS fabrication of nanofiber films. Like other spinnable polymers, nylon may increase the viscoelasticity of the gelatin-containing precursor solution, allowing for easier fiber formation [35].

3.2. FTIR Spectra Analysis

Figure 2a shows the FTIR spectra of the GA, PA66 films and GA/PA66 films with various weight ratios. For the GA film, the broad stretching band at 3000 to 3750 cm^{-1} (amide A) was related to N–H stretching vibrations and O–H, and three characteristic peaks of GA were observed at around 1647 cm^{-1} (amide I) corresponding to C=O and C–N stretching vibrations, 1542 cm^{-1} (amide II) corresponding to N–H bending and C–H stretching vibration, and 1259 cm^{-1} (amide III) [9,36]. For PA66 film, the adsorption peak appearing at 3305 cm^{-1} belonged to the N–H stretching band of amine group and the peaks at 1641 cm^{-1} and 1540 cm^{-1} [37]; and the characteristic peak at 2934 cm^{-1} corresponding to the stretching vibration of C–H also confirmed solid evidence of the existence of PA66 [38].

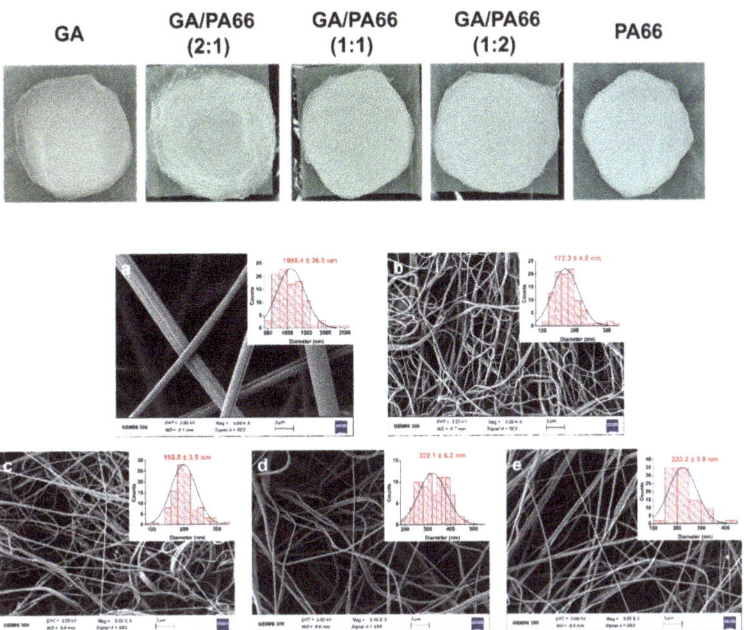

Figure 1. Pictures and scanning electron microscopy (SEM) images of the solution blowing spun gelatin (GA) (**a**), GA/PA66 (2:1) (**b**), GA/PA66 (1:1) (**c**), GA/PA66 (1:2) (**d**), PA66 (**e**) nanofibers.

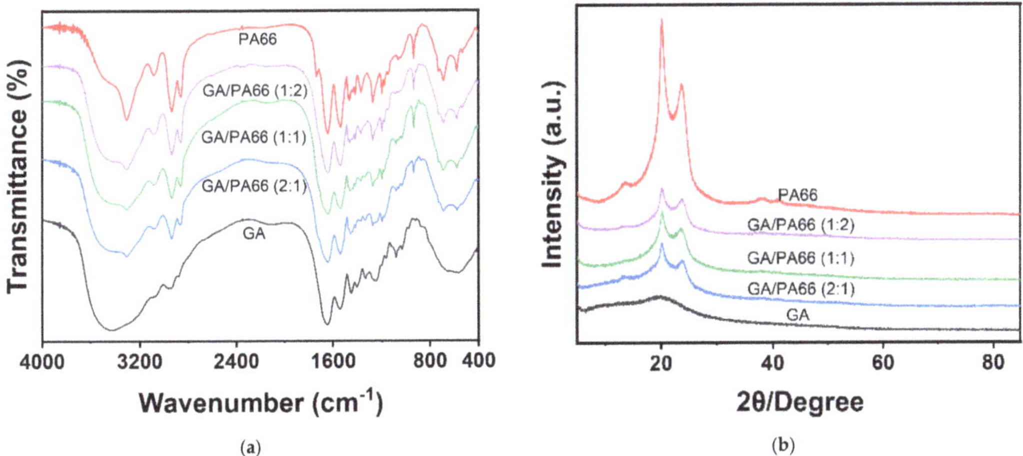

Figure 2. (**a**) Fourier transform infrared (FTIR) spectra and (**b**) X-ray diffraction (XRD) patterns of the GA, GA/PA66 (2:1), GA/PA66 (1:1), GA/PA66 (1:2), PA66 nanofibers.

The absence of peak splitting in all mixed nanofibers indicates that gelatin and PA66 were uniformly dispersed in the fibers. Overall, the GA/PA66 composite films exhibited similar major peaks, but with different amplitudes. The characteristic peaks of PA66 at around 3305 and 2934 cm^{-1}, and three characteristic peaks of at around 1647, 1542 and 1259 cm^{-1} belonging to GA, were observed in the GA/PA66 composite films, indicating that the introduction of PA66 did not disrupt the structure of GA and the homogeneous mixing in the GA/PA66 composite film. In addition, it can be observed that the peak around 3436 cm^{-1} in GA shifted to a higher band at about 3305 cm^{-1}. The lower intensity of the peak in the range 3125 cm^{-1} and 3400 cm^{-1} may be attributed to the interaction between the O–H and N–H groups of gelatin and PA66 molecules resulting in more intermolecular or intramolecular hydrogen bonding [20]. This result may also explain the phenomenon of improved GA spinnability mentioned above.

3.3. XRD Analysis

Figure 2b shows the XRD results of the GA, PA66 films and GA/PA66 films with various weight ratios. PA66 has two obvious diffraction peaks at around 2θ = 20.1° ascribed to the interchain hydrogen bond plane of the amide group [29] and 2θ = 23.7°, which is similar to the results of Noguchi's study [39]; whereas gelatin had a broad diffraction peak at 2θ = 19.2°, suggesting a low crystallinity of GA. For the GA/PA66 composite films, two diffraction peaks appeared between 20° and 25°, which were corresponding to the two characteristic peaks (α1 and α2) of the α-crystalline form of PA66 [38]. It was noticeable that the intensity of the peaks of the GA/PA66 composite film was enhanced compared to the pure GA film, suggesting that the crystallization of the GA/PA66 composite films was advanced with the addition of PA66.

3.4. Thermal Analysis

Figure 3a shows the DSC curves of the GA, PA66 films and GA/PA66 films with various weight ratios. The melting temperature (T_m) and melting enthalpy (ΔH_m) are summarized in (Table 1). The T_m of the pure GA and PA66 films were 126.50 °C and 260.43 °C, respectively. It should be noticed that the T_m and ΔH_m of the GA/PA66 composite films decreased with the addition of PA66. This might be because the GA nucleation was promoted to generate crystalline regions after the introduction of PA66.

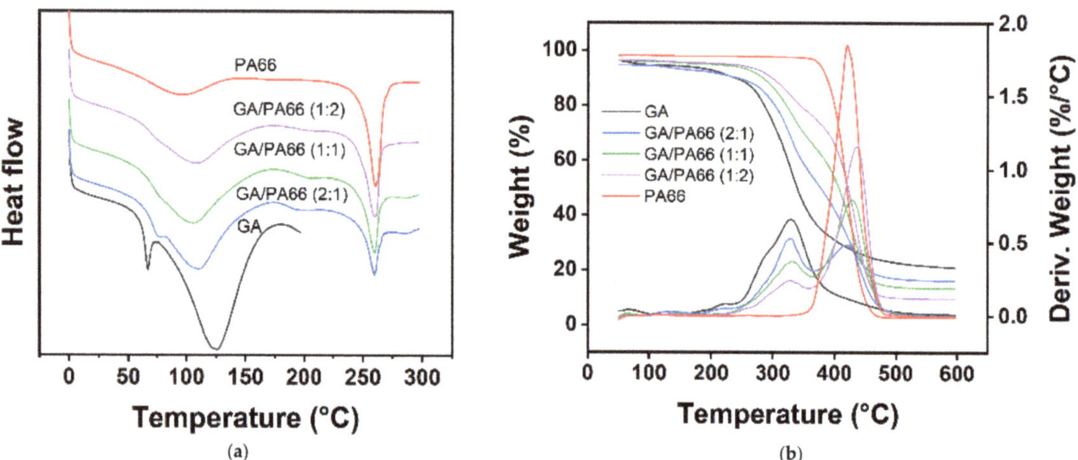

Figure 3. (**a**) Differential scanning calorimetry (DSC) and (**b**) thermogravimetric analysis (TGA) curves of the GA, GA/PA66 (2:1), GA/PA66 (1:1), GA/PA66 (1:2), PA66 nanofibers.

Table 1. Detailed data of the DSC and TGA thermograms for the nanofibrous films.

Sample	DSC Parameters				TGA Parameters		
	$T_{m, GA}$ (°C)	$\Delta H_{m, GA}$ (J/g)	$T_{m, PA66}$ (°C)	$\Delta H_{m, PA66}$ (J/g)	$T_{10wt\%}$ (°C) [1]	T_{max} (°C) [2]	W_{red} (%) [3]
GA	126.50	145.10	/	/	236.70	329.59	21.04
GA/PA66 (2:1)	110.20	148.7	259.71	21.87	240.17	330.22	16.20
GA/PA66 (1:1)	106.22	115.3	259.81	28.74	282.64	427.51	13.50
GA/PA66 (1:2)	108.28	102.9	260.39	38.52	298.13	435.59	9.70
PA66	/	/	260.43	65.61	387.09	420.18	2.95

[1] $T_{10wt\%}$ was the temperature at 10% mass loss. [2] T_{max} was the temperature at maximum weight loss rate. [3] W_{red} was the residual weight at 600 °C.

Figure 3b shows the TGA curves of the nanofibrous films, and the details were provided in (Table 1). The results indicate that there was a weight loss in the period from 50 to 260 °C, which was related to the vaporization of moisture [40]. After approximately 300 °C, a higher loss of nanofiber films resulting from polymer decomposition was recorded. The decomposition of GA occurs at about 300 °C, while the decomposition of PA66 at around 420 °C was observed. As the temperature hits 600 °C or higher, the mass of the films tended to be stable, indicating that the decomposition of the films was complete. At the completion of the TGA, the residual weights of the pure GA and PA66 films were 21.04% and 2.95%, respectively. The residual weights of the GA/PA66 composite films were 16.20%, 13.50% and 9.70%, corresponding to weight ratios of 2/1, 1/1 and 1/2, which were negatively correlated with the weight of PA66. The above results revealed that the residual amount of the composite films was reduced, while the rate of decomposition and the maximum decomposition temperature increased.

3.5. Mechanical Properties Analysis

Figure 4 shows the mechanical properties of the GA, PA66 films and GA/PA66 films with various weight ratios. The elastic modulus, elongation at break, and tensile strength of the GA film were 0.52 MPa, 7.98%, and 0.03 MPa, respectively. This result implied that the GA film was a material with poor ductility and mechanical strength. After blending with PA66, the elongation at break of the GA film increased substantially from 7.98% to 30.36% at the weight ratio of 1:1 (GA/PA66), and the tensile strength of the GA film were considerably enhanced from 0.03 MPa up to 1.42 MPa, which was 48 times higher, but there was a certain range of differences compared with pure PA66 nanofiber film. This might be due to the fact that the reduction of the diameter of nanofibers led to an increase in the density of inter-fiber alignment, and the uniformity of the nanofiber film might be improved [41]. In a similar study of Zhang et al. [21], they found that the mechanical properties of the GA films was also improved with the addition of poly(ε-caprolactone). In addition, the elongation at break of GA/PA66 was significantly improved compared to the pure GA film. These results indicate that the composite films had higher mechanical strength due to the addition of PA66.

3.6. Water Contact Angles Analysis

Figure 5 shows the WCAs of the films at the equilibration times of 0 s and 3 s. The WCA of the GA film was 111.73° at 0 s and 71.87° at 3 s, which implied the GA film had a comparatively hydrophilic surface. This was because GA was a hydrophilic material [42]. The WCA of the PA66 film was 136.83° at 0 s and 137.07° at 3 s, which meant the PA66 film had a comparatively hydrophobic surface. However, the hydrophilicity of GA/PA66 films was significantly increased since the composite films had a super hydrophilic surface of 0° at both 0 s and 3 s. Similar results were reported by Cao et al. [43], who found that compared to pure polyethersulfone film, the hydrophilicity of the polyethersulfone/PA66 filtration film surface was improved. This may be related to the fact that a considerable number of hydrophilic groups existed in PA66 [44], and the homogeneous dispersion of gelatin in the mixing system, thus exposing the hydrophilic groups more sufficiently to the surface. The hydrophilic surfaces of materials have been considered as an important factor in increasing the antimicrobial agent loading and antimicrobial activity. Karam et al. [45] found that hydrophilic surfaces could load more antimicrobial agents (nisin) and had higher antimicrobial activity than hydrophobic surfaces. Therefore, the GA/PA66 composite film presents promising potential in loading high doses of antimicrobial agents for food packaging.

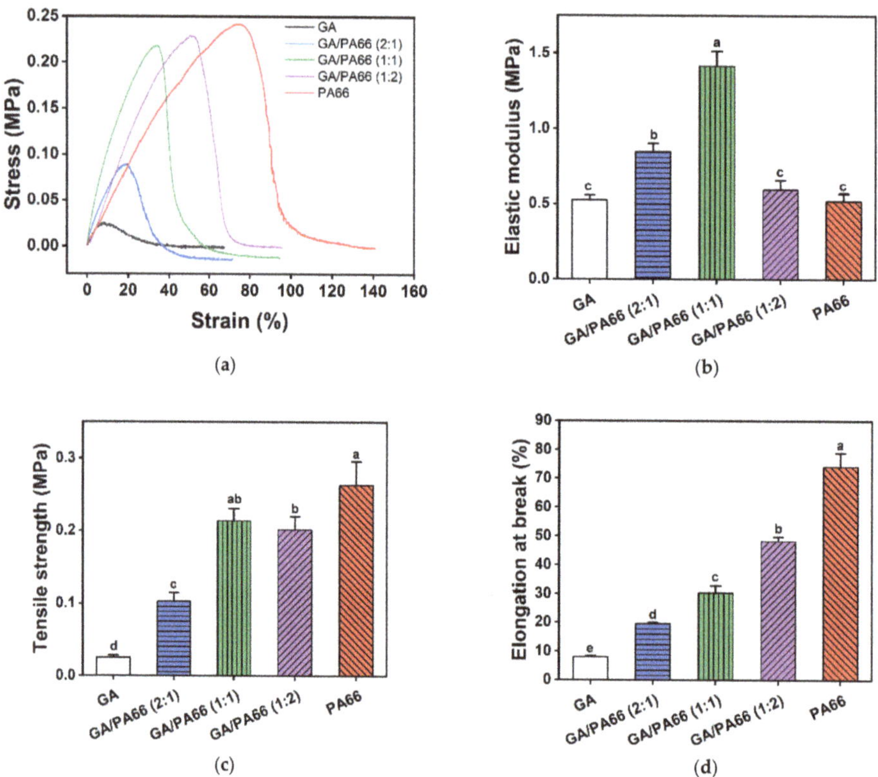

Figure 4. Mechanical properties of the nanofibrous films: (**a**) stress-strain curves; (**b**) elastic modulus; (**c**) elongation at break; (**d**) tensile strength. Values denoted with different letters (a–e) are significantly different ($p < 0.05$), where a is the highest value.

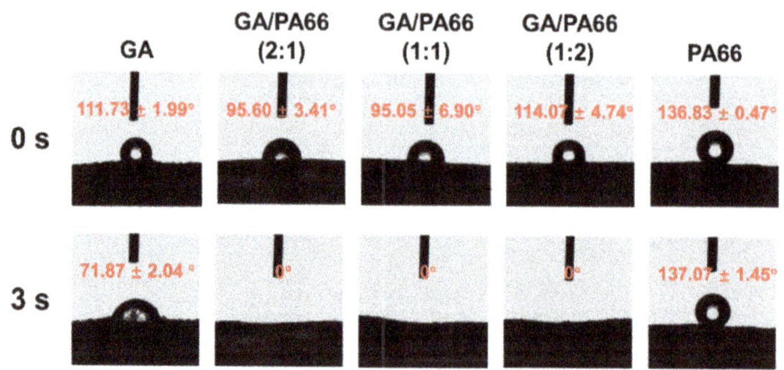

Figure 5. Water contact angles of the films at 0 s and 3 s (equilibration time).

3.7. Water Vapor Permeability Analysis

The rate of water vapor transmission was investigated to be proportional to the porosity of the nanofiber film due to the fiber diameter [46]. Figure 6 shows the WVP values of the GA, PA66 films and GA/PA66 films with various weight ratios. It is well acknowledged that WVP is an inevitable factor of packaging products, which is associated with the exchange of water with the food and the environment [34]. The WVP values of the pure GA and PA66 nanofibrous mat were 16.99 and 12.61 g mm/m^2 h kPa, respectively. When mixing with PA66, the WVP decreased significantly at weight ratios of 2:1 and 1:2 (GA/PA66), and the WVP for the GA/PA66 composite film declined from 16.99 to 9.93 g mm/m^2 h kPa at weight ratios of 2:1 (GA/PA66), indicating that the water vapor barrier performance was improved. This improved water vapor barrier performance could be attributed to a higher crystallization, which could function as a nucleating element, producing increased crystallinity and impermeable regions in the film [47]. In a similar study of the modification of gelatin by Mohammadi et al. [48], they also found that the WVP of GA/chitosan composite film significantly decreased after chitosan was added.

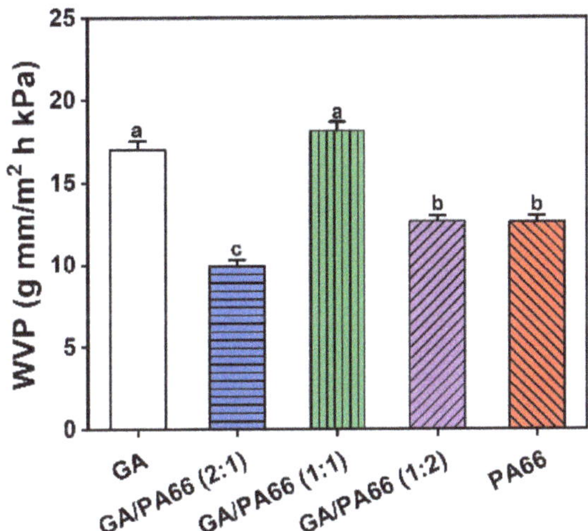

Figure 6. Water vapor permeability of the nanofibrous films. Values denoted with different letters (a–c) are significantly different ($p < 0.05$), where a is the highest value.

3.8. Solvent Resistance

Figure 7 shows the morphologies of the PA66 films and GA/PA66 films after immersing them in 1 × PBS for 24 h and drying. The pure gelatin film (Figure 7a) swelled in 1 × PBS solution, completely losing their fibrous structure and dried to form a transparent solid with a block-like structure, while the pure PA66 film (Figure 7e) remained intact in its fibrous form. With the addition of PA66, the fiber structure of the composite films (Figure 7b–d) was well maintained, indicating that the addition of PA66 could effectively improve the ability of GA to resist dissolution and thus maintained the various advantages brought by its nanometer size. The explanation for this could be that the PA66 network provided good support in an aqueous environment to maintain the basic structure of the gelatin. This intact fiber structure could improve the barrier performance of the fiber film in humid environments. The above results are consistent with the observation of Deng et al.'s work [16], which found that zein particles were distributed nicely in the gelatin network and helped preserve the compact structure by co-electrospinning. As reported by Ahammed et al. [49], the resistance to water of GA was also improved with the solubility in water decreased from 100% to 40% after the introduction of zein. The improved solvent resistance allows GA/PA66 composite films to be used for packaging of foods that generate water vapor due to respiration, such as fruits and vegetables.

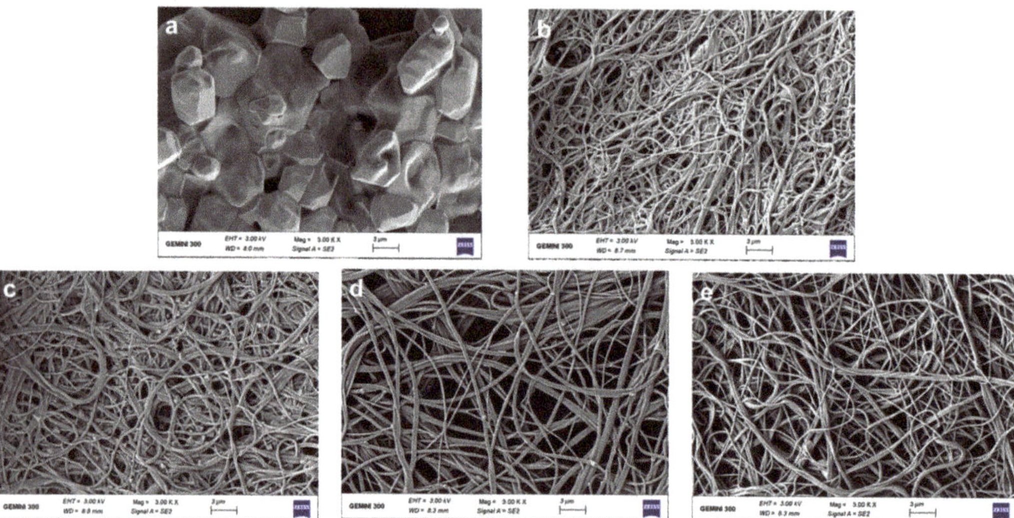

Figure 7. Morphologies of the GA (**a**), GA/PA66 (2:1) (**b**), GA/PA66 (1:1) (**c**), GA/PA66 (1:2) (**d**) and PA66 (**e**) films after immersed in 1 × phosphate-buffered saline (PBS) for 24 h.

4. Conclusions

In this work, GA/PA66 composite films were fabricated successfully by the SBS technique. The addition of PA66 resulted in a significant reduction in the diameter of the composite film. The SBS process mixed GA and PA66 into a homogeneous system by forming hydrogen bonding. Compared to the pure gelatin film, the GA/PA66 composite films had higher mechanical strength and the solvent resistance was improved. Overall, the comprehensive observation of GA/PA66 composite films indicated that PA66 can be used as a modified material for gelatin to improve the mechanical properties and solvent resistance, and SBS presents great promise for the rapid preparation of large area nanofiber film for food packaging.

Author Contributions: Data curation, methodology, formal analysis, investigation, writing—original draft, Z.Y.; methodology, formal analysis, C.S.; methodology, formal analysis, Y.Z.; conceptualization, supervision, funding acquisition, writing—review and editing, D.W.; conceptualization, Supervision, H.Z.; supervision, funding acquisition, K.C. All authors have read and agreed to the published version of the manuscript.

Funding: This research was funded by the Zhejiang Provincial Key R&D Program of China (2019C02074) and the Fundamental Research Funds for the Central Universities (K20210202).

Institutional Review Board Statement: Not applicable.

Informed Consent Statement: Not applicable.

Data Availability Statement: Not applicable.

Conflicts of Interest: The authors declare no conflict of interest.

References

1. Cui, T.; Yu, J.; Li, Q.; Wang, C.F.; Chen, S.; Li, W.; Wang, G. Large-scale fabrication of robust artificial skins from a biodegradable sealant-loaded nanofiber scaffold to skin tissue via microfluidic blow-spinning. *Adv. Mater.* **2020**, *32*, 2000982. [CrossRef] [PubMed]
2. Benavides, R.E.; Jana, S.C.; Reneker, D.H. Nanofibers from scalable gas jet process. *ACS Macro Lett.* **2012**, *1*, 1032–1036. [CrossRef]
3. Dadol, G.C.; Lim, K.J.A.; Cabatingan, L.K.; Tan, N.P.B. Solution blow spinning-polyacrylonitrile-assisted cellulose acetate nanofiber membrane. *Nanotechnology* **2020**, *31*, 345602. [CrossRef] [PubMed]
4. Dias, F.T.G.; Rempel, S.P.; Agnol, L.D.; Bianchi, O. The main blow spun polymer systems: Processing conditions and applications. *J. Polym. Res.* **2020**, *27*, 205. [CrossRef]
5. Zhang, C.; Li, Y.; Wang, P.; Zhang, H. Electrospinning of nanofibers: Potentials and perspectives for active food packaging. *Compr. Rev. Food Sci. Food Saf.* **2020**, *19*, 479–502. [CrossRef] [PubMed]
6. Deng, L.; Kang, X.; Liu, Y.; Feng, F.; Zhang, H. Characterization of gelatin/zein films fabricated by electrospinning vs solvent casting. *Food Hydrocoll.* **2018**, *74*, 324–332. [CrossRef]
7. Radusin, T.; Torres-Giner, S.; Stupar, A.; Ristic, I.; Miletic, A.; Novakovic, A.; Lagaron, J.M. Preparation, characterization and antimicrobial properties of electrospun polylactide films containing *Allium ursinum* L. extract. *Food Packag. Shelf Life* **2019**, *21*, 100357. [CrossRef]
8. Vargas Romero, E.; Lim, L.-T.; Suárez Mahecha, H.; Bohrer, B.M. The effect of electrospun polycaprolactone nonwovens containing chitosan and propolis extracts on fresh pork packaged in linear low-density polyethylene films. *Foods* **2021**, *10*, 1110. [CrossRef]
9. Wang, D.D.; Liu, Y.; Sun, J.; Sun, Z.; Liu, F.; Du, L.; Wang, D.D. Fabrication and characterization of gelatin/zein nanofiber films loading perillaldehyde for the preservation of chilled chicken. *Foods* **2021**, *10*, 1277. [CrossRef] [PubMed]
10. Tandon, B.; Kamble, P.; Olsson, R.T.; Blaker, J.J.; Cartmell, S.H. Fabrication and Characterisation of Stimuli Responsive Piezoelectric PVDF and Hydroxyapatite-Filled PVDF Fibrous Membranes. *Molecules* **2019**, *24*, 1903. [CrossRef] [PubMed]
11. Sett, S.; Stephansen, K.; Yarin, A.L. Solution-blown nanofiber mats from fish sarcoplasmic protein. *Polymer* **2016**, *93*, 78–87. [CrossRef]
12. Shen, C.; Cao, Y.; Rao, J.; Zou, Y.; Zhang, H.; Wu, D.; Chen, K. Application of solution blow spinning to rapidly fabricate natamycin-loaded gelatin/zein/polyurethane antimicrobial nanofibers for food packaging. *Food Packag. Shelf Life* **2021**, *29*, 100721. [CrossRef]
13. Van Vlierberghe, S.; Dubruel, P.; Schacht, E. Biopolymer-based hydrogels as scaffolds for tissue engineering applications: A review. *Biomacromolecules* **2011**, *12*, 1387–1408. [CrossRef]
14. Nair, L.S.; Laurencin, C.T. Biodegradable polymers as biomaterials. *Prog. Polym. Sci.* **2007**, *32*, 762–798. [CrossRef]
15. Yang, H.; Wang, Y. Effects of concentration on nanostructural images and physical properties of gelatin from channel catfish skins. *Food Hydrocoll.* **2009**, *23*, 577–584. [CrossRef]
16. Deng, L.; Zhang, X.; Li, Y.; Que, F.; Kang, X.; Liu, Y.; Feng, F.; Zhang, H. Characterization of gelatin/zein nanofibers by hybrid electrospinning. *Food Hydrocoll.* **2018**, *75*, 72–80. [CrossRef]
17. Nguyen, T.-H.; Lee, B.-T. Fabrication and characterization of cross-linked gelatin electro-spun nano-fibers. *J. Biomed. Sci. Eng.* **2010**, *3*, 1117–1124. [CrossRef]
18. Bigi, A.; Cojazzi, G.; Panzavolta, S.; Rubini, K.; Roveri, N. Mechanical and thermal properties of gelatin films at different degrees of glutaraldehyde crosslinking. *Biomaterials* **2001**, *22*, 763–768. [CrossRef]
19. Farris, S.; Song, J.; Huang, Q. Alternative reaction mechanism for the cross-linking of gelatin with glutaraldehyde. *J. Agric. Food Chem.* **2010**, *58*, 998–1003. [CrossRef] [PubMed]
20. Li, X.; Cui, R.; Sun, L.; Aifantis, K.E.; Fan, Y.; Feng, Q.; Cui, F.; Watari, F. 3D-printed biopolymers for tissue engineering application. *Int. J. Polym. Sci.* **2014**, *2014*, 829145. [CrossRef]
21. Zhang, Y.; Ouyang, H.; Chwee, T.L.; Ramakrishna, S.; Huang, Z.M. Electrospinning of gelatin fibers and gelatin/PCL composite fibrous scaffolds. *J. Biomed. Mater. Res.-Part B Appl. Biomater.* **2005**, *72*, 156–165. [CrossRef]
22. Meng, Z.X.; Wang, Y.S.; Ma, C.; Zheng, W.; Li, L.; Zheng, Y.F. Electrospinning of PLGA/gelatin randomly-oriented and aligned nanofibers as potential scaffold in tissue engineering. *Mater. Sci. Eng. C* **2010**, *30*, 1204–1210. [CrossRef]
23. Fallah, M.; Bahrami, S.H.; Ranjbar-Mohammadi, M. Fabrication and characterization of PCL/gelatin/curcumin nanofibers and their antibacterial properties. *J. Ind. Text.* **2016**, *46*, 562–577. [CrossRef]
24. Rose, J.B.; Sidney, L.E.; Patient, J.; White, L.J.; Dua, H.S.; El Haj, A.J.; Hopkinson, A.; Rose, F.R.A.J. In vitro evaluation of electrospun blends of gelatin and PCL for application as a partial thickness corneal graft. *J. Biomed. Mater. Res.-Part A* **2019**, *107*, 828–838. [CrossRef]
25. Han, B.; Ji, G.; Wu, S.; Shen, J. Preparation and characterization of nylon 66/montmorillonite nanocomposites with co-treated montmorillonites. *Eur. Polym. J.* **2003**, *39*, 1641–1646. [CrossRef]
26. Ahmadloo, E.; Gharehaghaji, A.A.; Latifi, M.; Saghafi, H.; Mohammadi, N. Effect of PA66 nanofiber yarn on tensile fracture toughness of reinforced epoxy nanocomposite. *Proc. Inst. Mech. Eng. Part C J. Mech. Eng. Sci.* **2019**, *233*, 2033–2043. [CrossRef]
27. Lee, J.E.; Han, D.H.; Kim, Y.M.; Choi, K.M. Comparison of mechanical and thermal properties of polyamide-66/rubber/fiber alloy composites. *Asian J. Chem.* **2013**, *25*, 5237–5244. [CrossRef]
28. Spina, R.; Cavalcante, B. Evaluation of grinding of unfilled and glass fiber reinforced polyamide 6,6. *Polymers* **2020**, *12*, 2288. [CrossRef] [PubMed]

29. Cakal Sarac, E.; Haghighi Poudeh, L.; Berktas, I.; Saner Okan, B. Scalable fabrication of high-performance graphene/polyamide 66 nanocomposites with controllable surface chemistry by melt compounding. *J. Appl. Polym. Sci.* **2021**, *138*, 49972. [CrossRef]
30. Zhang, J.; Gao, X.; Zhang, X.; Liu, H.; Zhang, H.; Zhang, X. Polyamide 66 and amino-functionalized multi-walled carbon nanotube composites and their melt-spun fibers. *J. Mater. Sci.* **2019**, *54*, 11056–11068. [CrossRef]
31. Xiong, Y.; Ren, C.; Zhang, B.; Yang, H.; Lang, Y.; Min, L.; Zhang, W.; Pei, F.; Yan, Y.; Li, H.; et al. Analyzing the behavior of a porous nano-hydroxyapatite/polyamide 66 (n-HA/PA66) composite for healing of bone defects. *Int. J. Nanomed.* **2014**, *9*, 485–494. [CrossRef] [PubMed]
32. Schuster, J.M.; Schvezov, C.E.; Rosenberger, M.R. Influence of Experimental Variables on the Measure of Contact Angle in Metals Using the Sessile Drop Method. *Procedia Mater. Sci.* **2015**, *8*, 742–751. [CrossRef]
33. Hasanpour Ardekani-Zadeh, A.; Hosseini, S.F. Electrospun essential oil-doped chitosan/poly(ε-caprolactone) hybrid nanofibrous mats for antimicrobial food biopackaging exploits. *Carbohydr. Polym.* **2019**, *223*, 115108. [CrossRef] [PubMed]
34. Cui, H.; Yuan, L.; Lin, L. Novel chitosan film embedded with liposome-encapsulated phage for biocontrol of *Escherichia coli* O157:H7 in beef. *Carbohydr. Polym.* **2017**, *177*, 156–164. [CrossRef] [PubMed]
35. Sinha-Ray, S.; Zhang, Y.; Yarin, A.L.; Davis, S.C.; Pourdeyhimi, B. Solution Blowing of Soy Protein Fibers. *Biomacromolecules* **2011**, *12*, 2357–2363. [CrossRef] [PubMed]
36. Sengor, M.; Ozgun, A.; Corapcioglu, G.; Ipekoglu, M.; Garipcan, B.; Ersoy, N.; Altintas, S. Core–shell PVA/gelatin nanofibrous scaffolds using co-solvent, aqueous electrospinning: Toward a green approach. *J. Appl. Polym. Sci.* **2018**, *135*, 46582. [CrossRef]
37. Lu, L.; Yang, B.; Liu, J. Flexible multifunctional graphite nanosheet/electrospun-polyamide 66 nanocomposite sensor for ECG, strain, temperature and gas measurements. *Chem. Eng. J.* **2020**, *400*, 125928. [CrossRef]
38. Gu, A.; Wu, J.; Shen, L.; Zhang, X.; Bao, N. High-strength GO/PA66 nanocomposite fibers via in situ precipitation and polymerization. *Polymers* **2021**, *13*, 1688. [CrossRef]
39. Noguchi, T.; Niihara, K.; Kawamoto, K.; Fukushi, M.; Jinnai, H.; Nakajima, K.; Endo, M. Preparation of high-performance carbon nanotube/polyamide composite materials by elastic high-shear kneading and improvement of properties by induction heating treatment. *J. Appl. Polym. Sci.* **2021**, *138*, 50512. [CrossRef]
40. Zou, Y.; Zhang, C.; Wang, P.; Zhang, Y.; Zhang, H. Electrospun chitosan/polycaprolactone nanofibers containing chlorogenic acid-loaded halloysite nanotube for active food packaging. *Carbohydr. Polym.* **2020**, *247*, 116711. [CrossRef]
41. Zhao, L.; Zhao, J.; Jiang, W.; Zhou, H.; He, J. Preparation and properties of composite phase-change nanofiber membrane by improved bubble electrospinning. *Mater. Res. Express* **2021**, *8*, 55011. [CrossRef]
42. Baseer, A.; Koenneke, A.; Zapp, J.; Khan, S.A.; Schneider, M. Design and characterization of surface-crosslinked gelatin nanoparticles for the delivery of hydrophilic macromolecular drugs. *Macromol. Chem. Phys.* **2019**, *220*, 1900260. [CrossRef]
43. Cao, J.; Cheng, Z.; Kang, L.; Zhang, Y.; Zhao, X.; Zhao, S.; Gao, B. Novel anti-fouling polyethersulfone/polyamide 66 membrane preparation for air filtration by electrospinning. *Mater. Lett.* **2017**, *192*, 12–16. [CrossRef]
44. Chi, E.; Tang, Y.; Wang, Z. In situ SAXS and WAXD investigations of polyamide 66/reduced graphene oxide nanocomposites during uniaxial deformation. *ACS Omega* **2021**, *6*, 11762–11771. [CrossRef] [PubMed]
45. Karam, L.; Jama, C.; Mamede, A.S.; Boukla, S.; Dhulster, P.; Chihib, N.E. Nisin-activated hydrophobic and hydrophilic surfaces: Assessment of peptide adsorption and antibacterial activity against some food pathogens. *Appl. Microbiol. Biotechnol.* **2013**, *97*, 10321–10328. [CrossRef] [PubMed]
46. Hosseini, S.F.; Nahvi, Z.; Zandi, M. Antioxidant peptide-loaded electrospun chitosan/poly(vinyl alcohol) nanofibrous mat intended for food biopackaging purposes. *Food Hydrocoll.* **2019**, *89*, 637–648. [CrossRef]
47. Karkhanis, S.S.; Stark, N.M.; Sabo, R.C.; Matuana, L.M. Water vapor and oxygen barrier properties of extrusion-blown poly(lactic acid)/cellulose nanocrystals nanocomposite films. *Compos. Part A Appl. Sci. Manuf.* **2018**, *114*, 204–211. [CrossRef]
48. Mohammadi, R.; Mohammadifar, M.A.; Rouhi, M.; Kariminejad, M.; Mortazavian, A.M.; Sadeghi, E.; Hasanvand, S. Physico-mechanical and structural properties of eggshell membrane gelatin- chitosan blend edible films. *Int. J. Biol. Macromol.* **2018**, *107*, 406–412. [CrossRef]
49. Ahammed, S.; Liu, F.; Khin, M.N.; Yokoyama, W.H.; Zhong, F. Improvement of the water resistance and ductility of gelatin film by zein. *Food Hydrocoll.* **2020**, *105*, 105804. [CrossRef]

Article

Development of *Nervilia fordii* Extract-Loaded Electrospun PVA/PVP Nanocomposite for Antioxidant Packaging

Peng Wen [1], Teng-Gen Hu [2,3], Yan Wen [2], Ke-Er Li [1], Wei-Peng Qiu [1], Zhi-Lin He [1], Hong Wang [1,*] and Hong Wu [2,*]

[1] College of Food Science, Guangdong Provincial Key Laboratory of Food Quality and Safety, South China Agricultural University, Guangzhou 510642, China; pwen@scau.edu.cn (P.W.); 201822010208@stu.scau.edu.cn (K.-E.L.); 201822010215@stu.scau.edu.cn (W.-P.Q.); 201822010206@stu.scau.edu.cn (Z.-L.H.)

[2] School of Food Science and Engineering, South China University of Technology/Guangdong Province Key Laboratory for Green Processing of Natural Products and Product Safety, Guangzhou 510640, China; hu.tenggen@gdaas.cn (T.-G.H.); ShakingW@aliyun.com (Y.W.)

[3] Sericultural & Agri-Food Research Institute, Guangdong Academy of Agricultural Sciences/Key Laboratory of Functional Foods, Ministry of Agriculture/Guangdong Key Laboratory of Agricultural Products Processing, Guangzhou 510640, China

* Correspondence: gzwhongd@163.com (H.W.); bbhwu@scut.edu.cn (H.W.)

Abstract: An ethyl acetate extract from of *Nervilia fordii* (NFE) with considerable suppression activity on lipid peroxidation (LPO) was first obtained with total phenolic and flavonoid contents and anti-LPO activity (IC_{50}) of 86.67 ± 2.5 mg GAE/g sample, 334.56 ± 4.7 mg RE/g extract and 0.307 mg/mL, respectively. In order to improve its stability and expand its application in antioxidant packaging, the nano-encapsulation of NFE within poly(vinyl alcohol) (PVA) and polyvinyl(pyrrolidone) (PVP) bio-composite film was then successfully developed using electrospinning. SEM analysis revealed that the NFE-loaded fibers exhibited similar morphology to the neat PVA/PVP fibers with a bead-free and smooth morphology. The encapsulation efficiency of NFE was higher than 90% and the encapsulated NFE still retained its antioxidant capacity. Fourier transform infrared spectroscopy (FTIR) and X-ray powder diffraction (XRD) analysis confirmed the successful encapsulation of NFE into fibers and their compatibility, and the thermal stability of which was also improved due to the intermolecular interaction demonstrated by thermo gravimetric analysis (TGA). The ability to preserve the fish oil's oxidation and extend its shelf-life was also demonstrated, suggesting the obtained PVA/PVP/NFE fiber mat has the potential as a promising antioxidant food packaging material.

Keywords: electrspinning; *Nervilia fordii* extract; nano-encapsulation; antioxidant activity

Citation: Wen, P.; Hu, T.-G.; Wen, Y.; Li, K.-E.; Qiu, W.-P.; He, Z.-L.; Wang, H.; Wu, H. Development of *Nervilia fordii* Extract-Loaded Electrospun PVA/PVP Nanocomposite for Antioxidant Packaging. *Foods* **2021**, *10*, 1728. https://doi.org/10.3390/foods10081728

Academic Editor: Pilar Montero

Received: 27 June 2021
Accepted: 23 July 2021
Published: 27 July 2021

1. Introduction

Antioxidant packing is a promising approach to avoiding oxidative damage of fatty foods by the incorporation of antioxidant agents into the packaging material [1]. This allows sustained release behavior of encapsulated antioxidants with high bioactivity, thus preserving food quality and extending its shelf-life. The increasing demand for healthy and safe food has promoted the study and the utilization of natural extracts, such as *allium ursinum L.* extract [2], *coptis chinensis* extract [3] and cinnamon essential oil [4]. *Nervilia fordii* (Hance) Schltr (Orchidaceae), also called "Qing Tian Kui", is one kind of *Nervilia* species that is mainly distributed in the Southwest of China and Southeast Asia [5]. It has been employed as a traditional Chinese herbal medicine for the treatment of tuberculosis, cough, and throat swelling. Phytochemical studies have focused on the isolation and identification of bioactive compounds [6]. This is the first report regarding the antioxidant activity of different solvent extracts from *Nervilia fordii* and its potential for antioxidant packaging.

Studies have shown that the nanostructure of packaging material with a high surface area is beneficial for the release of encapsulated compounds compared to that of a bulk material. Electrospinning is an emerging and promising technique to produce nanofibers due to the ease of operation, versatility at room temperature, controllable fiber morphology, and high entrapment capacity of different bioactive compounds [7]. It has been well established in food areas such as active food packaging, targeted delivery, and functional foods [8–10]. Recently, more attention has been paid to biodegradable polymers because of the serious environmental problem caused by traditional plastic waste. Polyvinylpyrrolidone (PVP), a nontoxic and biocompatible polymer with excellent film-forming ability, has been applied as a packaging material [11]. Previous study showed that electrospun PVP nanofibers can be used to increase the solubility and achieve the controlled release of curcumin [12]. However, the neat PVP film is too brittle and easily broken, which limits its application. Poly(vinyl alcohol) (PVA) is also an electrospinable and non-toxic polymer permitted for food contact materials [13]. It has been reported that PVA has been blended with other polymers to encapsulate bioactive compounds, including chitosan [14], polycaprolactone [15], and soy protein isolate [16]. To the best of our knowledge, studies on combinations of PVA and PVP for the preparation of electrospun antioxidant packaging material have not been reported yet.

Hence, in this study, different solvent extracts of *Nervilia fordii* were prepared and the antioxidant activity was measured using DPPH radical scavenging and lipid peroxidation assay. After that, the development of the PVA/PVP-based biodegradable polymer matrix for the encapsulation of ethyl acetate extract of *Nervilia fordii* (NFE) using electrospinning was performed to improve its stability and antioxidant bioactivity. The morphology, encapsulation efficiency and structural characterization of the obtained composite fibers were investigated. Finally, the applicability for lessening the fish oil oxidation was also explored.

2. Materials and Methods

2.1. Materials

The dried aerial part of *Nervilia fordii* with water content being around 11% was purchased from the Dashenlin Pharmaceutical Co. Ltd. (Guangzhou, China) with phenological stage being September in 2018; 2, 2-Diphenyl-1-picrylhydrazine (DPPH), ascorbic acid (Vc), rutin, gallic acid were obtained from Sigma-Aldrich Co. (Shanghai, China). PVP of molecular weight 1,300,000 was obtained from Aladdin Biological Technology Co., Ltd. (Shanghai, China); PVA (M w 85,000−124,000) was provided by Tianyi company (Guangzhou, China). All other reagents used were analytical grade.

2.2. Preparation of Different Solvent Extracts

The different *Nervilia fordii* extracts were prepared as follows: 10 g of air-dried aerial part of *Nervilia fordii* plant was mixed with 100 mL of a certain concentration of ethanol solution. The mixture was stirred at a certain temperature (50~90 °C) for 2 h, and the residues were re-extracted twice. Filtrates were pooled, and, after the evaporation of ethanol, the concentrated ethanol extract was then extracted with petroleum ether, ethyl acetate, and n-butanol for 4 h, respectively. The supernatants were combined and the residue was re-extracted by repeating the above procedure twice. The effects of variables like the ethanol concentration, temperature and solid-to-solvent ratio on DPPH activity were investigated. Finally, the obtained extracts were freeze-dried and kept in a desiccator for further analysis.

2.3. Determination of Antioxidant Contents in *Nervilia fordii* Extracts

(1) Total phenol content (TPC)

TPC was measured by Folin–Ciocalteu Reagent assay according to a previous study with some modifications [17]. In brief, 100 μL of different concentrations of solvent extracts were individually dissolved in 70% ethanol. Then, 0.5 mL of Folin–Ciocalteu Reagent and 7.9 mL of water were added, followed by shaking. After incubation in the dark for 5 min, the solution was reacted with 1.5 mL of 10% Na_2CO_3 solution and incubated in the dark for 2 h. After that, the absorbance at 765 nm against a blank was measured. A series of gallic acid standard solution was used to establish a calibration curve. The results were expressed as mg gallic acid equivalent (GAE)/g extract. For the NFE-loaded film, the sample was first immersed into 10 mL of 70% ethanol. The obtained extract solution from film was then determined by the above method. The TPC is expressed as mg of GAE/g film.

(2) Total flavonoid content (TFC)

TFC was calculated by $AlCl_3$-HAc-NaAc (pH 5.5) assay. Briefly, 200 µL different concentrations of solvent extracts were individually dissolved in 70% ethanol. Then 300 µL of 0.1 mol/L $AlCl_3$ solution and 200 µL of HAc-NaAc buffer solution (pH 5.5) were serially added and mixed with 4.5 mL of 70% ethanol. The absorbance at 405 nm was monitored. A series of rutin standard solution was used to establish a calibration curve. The results are expressed as mg rutin equivalent (RE)/g extract.

2.4. Antioxidant Activity Analysis

(1) DPPH assay

DPPH assay was used to evaluate the free radical scavenging activity of each sample as described by Neo with slight modifications [18]. Briefly, a certain amount of different extracts was dissolved in 70% ethanol aqueous solutions. One milliliter of the sample solution was then mixed with 3 mL of 0.1 mM DPPH that dissolved in ethanol in the dark for 30 min. The DPPH radical-scavenging rate was expressed as follows:

$$\text{DPPH radical-scavenging rate } \% = (A_{517,\text{ control}} - A_{517,\text{sample}})/A_{517,\text{control}} \times 100\%$$

$A_{517,sample}$ and $A_{517,control}$ are the absorbance of DPPH solution with or without samples.

(2) Lipid peroxidation (LPO) assay

First, the yolk lipoprotein solution was prepared as follows: albumen was removed from fresh eggs and the yolk was mixed with the same volume of 0.1 mol/L phosphate buffer (pH 7.4). After shaking for 10 min, the mixture was diluted into 1:25 using phosphate buffer. The obtained solutions were kept at 4 °C. Then, 0.2 mL of the above solutions was serially mixed with 0.1 mL of tested sample, 0.2 mL of $FeSO_4$ and 1.7 mL phosphate buffer. The mixtures were incubated at 37 °C for 1 h followed by adding 0.5 mL of trichloroacetic acid (TBA, 20%, *w*/*v*) solution. After 10 min incubation, the mixture was centrifuged at 4000 rpm for 10 min. The supernatant (2 mL) was taken out and mixed with 1 mL of TBA (0.8%) in a boiling water bath for 15 min. After cooling down to 25 °C, the absorbance was monitored at 532 nm. The LPO suppression ratio was determined as follows: LPO suppression ratio (%) = $((A_{control} - A_{sample})/A_{control}) \times 100\%$

A_{sample} and $A_{control}$ are the absorbance of solution with or without samples.

2.5. Encapsulation of *Nervilia fordii* Extract by Electrospinning

NFE was first prepared according to the optimized extract conditions described in Section 2.2. After that, the blending solution of PVA with PVP containing NFE was achieved as follows. First, 10%~15% (*w*/*v*) of PVA solution was prepared by stirring at 80 °C for 2 h. Then, the NFE-loaded PVP solution that was dissolved in 90% ethanol was added into the above PVA solution with a volume ratio of 1:1 to achieve a total polymer mass ratio and NFE of 12.5% and 4%, respectively. Finally, the electrospinning solution was injected into a syringe (21# needle), and the electrospinning process was conducted as a feed rate of 0.3 mL/h, voltage of 16 kV and distance of 14 cm. The encapsulation efficiency (EE) and loading capacity (LC) of NFE were determined based on the measurement of the amount of non-encapsulated extract. In brief, hexane was used to remove the free NFE from the electrospun nanofiber for 2 min, and the absorbance of obtained solution was measured at 472 nm. A series of NFE solution dissolved in hexane was used to establish a calibration curve. The EE and LC were determined as below:

$$\text{EE}\% = (\text{theoretical mass of NFE} - \text{free mass of NFE})/\text{theoretical mass of NFE} * 100$$

$$LC\% = (\text{theoretical mass of NFE} - \text{free mass of NFE})/\text{the mass of fiber}*100$$

2.6. Characterization of the Electrospun Fiber Mat

Scanning Electron Microscopy (SEM). The morphology of the obtained fiber was observed by a 3700 N scanning electron microscopy (SEM, Hitachi, Japan). The average diameter of fibers and the fibers size distribution was analyzed by Image-J software.

Fourier transform infrared Spectroscopy (FTIR). The interactions among PVA, PVP and NFE were investigated by FTIR spectroscopy (Bruker-VERTEX 70, Germany). The analysis was conducted under wave number of 3800–500 cm^{-1} and resolution of 4 cm^{-1}.

X-ray diffraction pattern (XRD). The XRD pattern of the different composite fibers was recorded to examine the crystallography of the prepared films using a MiniFlex 600 diffractometer (RigaKu, Tokyo, Japan) with Cu-Kα radiation. Data were collected in the 2θ range from 5° to 60° with a step of 0.02°.

Thermogravimetric Analysis (TGA). The thermal property of the fiber mats was characterized using a TGA Q500 (TA Instruments, New Castle, DE, USA). The sample was heated from 25 °C to 700 °C with a heating rate of 20 °C/min under nitrogen gas atmosphere.

2.7. Oxidative Stability in Accelerated Storage Test

The oxidative stability of encapsulated fish oil was analyzed under storage conditions of 45 °C since low and ambient temperatures require a relative long period of time. As described previously [19], 10 mL of fish oil sample was aliquoted into 20 mL brown glass vials, and 20 mg TA fibrous mat (2 cm × 3 cm) were added, followed by incubation at 45 °C in the dark for 30 days, as shown in Figure 1. This subject is of primary importance, since the goal of the paper is to evaluate the potential application of the antioxidant film as a pouch or direct contacting material. The un-encapsulated fish oil was used as the control, while oil samples with pure PVA/PVP film were used as a negative control. Peroxide value (PV) analysis was conducted by taking different amounts of samples out at different intervals as follows [20]. Briefly, 3 g of fish oil sample was dissolved in 50 mL of acetic acid and chloroform mixture (3:2 *v*/*v*). Then, 1 mL of saturated KI solution was added and the mixture was kept in the dark for 1 min. After adding distilled water (50 mL), the mixture was immediately titrated with 0.01 mol/L of sodium thiosulfate until the yellow color had almost disappeared. The PV value was determined as follows:

$$\text{PV value (mEq of } O_2/\text{kg sample)} = 12.69 \times 78.8 \times (V_S - V_B) \times C/m.$$

where V_S and V_B is the volume of titrant used in the titration of oil sample and a blank without any oil sample (mL), respectively, C is the concentration of sodium thiosulfate (mol/L) and m is the weight of oil sample (g).

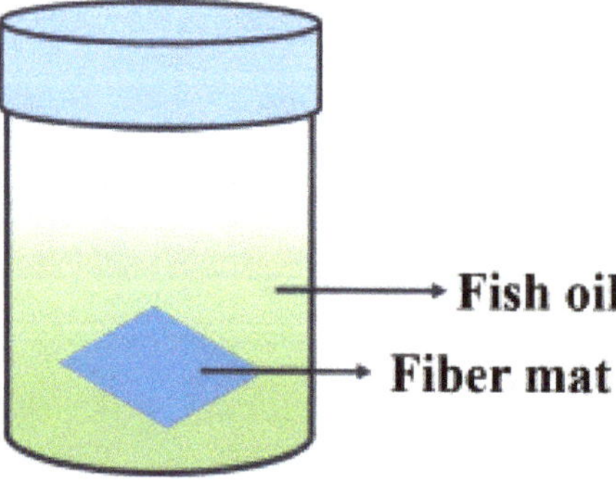

Figure 1. The schematic diagram for measuring the fiber mat's antioxidant activity on fish oil oxidation.

2.8. Statistical Analysis

All experiments were performed in triplicate. The obtained results were reported as the mean values ± standard deviations. Significant differences were carried out by SPSS 17 statistical software (SPSS Inc., Chicago, IL, USA), as $p < 0.05$ is regarded to be significantly different.

3. Results

3.1. Preparation of *Nervilia fordii* Ethanol Extract

The extraction parameters for *Nervilia fordii* ethanol extract was first optimized using single factor experiments by investigating the effects of variables such as the ethanol concentration, temperature and solid-to-solvent ratio on the antioxidant activity and yields of *Nervilia fordii* ethanol extract. Regarding the ethanol concentration, from Figure 2a, it can be seen that no significant difference ($p < 0.05$) in DPPH scavenging rate between the two ethanol concentrations (60% and 70%) was observed. For the extraction temperature, it was found that the increase of temperature would cause an enhanced DPPH scavenging activity (Figure 2(b1)), which may be ascribed to the accelerated molecular movement and decreased solvent viscidity. There was no statistical significance ($p > 0.05$) among the extraction temperatures with the value of 70 °C vs. 80 °C. In addition, higher temperature could cause the degradation of the flavonoids and phenolic compounds, as exhibited in the yields (Figure 2(b2)), a similar phenomenon was reported by Altemimi et al. [21]. Figure 2(c1) shows that DPPH activity was significantly increased ($p < 0.05$) with the increase of the solid-to-solvent ratio from 1:5 to 1:15, and no significant difference ($p < 0.05$) was observed between 1:10 and 1:15. This finding agreed with Prasad et al. who found that the permeation of compounds into the solvent can be enhanced under the higher solid-liquid ratio [22]. However, too high of a liquid-solid ratio can restrain the cavitation effect (Figure 2(c2)). Hence, from an economic point of view, the optimized process for *Nervilia fordii* ethanol extract was: ethanol concentration 60%, extraction temperature 70 °C and solid-to-solvent ratio 1:10.

After that, Box-behnken design (BBD) and response surface methodology (RSM) were then adopted for the optimization of extraction parameter. The levels for response surface design are shown in Table S1. The ANOVA results (Table S2) revealed that the model was remarkably significant ($p < 0.0001$), and the R-squared value obtained was 0.9968, indicating the model was very consistent with the experiment results. The fitted equation was: DPPH scavenging rate (%) = $38.95 - 0.75{*}A + 1.53{*}B + 1.69{*}C + 0.31{*}B{*}C + 0.28{*}A{*}C - 0.25{*}A{*}B - 0.43{*}A^2 - 0.81{*}B^2 - 1.04{*}C^2$ (A—Ethanol concentration; B—Temperature; C—Solid to liquid ratio). Besides that, two-dimensional contour plots and three-dimensional response surface for the correlation between any two variables are displayed in Figure S1. The optimal condition for extraction of *Nervilia fordii* ethanol extract was obtained as follows: ethanol concentration 55.55%, extraction temperature 75 °C and solid–liquid rate 14.2:1. Regarding the practical situation, the optimal condition was adjusted to an ethanol concentration of 56%, extraction temperature of 75 °C and solid–liquid rate of 14:1. Under which, the DPPH scavenging rate of *Nervilia fordii* ethanol extract obtained was 41.5%, which is closer to the theoretical prediction value of 40.9%, suggesting that the model was desirable.

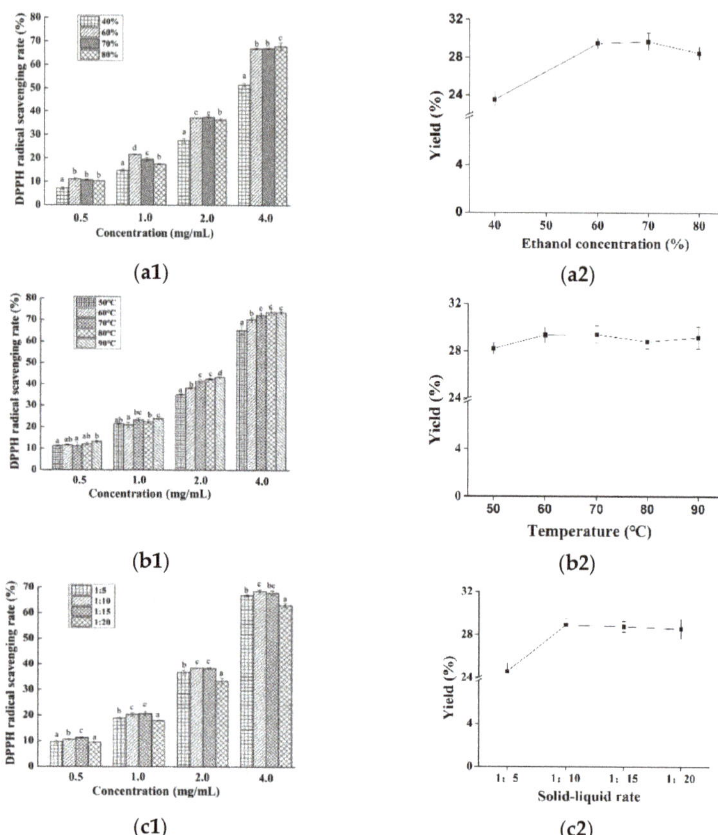

Figure 2. Effects of extraction condition on the extract's DPPH radical scavenging activity and yield. ((**a1**–**c1**) indicated the effects of ethanol concentration, temperature and solid-liquid rate on the DPPH radical scavenging rates of *Nervilia fordii* ethanol extract, respectively; (**a2**–**c2**) indicated the effects of ethanol concentration, temperature and solid-liquid rate on the yields of *Nervilia fordii* ethanol extract, respectively).

3.2. Antioxidant Capacities of Different Solvent Extracts

Then, different solvents (petroleum ether, ethyl acetate and n-butyl alcohol) were used to further extract the above obtained *Nervilia fordii* ethanol extract and their antioxidant activities were evaluated. From Figure 3a, it can be seen that the activity increased with the increase of extract concentration, and the scavenging abilities on DPPH radicals were in the order of ethyl acetate extract of *Nervilia fordii* (NFE) > n-butyl alcohol extract of *Nervilia fordii* (NFB) > petroleum ether extract of *Nervilia fordii* (NFP), which reached 94.5%, 60.5%, 33% and 18.4% at 3 mg/mL, respectively. However, a better scavenging rate was exhibited by Vc (96.3% at 0.1 mg/mL). As summarized in Table 1, the EC_{50} values of DPPH radical-quenching activity for NFE, NFB and NFP were 0.66, 2.43 and 4.25 mg/mL, respectively. Lipid peroxidation (LPO) is also considered to be another type of free radical oxidation, which is related to cellular damage. Similar to the tendency of DPPH scavenging activity of extracts, the LPO inhibition potency of different extracts were also concentration dependent and NFE possessed the highest inhibitory effect. Figure 3b shows that the inhibition rate rose from 24.9% to 77.4% for Vc and from 12.8% to 88.7% for NFE with the concentration increasing from 100 μg/mL to 500 μg/mL, respectively. In particular, a higher inhibition rate was achieved for NFE than that of Vc when the antioxidant concentration >300 μg/mL. From the IC_{50} values of LPO suppression activity in Table 1, NFE (0.307 mg/mL) has considerable activity in comparison to Vc (0.310 mg/mL), while NFB was 0.347 mg/mL and NFP was 0.436 mg/mL. These results clearly indicated that all of the solvent extracts had a noticeable effect on the inhibition of LPO, especially for the NFE, suggesting its potential in inhibiting the oxidation of fatty food or the application for functional foods.

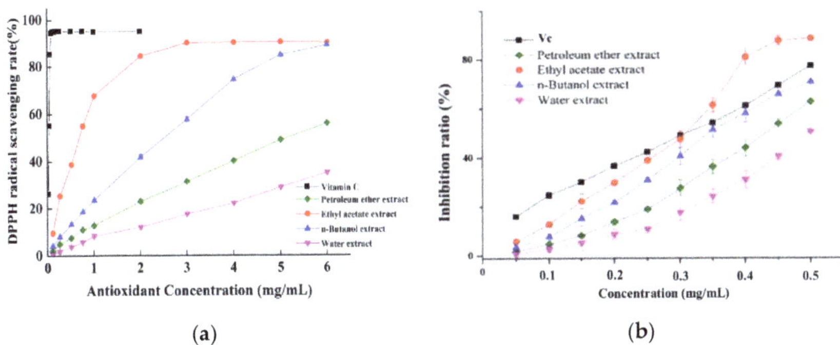

Figure 3. The DPPH radical scavenging ability (**a**) and the LPO suppression ratio (**b**) of different extracts and Vc.

3.3. Total Phenol and Flavonoid Contents

Being the main chain-breaking antioxidants that contributed to the antioxidant activity of extracts, the total phenolic and flavonoid contents in different extracts were examined using gallic acid equivalents (GAE) and rutin equivalents (RE), respectively. As presented in Table 1, the NFE had significantly higher contents of total phenols (86.67 ± 2.5 mg GAE/g extract) and total flavonoids (334.56 ± 4.7 mg RE/g extract), followed by NFB and NFP. It can be concluded that the polar solvent extracts (n-butanol and ethyl acetate) exhibited a higher content than nonpolar solvent extracts (petroleum ether) did, which may contribute to the stronger antioxidant activity of NFE and NFB than that of NFP. Taken the polarities of used solvents into consideration, it has been revealed that flavonoids from *Nervilia fordii* were more extractable by solvents with high polarity.

Table 1. Extraction yields, and total phenols, total flavonoid contents in different extracts.

Sample	Content		EC50 [c] of DPPH	IC50 [d] of LPO
	Total Phenols (mg GAE [a]/g Extract)	Total Flavonoids (mg RE [b]/g Extract)	Radical-Quenching Activity (mg Sample/mL)	Suppression Activity (mg Sample/mL)
Ethyl acetate extract (NFE)	86.67 ± 2.5	334.56 ± 4.7	0.66	0.307
n-Butyl alcohol extract (NFB)	41.85 ± 1.3	125.9 ± 2.6	2.43	0.347
Water extract	18.04 ± 1.2	46.95 ± 1.5	8.36	0.492
Petroleum ether extract (NFP)	15.17 ± 0.8	100.36 ± 3.1	4.25	0.436
Vc			0.087	0.310

Values (mean ± SD, n = 3) in the same column followed by a different letter are significantly different ($p < 0.05$). [a] GAE, Gallic acid equivalents. [b] RE, Rutin equivalents. [c] EC50 means the effective concentration of sample that can decrease 50% of DPPH radical scavenging rate. [d] IC50 means the effective concentration of sample that can inhibit 50% of lipid peroxidation.

3.4. Encapsulation of NFE into PVA/PVP Electrospun Nanofibers

As shown in Figure 4, the different NFE content-loaded PVA/PVP fibers did not show different morphology compared to the neat PVA/PVP fibers. However, the diameters of nanofibers significantly increased with the addition of NFE, which could be due to the solution characteristics as shown in Table 2. Although the conductivity decreased with the increase of NFE into the PVA/PVP solution, no significant decrease was observed. The viscosity of the solution increased from 1535 Pa·S to 1693 Pa·S because of the molecule entanglements, which was favorable for the formation of thick fibers [23]. A similar phenomenon was observed in tomato peel extract-loaded gelatin fibers [24]. In addition to fiber diameter, the DPPH radical scavenging rate also enhanced by increasing NFE concentration, suggesting the potential for antioxidant packaging (Table 2). When the NFE concentration was 4.0 mg/mL, the scavenging rate reached 78.4%, and there was no significant difference ($p > 0.05$) compared to that of 8 mg/mL. Besides that, the encapsulation efficiency values for 2 mg/mL, 4 mg/mL and 8 mg/mL NFE were 96.57 ± 1.46%, 94.32 ± 1.78%, and 91.42 ± 2.45%, respectively. These results suggested that almost no loss of NFE occurred during the electrospinning and more than 90% of NFE could be encapsulated into PVA/PVP fibers, indicating the efficient nano-encapsulation of NFE by electrospinning.

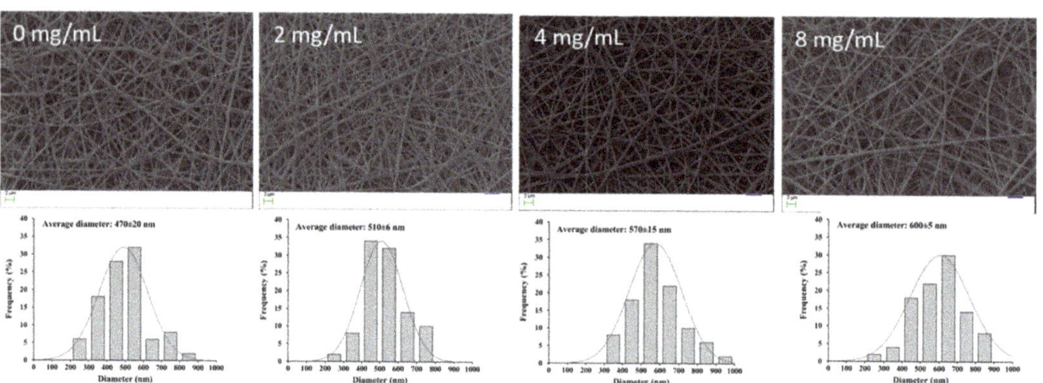

Figure 4. SEM images of different NFE content-loaded PVA/PVP fibers and its diameter distribution.

Table 2. Characteristics of electrospun solutions and the obtained fibers.

NFE Concentration (mg/mL)	Viscosity (Pa·S)	Conductivity (μS/cm)	Average Diameter (nm)	DPPH Radical Scavenging Rate (%)
0	1535	326	470	0 ^{a}
2.0	1572	315	510	38.7 ± 1.2 ^{b}
4.0	1604	307	570	78.4 ± 2.4 ^{c}
8.0	1693	302	600	80.2 ± 1.9 ^{c}

Note: different letters in the same column indicates the statistically different ($p < 0.05$).

3.5. Characterization of Electrospun Nanofibers

The interactions and compatibility among different components were examined by FTIR spectroscopy. Figure 5 shows that NFE exhibited peaks of O–H stretching (3330 cm^{-1}), C=O stretching (1705 cm^{-1}), and C–O–C) stretching (1242 and 1079 cm^{-1}). Compared to neat PVP and PVP fibers, almost no changes in the spectra of PVP/NFE and PVA/NFE were observed, suggesting that NFE could be compatibly entrapped into the PVP and PVA fibers. When NFE was added to the PVA/PVP matrices, the infrared spectra of NFE-loaded PVA/PVP fiber mat exhibited the characteristic peaks of both polymers, while the presence of NFE peaks almost disappeared, indicating the encapsulation of NFE into the fiber mat due to the formation of various kinds of intra/inter-molecular hydrogen bonds. In particular, the intensity and width of the peak broadening over 3380 cm^{-1} was in the following order of PVA/PVP/NFE > PVA/PVP, which indicated that a stronger hydrogen bond was formed in PVA/PVP/NFE.

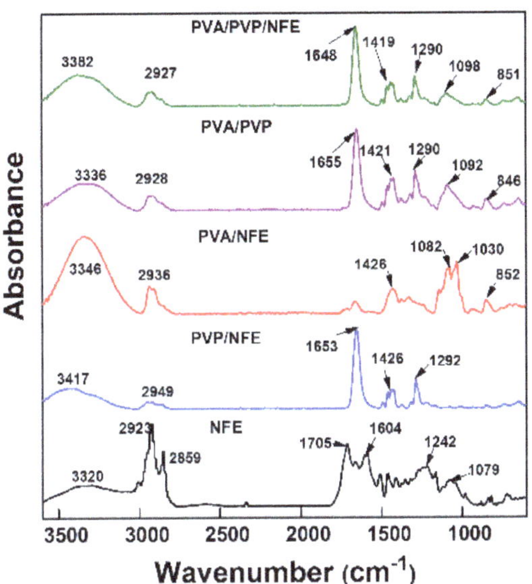

Figure 5. FTIR of different samples.

A crystallinity analysis by XRD pattern was carried out to confirm the inter-molecular interaction and biocompatibility among different components. As shown in Figure 6, pure NFE exhibited patterns of a crystalline state with diffraction peaks at 19.67°, 22.75° and 24.63°; however, these characteristic peaks for NFE were absent in the diffractograms of PVA/NFE, PVP/NFE and PVA/PVP/NFE. It can be suggested that NFE was converted into an amorphous state, which was ascribed to the electrospinning process and the interaction with amorphous PVP polymer. The rapid evaporation of solvent caused the quick transformation of polymer solution into solid form without enough time for crystallization [25]. Besides that, it can be seen that the first peak of PVP/NFE at 10.82° significantly diminished in PVP/PVA/NFE, and the intensity of the PVA/NFE peak at 19.47° was weakened and shifted to the higher 2θ position in PVP/PVA/NFE. These findings corresponded with the FTIR study where the intermolecular interaction existed among different components, particularly as hydrogen bonds.

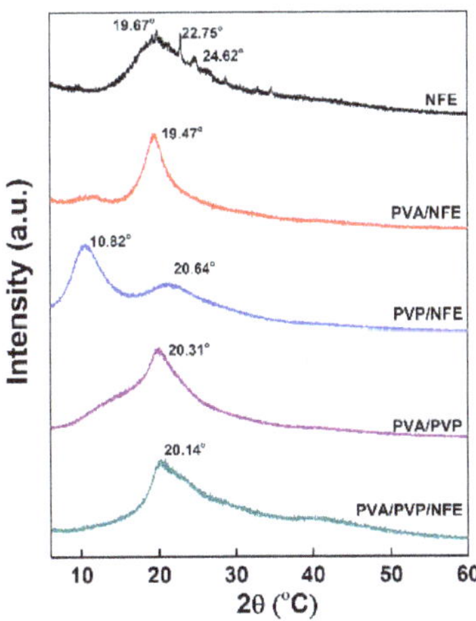

Figure 6. X-ray diffractograms of different samples.

TGA was used to evaluate the thermal stability of the obtained fiber mat. From Figure 7, the weight loss around 60~180 °C was attributed to the evaporation of free, freezing bound and chemical bound water for all curves. Free water is usually related to the absorption water and the freezing bound water is weakly interacted with the polymeric chain, while the chemical bound water is regarded as the water molecules bound to the polymeric chains via hydrogen bound. Thus, the mass loss in the first region of the thermograms related to different samples can be ascribed as evaporation of different types of water. For the PVA/NFE curve, another two main weight loss regions were displayed. The first region around 280~330 °C was due to the degradation of the side chain in the PVA matrix, like the C–O bond, and the second stage (400~460 °C) appeared as the cleavage of the C–C backbone in polymers, leading to so-called carbonization. In the case of PVP/NFE film, the region around 400~460 °C was attributed to the degradation of PVP. In particular, the higher degradation temperature is a result of increased number of hydroxyl groups among components. These findings are consistent with the previous results of FTIR and XRD analysis. Taking all results into consideration, it can be concluded that the obtained PVA/PVP/NFE fiber mat possesses adequate thermal stability and can be safely used for packaging application.

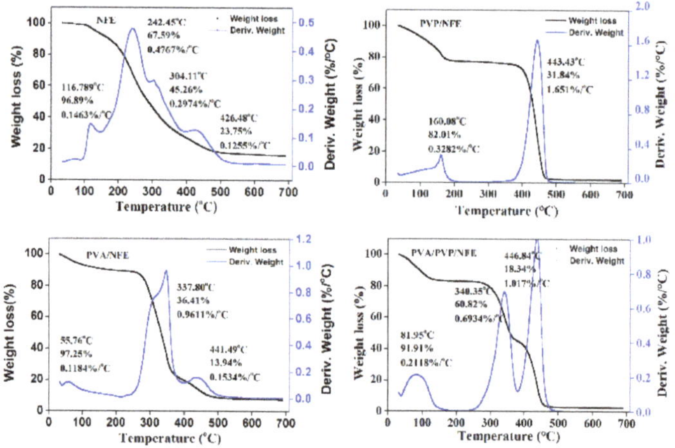

Figure 7. TGA curves of different samples.

3.6. *Effectiveness of the Active Films against Lipid Oxidation of Fish Oil*

Oxidation is one of the major causes of oil deterioration. For example, fish oil, a nutritive functional food, is highly susceptible to oxidation due to the presence of polyunsaturated fatty acids, thus limiting its application. Hydroperoxides are the primary products formed in the first stage of oxidation, which can be reflected by peroxide values (PV), an indicator that represents the extent of early phase lipid oxidation [26]. For an edible food product, the PV levels should be less than 30 meq peroxide/kg oil [27]. Herein, the effect of different antioxidant films on the oxidation extent of fish oil was determined by measuring the PV at 25 °C for 60 days. Unpacked fish oil was used as blank. As depicted in Figure 8, the initial raw fish oil presented a PV value of 3.03 meq peroxide/kg oil. It was increased markedly for non-packed fish oil after day 10, reaching a maximum level of 378 meq peroxide/kg oil at day 50, and then declined to 343 meq peroxide/kg oil at day 60 ($p > 0.05$), which may be due to the degradation of peroxides as secondary oxidation began [28]. A similar trend was also noted for fish oil packed with PVP/PVA film, although there was a slight suppression in the PV values compared with the control, which could be related to the excellent gas barrier property of PVA-based films [29]. On the contrary, the different forms of antioxidants resulted in different extents of inhibitory effects on oxidation, as the PV values of samples packed with PVP/PVA/NFE were always lower than that of the oil treated with free NFE on all sampling days. For instance, the PV value (30 meq peroxide/kg oil) occurred on the 10th, 30th and 40th day with respect to the unpacked oil and oil packed with PVA/PVP film and free NFE, respectively. The results indicated that the incorporation of natural antioxidants into an electrospun nanoscale delivery system was able to maintain its activity and was more effective than that of the non-encapsulated antioxidant due to the improved solubility. A similar trend was found in the edible guar gum-based nanofibrous mat for the encapsulation of tannic acid to inhibit the oxidation of flaxseed oil [19].

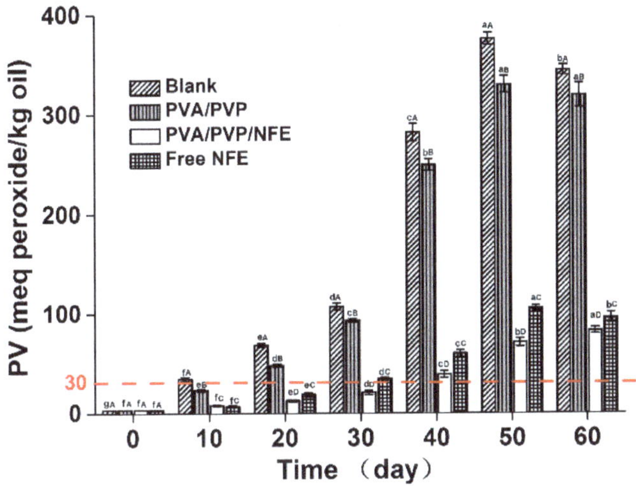

Figure 8. X-ray diffractograms of different samples. Different lowercase letters indicated a significant difference ($p < 0.05$) over different incubation time that treated with the same film. Different capital letters indicated a significant difference ($p < 0.05$) among different packaging film at a specific incubation time.

4. Conclusions

In this study, the ethyl acetate extract of *Nervilia fordii* (NFE) possessing excellent antioxidant activity was first obtained, followed by encapsulation into a PVA/PVP nanofiber with high encapsulation efficiency. FTIR-ATR, TGA and XRD analysis confirmed the presence of hydrogen bonding interactions among different components and their miscibility at the molecular level. The prepared nanofiber still exhibits good antioxidant activity by inhibiting the primary oxidation products of fish oil, in particular, it was more effective than the non-encapsulated NFE. These results suggested that nanoencapsulation by electrospinning is an effective way to stabilize NFE and improve its antioxidant activity. Hence, the PVA/PVP/NFE nanofiber could be a promising antioxidant packaging material for fatty food preservation.

Supplementary Materials: The following are available online at https://www.mdpi.com/article/10.3390/foods10081728/s1, Figure S1: Response surface and corresponding contour plots of three factors on the DPPH radical scavenging activity, Table S1: Factors and levels in BBD, Table S2: Analysis of variance (ANOVA) for the experimental results of BBD.

Author Contributions: P.W. experimental design, methodology, data curation, formal analysis, investigation, methodology, writing—original draft, writing—review and editing; T.-G.H. conceptualization, methodology, data curation, writing—review and editing; Y.W. experimental design, methodology, data curation, formal analysis; K.-E.L.—data curation; W.-P.Q.—data curation; Z.-L.H.—formal analysis; H.W. (Hong Wang) funding acquisition; H.W. (Hong Wu) supervision, project administration, writing—review and editing. All authors have read and agreed to the published version of the manuscript.

Funding: This research was funded by the National Natural Science Foundation of China, grant number 32001631, Guangdong Basic and Applied Basic Research Foundation, grant number 2020A1515110190 and the National Natural Science Foundation of China, grant number 31671852.

Acknowledgments: Authors acknowledge Ke-er Li, Wei-peng Qiu and Zhi-lin He for providing the electrospun fiber mat for experiments.

Conflicts of Interest: The authors declare no conflict of interest.

References

1. Zhang, C.; Li, Y.; Wang, P.; Zhang, H. Electrospinning of nanofibers: Potentials and perspectives for active food packaging. *Compr. Rev. Food Sci. Food Saf.* **2020**, *19*, 479–502. [CrossRef] [PubMed]
2. Radusin, T.; Torres-Giner, S.; Stupar, A.; Ristic, I.; Miletic, A.; Novakovic, A.; Lagaron, J.M. Preparation, characterization and antimicrobial properties of electrospun polylactide films containing Allium ursinum L. extract. *Food Packag. Shelf Life* **2019**, *21*, 9. [CrossRef]
3. Yang, S.B.; Kim, E.H.; Kim, S.H.; Kim, Y.H.; Oh, W.; Lee, J.T.; Jang, Y.A.; Sabina, Y.; Ji, B.C.; Yeum, J.H. Electrospinning fabrication of poly(vinyl alcohol)/coptis chinensis extract nanofibers for antimicrobial exploits. *Nanomaterials* **2018**, *8*, 14. [CrossRef]
4. Wen, P.; Zhu, D.H.; Wu, H.; Zong, M.H.; Jing, Y.R.; Han, S.Y. Encapsulation of cinnamon essential oil in electrospun nanofibrous film for active food packaging. *Food Control* **2016**, *59*, 366–376. [CrossRef]
5. Zhang, L.; Zhao, Z.X.; Lin, C.Z.; Zhu, C.C.; Gao, L. Three new flavonol glycosides from Nervilia fordii. *Phytochem. Lett.* **2012**, *5*, 104–107. [CrossRef]
6. Qiu, L.; Jiao, Y.; Xie, J.Z.; Huang, G.K.; Qiu, S.L.; Miao, J.H.; Yao, X.S. Five new flavonoid glycosides from Nervilia fordii. *J. Asian Nat. Prod. Res.* **2013**, *15*, 589–599. [CrossRef]
7. Wen, P.; Zong, M.H.; Linhardt, R.J.; Feng, K.; Wu, H. Electrospinning: A novel nano-encapsulation approach for bioactive compounds. *Trends Food Sci. Technol.* **2017**, *70*, 56–68. [CrossRef]
8. Fonseca, L.M.; dos Santos Cruxen, C.E.; Bruni, G.P.; Fiorentini, Â.M.; da Rosa Zavareze, E.; Lim, L.T.; Dias, A.R.G. Development of antimicrobial and antioxidant electrospun soluble potato starch nanofibers loaded with carvacrol. *Int. J. Biol. Macromol.* **2019**, *139*, 1182–1190. [CrossRef] [PubMed]
9. Kumar, T.S.M.; Kumar, K.S.; Rajini, N.; Siengchin, S.; Ayrilmis, N.; Rajulu, A.V. A comprehensive review of electrospun nanofibers: Food and packaging perspective. *Compos. Part B Eng.* **2019**, *175*, 107074. [CrossRef]
10. Zhang, C.; Li, Y.; Wang, P.; Li, J.W.; Weiss, J.; Zhang, H. Core-shell nanofibers electrospun from O/W emulsions stabilized by the mixed monolayer of gelatin-gum Arabic complexes. *Food Hydrocoll.* **2020**, *107*, 9. [CrossRef]
11. Nesic, A.; Ruzic, J.; Gordic, M.; Ostojic, S.; Micic, D.; Onjia, A. Pectin-polyvinylpyrrolidone films: A sustainable approach to the development of biobased packaging materials. *Compos. Part B Eng.* **2017**, *110*, 56–61. [CrossRef]
12. Rahma, A.; Munir, M.M.; Prasetyo, A.; Suendo, V.; Rachmawati, H. Intermolecular interactions and the release pattern of electrospun curcumin-polyvinyl(pyrrolidone) fiber. *Biol. Pharm. Bull.* **2016**, *39*, 163–173. [CrossRef] [PubMed]
13. Estevez-Areco, S.; Guz, L.; Candal, R.; Goyanes, S. Release kinetics of rosemary (*Rosmarinus officinalis*) polyphenols from polyvinyl alcohol (PVA) electrospun nanofibers in several food simulants. *Food Packag. Shelf Life* **2018**, *18*, 42–50. [CrossRef]
14. Haghju, S.; Bari, M.R.; Khaled-Abad, M.A. Affecting parameters on fabrication of beta-D-galactosidase immobilized chitosan/poly (vinyl alcohol) electrospun nanofibers. *Carbohyd. Polym.* **2018**, *200*, 137–143. [CrossRef] [PubMed]
15. Lan, X.Z.; Liu, Y.R.; Wang, Y.Q.; Tian, F.; Miao, X.M.; Wang, H.; Tang, Y.D. Coaxial electrospun PVA/PCL nanofibers with dual release of tea polyphenols and e-poly (L-lysine) as antioxidant and antibacterial wound dressing materials. *Int. J. Pharm.* **2021**, *601*, 10. [CrossRef]
16. Bruni, G.P.; de Oliveira, J.P.; Gomez-Mascaraque, L.G.; Fabra, M.J.; Martins, V.G.; Zavareze, E.D.; Lopez-Rubio, A. Electrospun beta-carotene-loaded SPI:PVA fiber mats produced by emulsion-electrospinning as bioactive coatings for food packaging. *Food Packag. Shelf Life* **2020**, *23*, 12.
17. Rybinski, W.; Karamac, M.; Sulewska, K.; Boerner, A.; Amarowicz, R. Antioxidant potential of grass pea seeds from european countries. *Foods* **2018**, *7*, 142. [CrossRef] [PubMed]
18. Neo, Y.P.; Ray, S.; Jin, J.; Gizdavic-Nikolaidis, M.; Nieuwoudt, M.K.; Liu, D.; Quek, S.Y. Encapsulation of food grade antioxidant in natural biopolymer by electrospinning technique: A physicochemical study based on zein–gallic acid system. *Food Chem.* **2013**, *136*, 1013–1021. [CrossRef] [PubMed]
19. Yang, W.; Li, X.; Jiang, J.; Fan, X.; Du, M.; Shi, X.; Cao, R. Improvement in the oxidative stability of flaxseed oil using an edible guar gum-tannic acid nanofibrous mat. *Eur. J. Lipid Sci. Technol.* **2019**, *121*, 1800438. [CrossRef]
20. Nand, A.V.; Swift, S.; Uy, B.; Kilmartin, P.A. Evaluation of antioxidant and antimicrobial properties of biocompatible low density polyethylene/polyaniline blends. *J. Food Eng.* **2013**, *116*, 422–429. [CrossRef]
21. Altemimi, A.; Watson, D.G.; Kinsel, M.; Lightfoot, D.A. Simultaneous extraction, optimization, and analysis of flavonoids and polyphenols from peach and pumpkin extracts using a TLC-densitometric method. *Chem. Cent. J.* **2015**, *9*, 39. [CrossRef]

22. Prasad, K.N.; Yang, E.; Yi, C.; Zhao, M.; Jiang, Y. Effects of high pressure extraction on the extraction yield, total phenolic content and antioxidant activity of longan fruit pericarp. *Innov. Food Sci. Emerg.* **2009**, *10*, 155–159. [CrossRef]
23. Aslaner, G.; Sumnu, G.; Sahin, S. Encapsulation of grape seed extract in rye flour and whey protein-based electrospun nanofibers. *Food Bioprocess Technol.* **2021**, *14*, 1118–1131. [CrossRef]
24. İnanç Horuz, T.; Belibağlı, K.B. Nanoencapsulation by electrospinning to improve stability and water solubility of carotenoids extracted from tomato peels. *Food Chem.* **2018**, *268*, 86–93. [CrossRef]
25. Wen, P.; Zong, M.H.; Hu, T.G.; Li, L.; Wu, H. Preparation and characterization of electrospun colon-specific delivery system for quercetin and its antiproliferative effect on cancer cells. *J. Agric. Food Chem.* **2018**, *66*, 11550–11559. [CrossRef] [PubMed]
26. Wang, R.; Tian, Z.; Chen, L. A novel process for microencapsulation of fish oil with barley protein. *Food Res. Int.* **2011**, *44*, 2735–2741. [CrossRef]
27. Gotoh, N.; Wada, S. The importance of peroxide value in assessing food quality and food safety. *J. Am. Oil Chem. Soc.* **2006**, *83*, 473–474. [CrossRef]
28. Nilsuwan, K.; Benjakul, S.; Prodpran, T.; de la Caba, K. Fish gelatin monolayer and bilayer films incorporated with epigallocatechin gallate: Properties and their use as pouches for storage of chicken skin oil. *Food Hydrocoll.* **2019**, *89*, 783–791. [CrossRef]
29. Jang, J.; Lee, D.K. Oxygen barrier properties of biaxially oriented polypropylene/polyvinyl alcohol blend films. *Polymer* **2004**, *45*, 1599–1607. [CrossRef]

Article

Portable, Rapid, and Sensitive Time-Resolved Fluorescence Immunochromatography for On-Site Detection of Dexamethasone in Milk and Pork

Xiangmei Li [1], Xiaomin Chen [1], Jinxiao Wu [2], Zhiwei Liu [1], Jin Wang [1], Cuiping Song [3], Sijun Zhao [3], Hongtao Lei [1,*] and Yuanming Sun [1,*]

1 Guangdong Provincial Key Laboratory of Food Quality and Safety, College of Food Science, South China Agricultural University, Guangzhou 510642, China; lixiangmei12@163.com (X.L.); chenxiaomin2021@163.com (X.C.); liuzhiwei2021888@163.com (Z.L.); wj646607936@163.com (J.W.)

2 Shanxi Institute of Feed and Veterinary Drug control, No. 5 Shengli West Street, Jiancaoping District, Taiyuan 030000, China; wdwk030@163.com

3 China Animal Health and Epidemiology Center, 369 Nanjing Rd, Si Fang Qu, Qingdao 266032, China; wdwk045@163.com (C.S.); lixiangmei12@scau.edu.cn (S.Z.)

* Correspondence: hongtao@scau.edu.cn (H.L.); gzsyming@163.com (Y.S.); Tel.: +86-20-8528-3925 (H.L.); +86-20-8528-3448 (Y.S.); Fax: +86-20-8528-0270 (H.L.); +86-20-8528-3448 (Y.S.)

Abstract: Dexamethasone (DEX) is widely used because of its anti-inflammatory, anti-endotoxin, anti-shock, and stress-enhancing response activities. It can increase the risk of diabetes and hypertension if it is abused or used improperly. However, there is a lack of sensitive and rapid screening methods for DEX in food. In this study, a time-resolved fluorescent microspheres immunochromatographic assay (TRFM-ICA) integrated with a portable fluorescence reader was developed for the quantitative detection of DEX in milk and pork. The cut-off values of the TRFM-ICA were 0.25 ng/mL and 0.7 μg/kg, respectively. The limits of quantitation (LOQs) were 0.003 ng/mL and 0.062 μg/kg, respectively. The recovery rates were 80.0–106.7%, and 78.6–83.6%, respectively, with the coefficients of variation ranging 6.3–12.5%, and 7.5–10.3%, respectively. A parallel experiment for 20 milk and 10 pork samples with LC-MS/MS was carried out to confirm the performance of the on-site application of the developed TRFM-ICA. The results of the two methods are basically the same; the correlation (R^2) was >0.98. The establishment of TRFM-ICA will provide a new sensitive and efficient technical support for the rapid screening of DEX in food.

Keywords: dexamethasone; time-resolved fluorescent microspheres; immunochromatographic assay; milk; pork

Citation: Li, X.; Chen, X.; Wu, J.; Liu, Z.; Wang, J.; Song, C.; Zhao, S.; Lei, H.; Sun, Y. Portable, Rapid, and Sensitive Time-Resolved Fluorescence Immunochromatography for On-Site Detection of Dexamethasone in Milk and Pork. *Foods* **2021**, *10*, 1339. https://doi.org/10.3390/foods10061339

Academic Editors: Hong Wu and Hui Zhang

Received: 28 April 2021
Accepted: 8 June 2021
Published: 10 June 2021

1. Introduction

Dexamethasone (DEX) is a synthetic corticosteroid and has pharmacological effects including anti-inflammatory, anti-toxic, anti-allergic, and anti-rheumatic activities [1]. Therefore, it is widely used in veterinary clinical treatment of maternal metabolic diseases or in combination with antibiotics to treat infectious diseases, and it is also one of the commonly used drugs in livestock and poultry breeding [2]. However, DEX can also cause certain adverse reactions to animals, such as gastrointestinal reactions, allergic reactions, liver dysfunction, skin and mucosal symptoms, etc. Long-term consumption of animal products with excessive DEX will cause diabetes, hypertension, myocardial infarction, gastrointestinal ulcer, and other symptoms [3]. Therefore, DEX is strictly forbidden to be used as a growth hormone in animal-derived food globally [4]. Many countries and organizations have established the maximum residue limits (MRLs) for DEX in animal foods. For example, China and Codex Alimentarius Commission set MRLs of 0.3 and 0.75 μg/kg in milk and pork, respectively [5,6]. Based on the wide application, serious side effects, and trace detection requirements of DEX, it is imperative to establish a rapid and sensitive detection method for DEX in food to ensure the health of humans and animals.

Since the emergence of immunochromatographic assay (ICA), it has become the most popular rapid detection method for food safety testing due to its outstanding advantages such as simple sample preparation, fast acquisition of test results, no professional training, low cost, and being suitable for screening large quantities of samples [7–9]. ICA based on gold nanoparticle (GNP) is the most widely used method on the market. However, with the increasing demand for food safety detection, the sensitivity of GNP-ICA has become a bottleneck restricting its development [10]. Therefore, how to improve the sensitivity of ICA has become the focus of research. There are mainly two ways to overcome this deficiency from the published literature. On the one hand, novel Ab-labeled tracers are synthesized for signal amplification, such as quantum dots (QD) [11,12], fluorescent microspheres (FM) [8,13], chemiluminescent materials [14,15], up conversion phosphorescence (UCP) [16], biotin-affinity [17], and metal–organic frameworks [18]. On the other hand, external analytical instruments are developed, such as desktop [19,20], hand-held [21,22], or smartphone-based [23,24] reading platforms. Therefore, with the continuous development of material technology and equipment, the problem of ICG restricted by sensitivity will be alleviated, and its advantages in rapid detection will become increasingly prominent. However, up to now, there have been only three reports on the detection of DEX in animal-derived foods by ICA. One used GNP-ICA for milk detection, where the cut off value was 0.5 μg/kg, which could not meet the requirement of residue detection [25]. The second used UCP-ICA for the animal tissue detection, and the stability of UCNP is controversial [26]. The third use latex microspheres (LMs)-ICA for milk and pork detection; based on the color diversification of LMs, different samples could be distinguished by color. However, the sensitivity was not as good as our present work [27].

Therefore, in order to provide a stable, sensitive, reliable, and rapid detection method for DEX residue detection, ICG based on time-resolved fluorescent microspheres (TRFM) was established. TRFM was employed as the signal-labeled tracer with several advantageous features: (a) Rare earth ions with longer fluorescence half-life (103–106 times the traditional fluorescence) are used as labels, which have extremely wide stokes shift (he excitation wavelength and emission wavelength are 365 and 610 nm, respectively, and the stoke shift is more than 200 nm) and long fluorescence quenching time, thus effectively eliminating the interference of various non-specific fluorescence and improving the accuracy and sensitivity [24]. (b) There are thousands of fluorescent molecules in the TRFM, which greatly improves the labeling efficiency of fluorescence and analytical sensitivity [28]. (c) The surface of TRFM is modified with carboxyl or other functional groups, which are used for covalent coupling with proteins or Ab, improving the stability of the conjugates. These features are ideal for the development of ICG. Meanwhile, a portable, compact desk reader was used to quantify results. This integrated strategy could provide valuable technical support for the on-site detection of DEX in animal-derived food.

2. Materials and Methods

2.1. Reagents and Instruments

DEX, hydrocortisone, prednisone, triamcinolone, betamethasone, 1-ethyl-3-(3-dimethy laminopropyl) carbodiimide hydrochloride (EDC), N-hydroxysuccinimide (NHS), 2-(N-morpholino) ethanesulfonic acid (MES), ovalbumin (OVA), and bovine serum albumin (BSA) were purchased from Sigma-Aldrich (St. Louis, MO, USA). TRFM, europium chelates (365/610), with 1% solid content (*w/v*) and 0.2 μm particle size, was purchased from Bangs Laboratories, Inc (Fishers, IN, USA). Anti-DEX monoclonal A and DEX-OVA coating antigen (Ag) were prepared in our laboratory. The nitrocellulose filter (NC) membrane (Sartorius, UniSart CN95) was purchased from Sartorius Stedim Biotech GmbH (Goettingen, Germany). The microtiter plates were supplied by the JET BIOFIL Co. (Guangzhou, China). The polyvinylchloride (PVC) backing plate (SMA31-40), sample pad (GF-2), and absorbent pad (CH37) were purchased from Shanghai Kinbio Tech. Co., Ltd. (Shanghai, China).

Sucrose, sodium chloride, and other chemicals reagents were bought from Sinopharm Chemical Reagent Co., Ltd. (Shanghai, China).

The strip cutter ZQ-2000 and the slitting machine SPT300 were purchased from Shanghai kinbio Tech. Co., Ltd. (Shanghai, China). The XYZ-3060 Dispensing Platform was bought from BioDot, Inc. (Irvine, CA, USA). The UV spectrometer was provided by Thermo Fisher Scientific Co. (Waltham, Massachusetts, USA). The Lynx-4000 centrifuge was obtained from Thermo Fisher Scientific GmbH (Berlin, Germany). The time-resolved fluorescence quantitative analysis reader (FQ-S2, 254 nm, 365 nm) was purchased from WDWK Biotechnology Co., Ltd. (Beijing, China).

2.2. Preparation of TRFM Immunoprobe

The preparation of TRFM immunoprobe mainly includes two steps: activation of carboxyl groups on the surface of TRFM and covalent coupling with DEX Ab [28] (Figure 1A). Briefly, with constant stirring (300 rpm/min), 0.1 mg of TRFM was added in 1 mL of MES buffer (50 mM, pH 5.5). Then, 15 μL of freshly prepared 0.5 mg/mL EDC and NHS solution were sequentially added. The mixture was centrifuged at 14,000× g for 15 min at 4 °C after reaction for 15 min. The supernatant was discarded, and the bottom sediment was redissolved with 1 mL of borate buffer (BB, 50 mM, pH 8.0). Anti-DEX Ab (1 μL, 1.0 mg/mL), which was dissolved in 60 μL of BB (2 mM, pH 8.0), was added. The reaction solution was well mixed and incubated at room temperature for 45 min, and then 20 μL of 20% BSA (*w/v*) were added for blocking. The above solution was centrifuged at 14,000× g for 15 min at 4 °C after another 60 min of blocking reaction. The bottom sediment was dissolved in 200 μL of resuspension, which was stored at 4 °C for later use. The key technical parameters are shown in Table S1.

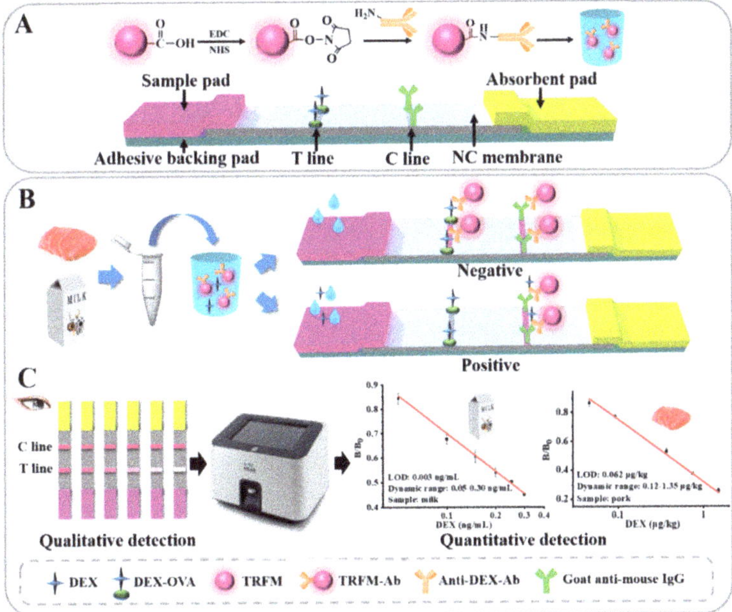

Figure 1. Schematic diagram of TRFM-ICG for DEX detection in milk and pork: (**A**) the preparation principle of TRFM-DEX Ab immunoprobe and structure of test strip; (**B**) the detection principle of test strip; and (**C**) the qualitative and quantitative test results.

2.3. Preparation of the Test Strips

The goat anti-mouse secondary Ab and coating Ag (DEX-OVA) were diluted to the optimal concentration, and then sprayed on the NC film to form the control (C) and test (T) lines, respectively. The technical parameters of the spray film were as follows: spray length, 30 cm; distance between T-C line, 8 mm; and spray volume, 0.8 μL/cm. The processed NC films were placed in a 37 °C oven to dry overnight. The sample pads were immersed in the designed sample pad treatment solution for 30 s, and then dried for 2 h at 60 °C. Finally, the dried NC film, sample pad, and absorbent pad were pasted on the PVC backing pad with 2 mm overlap each other (Figure 1A). The PVC sheet was cut into 3.5 mm strips for use. The working principle is that the analyte competes with coating Ag for the limited Abs. The binding of the analyte to the Abs inhibited that of the coating Ag to the Abs, which was judged by the fluorescence intensity of T-line. The greater is the amount analyte, the weaker is the T-line signal, or there is even no color (Figure 1B). The working parameters of the goat anti-mouse secondary Ab and coating Ag are shown in Table S1.

2.4. Sample Pretreatment

Milk: Milk samples can be detected directly without pretreatment. If the sample is rich in fat, the fat can be removed by centrifugation.

Pork: After the pork was chopped with scissors and homogenized, 4 g were accurately weighed into a centrifuge tube, and 4 mL of 0.2 M phosphate buffer (PB, pH 7.4)-methanol solution (*v/v*) were added. The sample was vortexed vigorously for 3 min, sonicated for 3 min, and then centrifuged at 4000× *g* for 10 min at room temperature. The supernatant was transferred out for use.

2.5. Test Procedure

Five microliters of TRFM-DEX Ab immunoprobe were added to the microwell, and then 150 μL of standard solution or sample solution were added. After reaction for 4 min, the test strip was inserted into the microwell for 5 min for chromatographic reaction. The test strip was taken out of the microwell, and the sample pad was quickly peeled off. The qualitative result was observed under an ultraviolet lamp (Figure 1C). The quantitative results were obtained by inserting the test strip into the fluorescence quantitative analysis reader (Figure 1C). The fluorescence intensity of the T/C lines were converted into the corresponding peak. The stronger is the signal, the larger is the peak area.

2.6. Method Performance Evaluation

2.6.1. Sensitivity

In this study, the spiked milk and pork samples with a series of DEX concentrations were used to assess the sensitivity of the developed TRFM-ICG; each spiked concentration was detected in triplicate. The sensitivity was expressed by the cut-off value, calibration curve, and limit of quantitation (LOQ). The cut-off value was defined as the lowest DEX concentration that can cause the T-line signal to disappear. The fluorescence intensity of the T/C line can be achieved by the fluorescence reader. The DEX concentration in milk or pork sample was quantified by the calibration curve, taking the concentration of DEX as the x-axis and B/B_0 (T/C line fluorescence signal ratio, where B is with analyte in the reaction system and B_0 is without analyte in the reaction system) as the y-axis. The LOQ was defined as the concentration corresponding to 80% of B/B_0 value on the calibration curve [29].

2.6.2. Specificity

Method specificity is expressed by cross-reactivity (CR). DEX and several structural analogs, such as hydrocortisone, prednisone, triamcinolone, and betamethasone, were analyzed at different concentrations for the CR by the indirect competitive enzyme-linked immunosorbent assay (icELISA) and TRFM-ICG. The CR is calculated by the ratio of the IC_{50} of the target analyte/analogs.

2.6.3. Accuracy and Precision

Milk and pork samples, identified by LC-MS/MS as DEX-free, were spiked with three known concentrations of DEX, respectively. All spiked samples were processed according to the previous description. Each spiked level was tested in triplicate on three different days using the fluorescence reader. The precision and accuracy of the TRFM-ICG were assessed by the coefficient of variation (CV) and recovery, respectively.

2.7. Blind Sample Detection

Twenty milk and ten pork blind samples were provided by Guangdong Provincial Key Laboratory of Food Quality and Safety, which were tested by the developed TRFM-ICG and LC-MS/MS as the confirmatory method. It should be noted that we did not know the concentration of DEX in each sample. Each sample was tested in triplicate; the correlation between the two methods was compared; and the consistency of the results reflects the reliability of the method.

The working parameters of the LC-MS/MS method are given in the Supplementary Materials. The sample pretreatment of milk and pork was consistent with the national standards [30,31].

3. Results and Discussion

3.1. Optimization of the TRFM-ICG

Several important technical parameters were optimized to achieve the best detection performance of the established TRFM-ICG, including the particle size and activated pH value of TRFM, the optimal pH value and ion concentration of Ab coupling with TRFM, Ab amount, sample pad treatment solution formula, etc.

3.1.1. Particle Size of TRFM

The particle size of the microsphere determines the specific surface area, which affects not only the binding efficiency of the carboxyl groups on the surface of the microsphere with Ab but also the chromatographic release of the immunoprobe [8,28]. The detection results of 200 and 300 nm microspheres were compared. The results show that the release effect of the 300 nm microsphere immunoprobe was not as good as that of the 200 nm microsphere immunoprobe. There were many residues in the reaction zone, which led to the background being very red and fuzzy (Figure S1). However, high background values will have a negative impact on the subsequent visual judgment and quantitative detection. Therefore, 200 nm microspheres were selected as DEX-Ab labeling tracers.

3.1.2. Activation pH Value of TRFM

TRFM are modified with carboxyl groups and need to be activated before coupling with Ab. The appropriate pH value can improve the activation efficiency of carboxyl group, and thus improve the binding efficiency with Ab [24]. The carbodiimide method was applied to activate the carboxyl groups under four different pH conditions, including pH 5.0, 5.5, 6.0, and 6.5. The results show that the fluorescence signal intensity of T-line increased with the increase of pH value, but the inhibitory effect was difficult to distinguish with the naked eyes. With the help of quantitative analysis results, the inhibition rates were the highest in the detection of milk and pork samples under pH 5.5 (Figure S2). Therefore, the optimal activation pH was selected as 5.5.

3.1.3. The Ab Dilution Buffer

The Ab dilution buffer can not only maintain the biological activity of Ab but also prevent non-specific reactions and improve the sensitivity of the method [7,32]. Five different Ab dilution buffers, namely ultrapure water, 0.01 M PB (pH 7.4), 0.01 M PB (pH 7.4, 0.5% BSA), 0.5% BSA, and 0.002 M BB (pH 8.0), were used to dilute DEX Ab, adding a control group with no Ab diluent. The photos of the test strips and quantitative detection results show that 0.002 M BB (pH 8.0) as an Ab diluent had outstanding advantages in inhibition effect, and the fluorescence intensity was also satisfactory (Figure 2). This result shows that pH was a key factor affecting the binding of Ab-FMs, and BSA would hinder the coupling of Ab-FMs. Therefore, BSA should not be contained in the Ab diluent. Interestingly, this result is exactly the opposite of our previous research [27]. Therefore, there was no doubt that 0.002 M BB (pH 8.0) was our target Ab dilution.

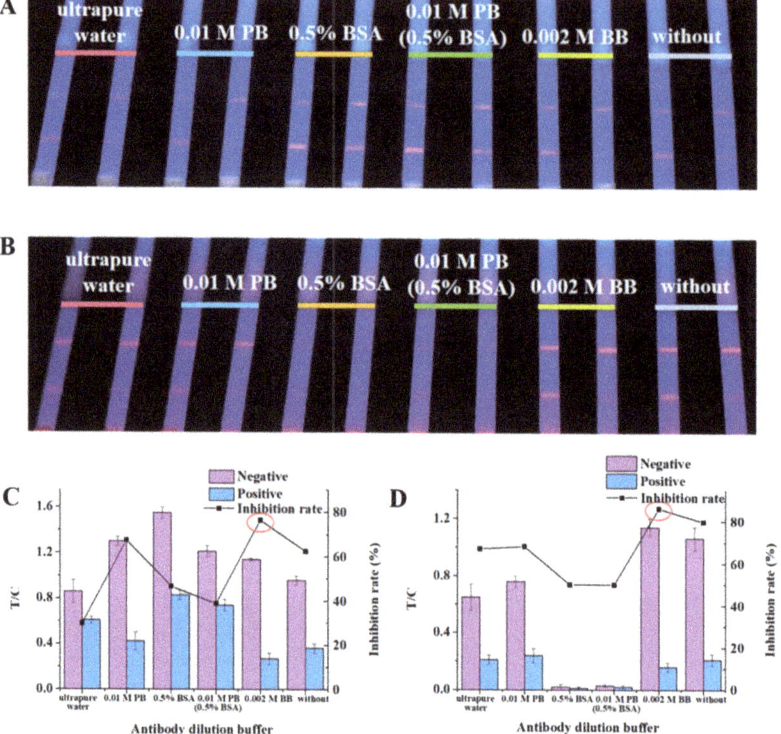

Figure 2. The fluorescence intensity, inhibition effect, and inhibition rate results of the Ab dilution buffer: (**A**) ultraviolet lamp results for milk detection; (**B**) ultraviolet lamp results for pork detection; (**C**) fluorescence quantitative results for milk detection; and (**D**) fluorescence quantitative results for pork detection.

3.1.4. The Ab Amount

The Ab amount plays a decisive role in the signal intensity and inhibitory effect of the immunoassay method [10,33]. The detection performance of four different Ab amounts (0.8, 1.0, 1.2, and 1.4 μg for milk and 0.6, 0.8, 1.0, and 1.2 μg for pork) were investigated. As shown in Figure S3, with the increased of Ab amount, the fluorescent signal of T-line increased gradually, but the inhibition effect became worse. Combined with the results of quantitative detection, it was not difficult to find that, when the amount of Ab was 0.8 μg, which was equivalent to adding 80 μL of Ab solution, the inhibition rate of the test strip was the best. Therefore, the optimal amount of Ab was 0.8 μg for both milk and pork detection.

3.1.5. Key Reagents of Sample Pad Treatment Solution

The main function of the sample pad is the carrier of the sample, which can promote the release of Ab probe and eliminate the matrix interference of the sample [24,28]. Therefore, the handling of the sample pad is very important.

Surfactant. Surfactant can promote the release of immunoprobe, affect the behavior of immunoprobe on chromatographic pads, and reduce non-specific reaction [28,34]. In this study, we employed Tween-20 as a surfactant and studied the effect of its concentration changes on the performance of test strips. As shown in Figure S4, we prepared a series of different concentrations of Tween-20 in the sample pad treatment solution and found that, as the concentration of Tween-20 increased, the release rate of immunoprobe accelerated. The results of quantitative analysis show that the inhibition rate was the best when the concentration of Tween-20 was 0.5% for milk samples and 0.2% for pork samples. The reason for the inconsistency of the optimal concentration should be closely related to the viscosity and fluidity of the samples.

Sample pad treatment buffer. It is particularly important for the sample pad treatment buffer to protect Ab activity and eliminate the interference of the sample matrix [24,29]. In this study, the effects of three kinds of buffers with different ionic strength and pH values on the detection performance of test strips were compared. The results (Figure 3) show that pH value > 9.0 was not conducive to the performance of Ab activity because the fluorescence intensity of test strip was not ideal. However, high ionic strength was conducive to inhibition. Considering the fluorescence signal intensity, inhibition effect, and inhibition rate, 0.05 M PB was finally determined as the buffer of the sample pad treatment solution.

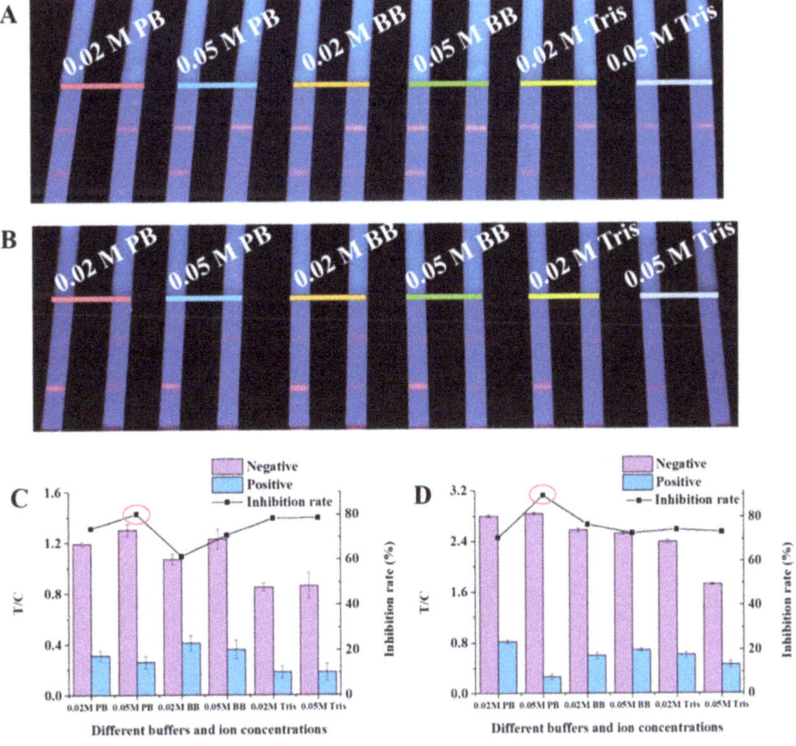

Figure 3. The fluorescence intensity, inhibition effect, and inhibition rate results of the sample pad treatment buffer: (**A**) ultraviolet lamp results for milk detection; (**B**) ultraviolet lamp results for pork detection; (**C**) fluorescence quantitative results for milk detection; and (**D**) fluorescence quantitative results for pork detection.

To sum up, the formula of the sample pad treatment solution for milk detection was 0.05 M PB (pH 7.4, 0.5% Tween-20, 0.3% PVP) and for pork detection was 0.05 M PB (pH 7.4, 0.2% Tween-20, 0.3% PVP).

3.2. *Method Performance Evaluation*

3.2.1. Sensitivity

Based on the above optimization conditions, negative milk and pork samples were spiked with DEX at different concentrations and tested by the optimized TRFM-ICG. As shown in Figure 4, the cut-off values of TRFM-ICG for DEX in milk and pork were 0.25 ng/mL and 0.7 μg/kg, respectively. The LOQ were 0.003 ng/mL and 0.062 μg/kg, respectively. The dynamics ranged 0.05–0.3 ng/mL and 0.12–1.35 μg/kg, respectively.

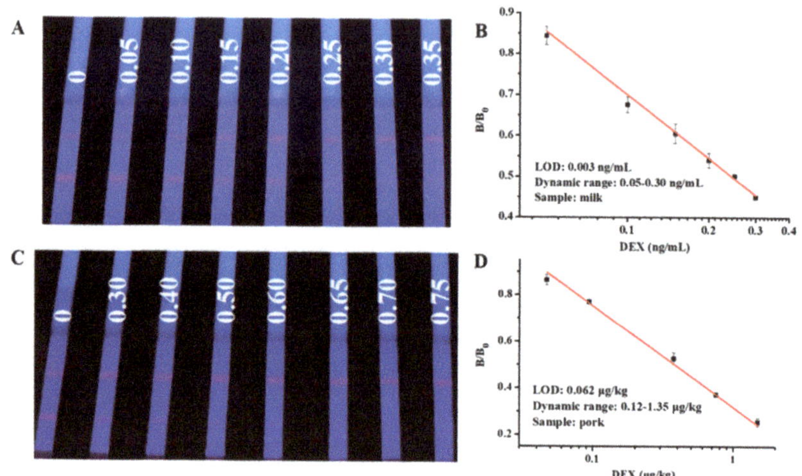

Figure 4. The results of TRFM-ICG for milk and pork detection: (**A**) the cut-off value for milk was 0.25 ng/mL; (**B**) calibration curve for quantitative detection of DEX in milk; (**C**) the cut-off value for pork was 0.7 μg/kg; and (**D**) calibration curve for quantitative detection of DEX in pork.

3.2.2. Specificity

The specific results are shown in Table S2. DEX Ab exhibited a strong CR to triamcinolone, with a CR rate of 54.5%, which may be due to the similar structure of the Ag recognition site. The CR rates to betamethasone, prednisolone, and hydrocortisone were 24.0%, 14.0%, and 1.5%, respectively. The specificity results of TRFM-ICG were consistent with that of icELISA, and the CR rates to DEX, triamcinolone, betamethasone, prednisolone, and hydrocortisone were 100%, 62.5%, 31.3%, 20.8%, and 4.2%, respectively. The specificity results show that the established TRFM-ICG method can be used for the multiple targets' detection of five glucocorticoids.

3.2.3. Accuracy and Precision

The DEX-free milk and pork samples were spiked with DEX standard working solution so that the concentrations of DEX in milk were 0.075, 0.15, and 0.3 ng/mL and in pork were 0.35, 0.7, and 1.4 μg/kg. The recovery rates of DEX in milk and pork samples were 80.0–106.7%, and 78.6–83.6%, respectively, with the CVs of 6.3–12.5%, and 7.5–10.3%, respectively (Table 1). The recoveries and CVs meet the requirement for residue detection. These results indicate that the established TRFM-ICG method has good accuracy and reproducibility.

Table 1. Recovery of the TRFM-ICG for the determination of DEX in milk and pork samples (n = 3).

Sample	Spiked Level (ng/mL or μg/kg)	Measured Level (ng/mL or μg/kg)	Recovery (%)	CV (%)
Milk	0.075	0.08 ± 0.01	106.7	12.5
	0.15	0.16 ± 0.01	106.6	6.3
	0.3	0.24 ± 0.03	80.0	12.5
Pork	0.35	0.28 ± 0.02	80.4	7.5
	0.7	0.55 ± 0.05	78.6	9.1
	1.4	1.17 ± 0.12	83.6	10.3

3.3. Blind Sample Detection

Twenty milk and ten pork blind samples were analyzed simultaneously with our established TRFM-ICG method and the national standard method (LC-MS/MS). The test results are shown in Table 2 and Figure 5. Eighteen milk and eight pork samples were detected to contain DEX using LC-MS/MS, and the same results were obtained by TRFM-ICG. The detection results of the two methods were basically consistent; the correlation coefficient was greater than 0.98 (R^2 > 0.98). These results indicate that the established TRFM-ICG method was accurate and reliable, and it can be used in the actual detection.

Table 2. Determination of DEX in blind milk and pork samples by LC-MS/MS and TRFM-ICG (n = 3).

Sample	TRFM-ICG (ng/mL or μg/kg)	CV (%)	LC-MS/MS (ng/mL or μg/kg)	CV (%)
Milk 1	0.09 ± 0.01	11.1	0.14 ± 0.02	14.3
Milk 2	0.55 ± 0.02	3.6	0.58 ± 0.04	6.9
Milk 3	0.10 ± 0.01	10.0	0.13 ± 0.01	7.7
Milk 4	0.24 ± 0.03	12.5	0.22 ± 0.01	4.6
Milk 5	0.29 ± 0.01	3.5	0.33 ± 0.04	12.1
Milk 6	0.81 ± 0.06	7.4	0.91 ± 0.05	5.5
Milk 7	0.27 ± 0.01	3.7	0.30 ± 0.02	6.7
Milk 8	0.24 ± 0.01	4.2	0.24 ± 0.05	16.7
Milk 9	0.12 ± 0.01	8.3	0.13 ± 0.04	15.4
Milk 10	0.80 ± 0.07	8.8	0.83 ± 0.05	6.0
Milk 11	0.55 ± 0.05	9.1	0.52 ± 0.02	3.9
Milk 12	ND	-	ND	-
Milk 13	0.11 ± 0.02	18.2	0.15 ± 0.03	13.3
Milk 14	0.30 ± 0.04	13.3	0.32 ± 0.05	15.6
Milk 15	0.16 ± 0.02	12.5	0.19 ± 0.03	15.8
Milk 16	0.46 ± 0.06	13.0	0.58 ± 0.04	6.9
Milk 17	ND	-	ND	-
Milk 18	0.22 ± 0.01	4.6	0.22 ± 0.01	4.6
Milk 19	0.85 ± 0.03	3.5	0.96 ± 0.01	1.0
Milk 20	0.30 ± 0.04	13.3	0.33 ± 0.02	6.1
Pork 1	ND	-	ND	-
Pork 2	0.52 ± 0.02	3.9	0.49 ± 0.01	2.0
Pork 3	2.17 ± 0.07	3.2	2.28 ± 0.04	1.8
Pork 4	0.21 ± 0.01	4.8	0.18 ± 0.03	16.7
Pork 5	1.17 ± 0.01	0.9	0.97 ± 0.03	3.1
Pork 6	2.78 ± 0.15	5.4	2.57 ± 0.08	3.1
Pork 7	0.43 ± 0.01	2.3	0.37 ± 0.03	8.1
Pork 8	ND	-	ND	-
Pork 9	0.58 ± 0.01	1.7	0.50 ± 0.04	8.0
Pork 10	0.39 ± 0.03	7.7	0.36 ± 0.04	11.1

ND, not detected; -, unavailable.

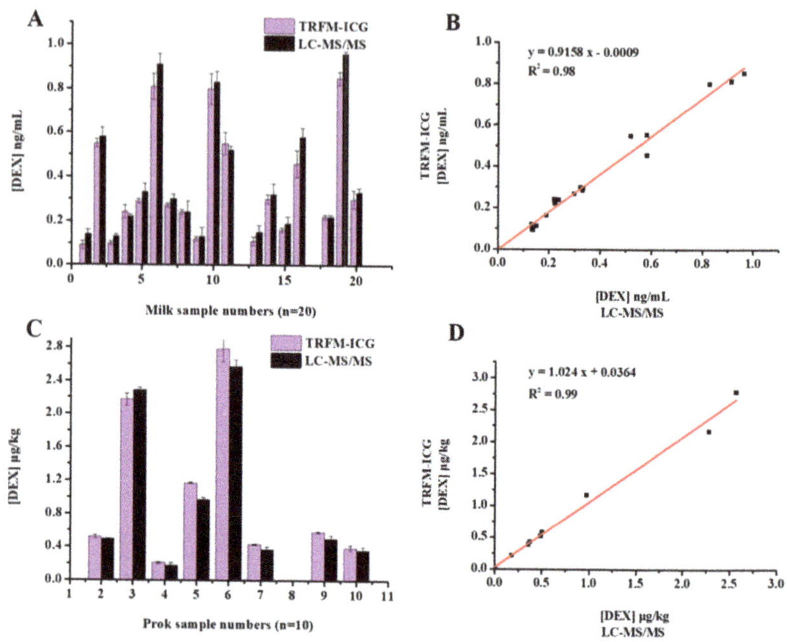

Figure 5. The correlation diagram of DEX detection results of the LC-MS/MS and TRFM-ICG: (**A**,**B**) in 20 milk samples; and (**C**,**D**) in 10 pork samples.

3.4. Comparison of DEX Immunoassay

The European Union, Japan, China, and many other countries and organizations have clearly stipulated the MRLs of DEX in animal-derived food [5,6,35,36]. Unfortunately, the detection methods of DEX residue in animal-derived food are rare, and there are even fewer rapid immunoassay methods. Until now, there are only six reports on immunoassay methods for DEX in foods (Table 3), of which three are ELISA methods [37–39] and the other three are ICG methods [25–27]. As is known, compared to ICG, the operation process and sample pretreatment of the ELISA method are relatively cumbersome. The outstanding advantage of ICG is that it is simple and fast, and the results can be achieved within 5–10 min. Therefore, the development of ICG can greatly improve the screening efficiency of DEX, and it is a useful supplement to monitoring methods.

Table 3. Comparison of immunological methods for detecting DEX in animal-derived foods.

Method	Sample	Cut-Off Value (ng/mL or μg/kg)	LOQ (ng/mL or μg/kg)	Reference
CL-ELISA	Milk	-	0.3 (sample)	[36]
ELISA	Chicken muscle, liver	-	0.3, 0.5 (sample)	[38]
ELISA	Milk, liver	-	0.2, 0.6 (sample)	[37]
CG-ICG	Milk	0.5 (sample)	0.017 (buffer)	[25]
UCNP-ICG	Animal tissue	0.3 (sample)	0.05 (buffer)	[26]
LM-ICG	Milk, pork	0.3, 0.7 (sample)	0.047, 0.087 (sample)	[27]
TRFM-ICG	Milk, pork	0.25, 0.7 (sample)	0.003, 0.062 (sample)	This work

-, unavailable.

Among the three reported ICG methods, one used traditional colloidal gold as the Ab tracer, the detection sample was milk, and the sensitivity could not meet the requirements for the detection of DEX residues [25]. The second used UCP as the Ab tracer, only qualitative detection was carried out on animal tissue sample, and the stability of

UCP is controversial [26]. The third was the work of our team. The advantage of that work was that we could use LMs with different colors to distinguish different samples. However, the sensitivity of that method was not as good as the one in our current work [27]. This may be because there are thousands of fluorescent molecules in TRFM, which greatly improves the labeling efficiency of fluorescence and analytical sensitivity.

4. Conclusions

In this study, a rapid, sensitive TRFM-ICG method based on a portable fluorescence reader was firstly established and confirmed for screening detection of DEX in milk and pork. TRFM was employed as the Ab tracer, and the cut-off values for DEX in milk and pork were 0.25 ng/mL and 0.7 μg/kg, respectively. The LOQs were 0.003 ng/mL and 0.062 μg/kg, respectively. The recovery rates ranged 80.0–106.7%, and 78.6–83.6%, respectively, with the CVs ranging 6.3–12.5% and 7.5–10.3%, respectively. The results could be obtained within 10 min. Parallel testing of blind milk and pork samples with LC-MS/MS demonstrated that the developed quantitative TRFM-ICG method was accurate, reliable, and user-friendly. The establishment of the TRFM-ICG method can provide a new, efficient, and useful technical support for the rapid screening of DEX in food.

Supplementary Materials: The following are available online at https://www.mdpi.com/article/10.3390/foods10061339/s1, Figure S1: The results of particle size of TRFM, Figure S2: The fluorescence intensity, inhibition effect, and inhibition rate results of the activation pH value. (A) Ultraviolet lamp results for milk detection; (B) Fluorescence quantitative results for milk detection; (C) Ultraviolet lamp results for pork detection; (D) Fluorescence quantitative results for pork detection, Figure S3: The fluorescence intensity, inhibition effect, and inhibition rate results of the Ab amount. (A) Ultraviolet lamp results for milk detection; (B) Fluorescence quantitative results for milk detection; (C) Ultraviolet lamp results for pork detection; (D) Fluorescence quantitative results for pork detection, Figure S4: The fluorescence intensity, inhibition effect, and inhibition rate results of the surfactant concentration. (A) Ultraviolet lamp results for milk detection; (B) Ultraviolet lamp results for pork detection; (C) Fluorescence quantitative results for milk detection; (D) Fluorescence quantitative results for pork detection, Table S1: Working parameters of the TRFM-ICG for milk and pork detection, Table S2: The IC50 and CR values of anti-DEX Ab determined by icELISA, and TRFM-ICG.

Author Contributions: Conceptualization, writing—original draft preparation, and methodology, X.L. and X.C.; methodology, main data analysis, and writing—review and editing, J.W. (Jinxiao Wu) and Z.L.; validation, data curation, and writing—review and editing, J.W. (Jin Wang), C.S. and S.Z.; and conceptualization, supervision, project administration, and writing—review and editing, H.L. and Y.S. All authors have read and agreed to the published version of the manuscript.

Funding: This research was funded by National Key Research and Development Program of China (2017YFC1601205), the Natural Science Foundation of China (31701703 and 31871883), Guangzhou Planned Program in Science and Technology (201803020026, 201807010109, and 2017B020207010), and Shanxi Provincial Key Research and Development Program (201803D221028-1).

Institutional Review Board Statement: Not applicable.

Informed Consent Statement: Not applicable.

Conflicts of Interest: The authors declare no conflict of interest.

Abbreviations

DEX, dexamethasone; MRLs, maximum residue limits; ICA, immunochromatographic assay; GNP, gold nanoparticle; QD, quantum dots; FM, fluorescent microspheres; UCP, up conversion phosphorescence; LMs, latex microspheres; TRFM, time-resolved fluorescent microspheres; EDC, 1-ethyl-3-(3-dimethylaminopropyl) carbodiimide hydrochloride; NHS, N-hydroxysuccinimide; MES, 2-(N-morpholino) ethanesulfonic acid; OVA, ovalbumin; BSA, bovine serum albumin; Ag, antigen; Ab, antibody; BB, borate buffer; PB, phosphate buffer; LOQ, limit of quantitation; icELISA, indirect competitive enzyme-linked immunosorbent assay; CR, cross-reactivity; CV, coefficient of variation.

References

1. Jeon, M.-Y.; Woo, S.M.; Seo, S.U.; Kim, S.H.; Nam, J.-O.; Kim, S.; Park, J.-W.; Kubatka, P.; Min, K.-J.; Kwon, T.K. Dexamethasone inhibits TRAIL-induced apoptosis through c-FLIP(L) upregulation and DR5 downregulation by GSK3beta activation in cancer cells. *Cancers* **2020**, *12*, 2901. [CrossRef] [PubMed]
2. Villar, J.; Ferrando, C.; Martinez, D.; Ambros, A.; Munoz, T.; Soler, J.A.; Aguilar, G.; Alba, F.; Gonzalez-Higueras, E.; Conesa, L.A.; et al. Dexamethasone treatment for the acute respiratory distress syndrome: A multicentre, randomised controlled trial. *Lancet Respir. Med.* **2020**, *8*, 267–276. [CrossRef]
3. Choi, M.Y.; Jee, D.; Kwon, J.-W. Characteristics of diabetic macular edema patients refractory to anti-VEGF treatments and a dexamethasone implant. *PLoS ONE* **2019**, *14*, e0222364. [CrossRef] [PubMed]
4. Yuan, Y.; Xu, C.; Peng, C.; Jin, Z.; Chen, W.; Liu, L. Analytical methods for the detection of corticosteroids-residues in animal-derived foodstuffs. *Crit. Rev. Anal. Chem.* **2008**, *38*, 227–241. [CrossRef]
5. Ministry of Agriculture and Rural Affairs of the People's Republic of China. *No. 235 Announcement on Maximum Residue Limit for Veterinary Drugs in Animal Foods*; Ministry of Agriculture and Rural Affairs of the People's Republic of China: Beijing, China, 2002.
6. Codex Alimentarius Commission (CAC). In Proceedings of the 32nd Session: Compendium of Methods of Analysis Identified as Suitable to Support Codex MRLs, Rome, Italy, 29 June–4 July 2009.
7. Li, X.; Luo, P.; Tang, S.; Beier, R.C.; Wu, X.; Yang, L.; Li, Y.; Xiao, X. Development of an immunochromatographic strip test for rapid detection of melamine in raw milk, milk products and animal feed. *J. Agric. Food Chem.* **2011**, *59*, 6064–6070. [CrossRef]
8. Li, X.; Shen, J.; Wang, Q.; Gao, S.; Pei, X.; Jiang, H.; Wen, K. Multi-residue fluorescent microspheres immunochromatographic assay for simultaneous determination of macrolides in raw milk. *Anal. Bioanal. Chem.* **2015**, *407*, 9125–9133. [CrossRef]
9. Li, X.; Wen, K.; Chen, Y.; Wu, X.; Pei, X.; Wang, Q.; Liu, A.; Shen, J. Multiplex immunogold chromatographic assay for simultaneous determination of macrolide antibiotics in raw milk. *Food Anal. Methods* **2015**, *8*, 2368–2375. [CrossRef]
10. Li, X.; Wang, W.; Wang, L.; Wang, Q.; Pei, X.; Jiang, H. Rapid determination of phenylethanolamine A in biological samples by enzyme-linked immunosorbent assay and lateral-flow immunoassay. *Anal. Bioanal. Chem.* **2015**, *407*, 7615–7624. [CrossRef]
11. Duan, H.; Li, Y.; Shao, Y.; Huang, X.; Xiong, Y. Multicolor quantum dot nanobeads for simultaneous multiplex immunochromatographic detection of mycotoxins in maize. *Sens. Actuators B-Chem.* **2019**, *291*, 411–417. [CrossRef]
12. Wang, Y.; Xu, J.; Qiu, Y.; Li, P.; Liu, B.; Yang, L.; Barnych, B.; Hammock, B.D.; Zhang, C. Highly specific monoclonal antibody and sensitive quantum dot beads-based fluorescence immunochromatographic test strip for tebuconazole assay in agricultural products. *J. Agric. Food Chem.* **2019**, *67*, 9096–9103. [CrossRef]
13. Liu, Y.; Ji, J.; Cui, F.; Sun, J.; Wu, H.; Pi, F.; Zhang, Y.; Sun, X. Development of a two-step immunochromatographic assay for microcystin-LR based on fluorescent microspheres. *Food Control* **2019**, *95*, 34–40. [CrossRef]
14. Deng, J.; Yang, M.; Wu, J.; Zhang, W.; Jiang, X. A self-contained chemiluminescent lateral flow assay for point-of-care testing. *Anal. Chem.* **2018**, *90*, 9132–9137. [CrossRef]
15. Ouyang, H.; Wang, M.; Wang, W.; Fu, Z. Colorimetric/chemiluminescent immunochromatographic test strip by using luminol-reduced gold nanoparticles as dual-response probes. *Sens. Actuators B-Chem.* **2018**, *266*, 318–322. [CrossRef]
16. Wang, P.; Wang, R.; Zhang, W.; Su, X.; Luo, H. Novel fabrication of immunochromatographic assay based on up conversion phosphors for sensitive detection of clenbuterol. *Biosens. Bioelectron.* **2016**, *77*, 866–870. [CrossRef]
17. Shao, Y.; Duan, H.; Zhou, S.; Ma, T.; Guo, L.; Huang, X.; Xiong, Y. Biotin-streptavidin system-mediated ratiometric multiplex immunochromatographic assay for simultaneous and accurate quantification of three mycotoxins. *J. Agric. Food Chem.* **2019**, *67*, 9022–9031. [CrossRef]
18. Li, R.; Bu, T.; Zhao, Y.; Sun, X.; Wang, Q.; Tian, Y.; Bai, F.; Wang, L. Polydopamine coated zirconium metal-organic frameworks-based immunochromatographic assay for highly sensitive detection of deoxynivalenol. *Anal. Chim. Acta* **2020**, *1131*, 109–117. [CrossRef]
19. Chen, C.; Yu, X.; Han, D.; Ai, J.; Ke, Y.; Wang, Z.; Meng, G. Non-CTAB synthesized gold nanorods-based immunochromatographic assay for dual color and on-site detection of aflatoxins and zearalenones in maize. *Food Control* **2020**, *118*, 107418. [CrossRef]
20. Hou, S.; Ma, J.; Cheng, Y.; Wang, H.; Sun, J.; Yan, Y. Quantum dot nanobead-based fluorescent immunochromatographic assay for simultaneous quantitative detection of fumonisin B-1, dexyonivalenol, and zearalenone in grains. *Food Control* **2020**, *117*, 107331. [CrossRef]
21. Guo, L.; Liu, L.; Cui, G.; Ma, S.; Wu, X.; Kuang, H. Gold immunochromatographic assay for kitasamycin and josamycin residues screening in milk and egg samples. *Food Agric. Immunol.* **2019**, *30*, 1189–1201. [CrossRef]
22. Kong, D.; Liu, L.; Song, S.; Suryoprabowo, S.; Li, A.; Kuang, H.; Wang, L.; Xu, C. A gold nanoparticle-based semi-quantitative and quantitative ultrasensitive paper sensor for the detection of twenty mycotoxins. *Nanoscale* **2016**, *8*, 5245–5253. [CrossRef]
23. Cheng, N.; Shi, Q.; Zhu, C.; Li, S.; Lin, Y.; Du, D. Pt-Ni(OH)(2) nanosheets amplified two-way lateral flow immunoassays with smartphone readout for quantification of pesticides. *Biosens. Bioelectron.* **2019**, *142*, 111498. [CrossRef]
24. Liu, Z.; Hua, Q.; Wang, J.; Liang, Z.; Li, J.; Wu, J.; Shen, X.; Lei, H.; Li, X. A smartphone-based dual detection mode device integrated with two lateral flow immunoassays for multiplex mycotoxins in cereals. *Biosens. Bioelectron.* **2020**, *158*, 112178. [CrossRef]
25. Wang, Z.; Zheng, Q.; Guo, L.; Suryoprabowo, S.; Liu, L.; Kuang, H. Preparation of an anti-dexamethasone monoclonal antibody and its use in development of a colloidal gold immunoassay. *Food Agric. Immunol.* **2017**, *28*, 958–968. [CrossRef]

26. Zhang, S.; Yao, T.; Wang, S.; Feng, R.; Chen, L.; Zhu, V.; Hu, G.; Zhang, H.; Yang, G. Development of an upconversion luminescence nanoparticles-based immunochromatographic assay for the rapid detection of dexamethasone in animal tissue. *Food Anal. Methods* **2019**, *12*, 752–760. [CrossRef]
27. Li, X.; Chen, X.; Liu, Z.; Wang, J.; Hua, Q.; Liang, J.; Shen, X.; Xu, Z.; Lei, H.; Sun, Y. Latex microsphere immunochromatography for quantitative detection of dexamethasone in milk and pork. *Food Chem.* **2021**, *345*, 128607. [CrossRef]
28. Li, X.; Wu, X.; Wang, J.; Hua, Q.; Wu, J.; Shen, X.; Sun, Y.; Lei, H. Three lateral flow immunochromatographic assays based on different nanoparticle probes for on-site detection of tylosin and tilmicosin in milk and pork. *Sens. Actuators B-Chem.* **2019**, *301*, 127059. [CrossRef]
29. Zhang, X.; Yu, X.; Wen, K.; Li, C.; Marti, G.M.; Jiang, H.; Shi, W.; Shen, J.; Wang, Z. Multiplex Lateral Flow Immunoassays Based on Amorphous Carbon Nanoparticles for Detecting Three Fusarium Mycotoxins in Maize. *J. Agric. Food Chem.* **2017**, *65*, 8063–8071. [CrossRef]
30. GB-T20741-2006. *Determination of Dexamethasone Residue in Livestock and Poultry Meat by Liquid Chromatography-Tandem Mass Spectrometry;* State Standard of the People's Republic of China: Beijing, China, 2006.
31. GB-T22978-2008. *Determination of Dexamethasone in Milk and Milk Powder by Liquid Chromatography-Tandem Mass Spectrometry;* State Standard of the People's Republic of China: Beijing, China, 2008.
32. Wang, C.; Li, X.; Peng, T.; Wang, Z.; Wen, K.; Jiang, H. Latex bead and colloidal gold applied in a multiplex immunochromatographic assay for high-throughput detection of three classes of antibiotic residues in milk. *Food Control* **2017**, *77*, 1–7. [CrossRef]
33. Li, X.; Wang, J.; Yi, C.; Jiang, L.; Wu, J.; Chen, X.; Shen, X.; Sun, Y.; Lei, H. A smartphone-based quantitative detection device integrated with latex microsphere immunochromatography for on-site detection of zearalenone in cereals and feed. *Sens. Actuators B-Chem.* **2019**, *290*, 170–179. [CrossRef]
34. Li, X.; Chen, X.; Wu, X.; Wang, J.; Liu, Z.; Sun, Y.; Shen, X.; Lei, H. Rapid detection of adulteration of dehydroepiandrosterone in slimming products by competitive indirect enzyme-linked immunosorbent assay and lateral flow immunochromatography. *Food Agric. Immunol.* **2019**, *30*, 123–139. [CrossRef]
35. The Japan Food Chemical Research Foundation. *Maximum Reside Limits (MRLs) List of Agricultural Chemicals in Foods;* The Japan Food Chemical Research Foundation: Tokyo, Japan, 2015.
36. European Union (EU) Commission Regulation. No 37/2010 of 22 December 2009 on pharmacologically active substances and their classification regarding maximum residue limits in foodstuffs of animal origin. *Off. J. Eur. Union* **2010**, *15*, 1–72.
37. Vdovenko, M.M.; Gribas, A.V.; Vylegzhanina, A.V.; Sakharov, I.Y. Development of a chemiluminescent enzyme immunoassay for the determination of dexamethasone in milk. *Anal. Methods* **2012**, *4*, 2550–2554. [CrossRef]
38. Kolanovic, B.S.; Bilandzic, N.; Varenina, I. Validation of a multi-residue enzyme-linked immunosorbent assay for qualitative screening of corticosteroids in liver, urine and milk. *Food Addit. Contam. Part A-Chem. Anal. Control Expo. Risk Assess.* **2011**, *28*, 1175–1186. [CrossRef] [PubMed]
39. Wang, W.; He, J.; Han, X.; Zheng, X.; Li, L. One-step indirect competitive ELISA development for dexamethasone determination in chicken. *J. Food Saf. Hyg.* **2016**, *2*, 75–83.

 foods

MDPI

Article

Development of a Sodium Alginate-Based Active Package with Controlled Release of Cinnamaldehyde Loaded on Halloysite Nanotubes

Rui Cui [1], Bifen Zhu [1], Jiatong Yan [1], Yuyue Qin [1,*], Mingwei Yuan [2,*], Guiguang Cheng [1] and Minglong Yuan [2]

1 Institute of Agriculture and Food Engineering, Kunming University of Science and Technology, Kunming 650550, China; cuirui910@163.com (R.C.); yi199602@163.com (B.Z.); 17854117210@163.com (J.Y.); ggcheng@kmust.edu.cn (G.C.)

2 Engineering Research Center of Biopolymer Functional Materials of Yunnan, Yunnan Nationalities University, Kunming 650550, China; yml@vip.163.com

* Correspondence: rabbqy@163.com (Y.Q.); yuanmingwei@163.com (M.Y.)

Abstract: The worsening environment and the demand for safer food have accelerated the development of new food packaging materials. The objective of this research is to prepare antimicrobial food packaging film with controlled release by loading cinnamaldehyde (CIN) on etched halloysite nanotubes (T-HNTs) and adding it to sodium alginate (SA) matrix. The effects of T-HNTs-CIN on the physical functional properties and antibacterial activity of the film were systematically evaluated, and the release of CIN in the film was also quantified. Transmission electron microscopy and nitrogen adsorption experiments showed that the halloysite nanotubes had been etched and CIN was successfully loaded into the T-HNTs. The addition of T-HNTs-CIN significantly improved the water vapor barrier properties and tensile strength of the film. Similarly, the presence of T-HNTs-CIN in the film greatly reduced the negative effects of ultraviolet rays. The release experiment showed that the diffusion time of CIN in SA/T-HNTs-CIN film to fatty food simulation solution was delayed 144 h compared with that of SA/CIN film. Herein, the antibacterial experiment also confirmed the controlled release effect of T-HNTs on CIN. In conclusion, SA/T-HNTs-CIN film might have broad application prospects in fatty food packaging.

Keywords: cinnamaldehyde; halloysite nanotubes; sodium alginate; release

Citation: Cui, R.; Zhu, B.; Yan, J.; Qin, Y.; Yuan, M.; Cheng, G.; Yuan, M. Development of a Sodium Alginate-Based Active Package with Controlled Release of Cinnamaldehyde Loaded on Halloysite Nanotubes. *Foods* **2021**, *10*, 1150. https://doi.org/10.3390/foods10061150

Academic Editors: Hong Wu and Hui Zhang

Received: 23 March 2021
Accepted: 11 May 2021
Published: 21 May 2021

1. Introduction

In recent years, the need to coordinate the growing environmental pollution problems with technological progress has become very urgent [1]. Hence, the use of biodegradable polymer materials to develop functional food packaging materials has been widely concerned and studied, in order to reduce environmental pollution and meet people's needs for active food packaging [2,3]. Among many biopolymers, alginate, as a carbohydrate, is often used as a raw material for biodegradable food packaging, which has the characteristics of low cost, good biocompatibility, and excellent film-forming properties [4,5]. Sodium alginate (SA) is a naturally linear water-soluble polysaccharide extracted from brown algae, which is composed of β-D-mannuronic acid and α-l-guluronic acid (1–4) linking units [6,7]. SA has been widely used in the field of food packaging and biomedical fields because of its low price, easy access, and easy processing. As a food packaging material, SA has advantages, such as good mechanical properties and high transparency [8,9].

Pathogenic microorganisms can cause food spoilage and food-borne diseases, as well as consumers' concerns about chemical residues in food, making the food industry more and more concerned about the research of natural antibacterial agents in food packaging [10,11]. In order to maximize the function of packaging, extend the use period of food and protect consumers from the threat of food-borne disease outbreaks, antibacterial active substances (such as essential oils) are usually added to increase its function [12]. Cinnamon essential oil is widely used in the field of food, which can be used to protect food without causing harm to human health. There have been many studies that add cinnamon essential oil to food packaging as an antimicrobial agent for food preservation [13,14]. Cinnamaldehyde (CIN) has been approved by the Food and Drug Administration (FDA) and can be used in food products. It is the main ingredient extracted from cinnamon essential oil, which is not only harmless, but also has a broad spectrum of antibacterial and antifungal activities [15]. However, it must be taken into account that CIN is liquid and volatile at room temperature, which is a severe challenge we are currently facing. In this regard, there have been many solutions, such as using porous and other special structure carriers to load active compounds or encapsulating the active substance in chitosan or cyclodextrin [16,17]. In this context, it is one of the most convenient strategies to load CIN with nanoparticles of special structure, which can not only reduce CIN volatilization losses, but also impart the characteristics of nanoparticles to the food packaging system.

Among many nanoparticle carriers, halloysite nanotubes (HNTs) have aroused great interest of researchers due to their ability to trap, protect, and control the release of active substances [18]. HNTs, whose chemical formula is $Al_2Si_2O_5(OH)_4.nH_2O$, which is a subgroup of kaolin. In addition, HNTs have been listed by the FDA as a safe food packaging material [19]. HNTs can be used as a carrier for drug delivery in the medical field because of their unique hollow tubular structure, low cost, and good biocompatibility [12,20]. In particular, HNTs have been shown to have higher adsorption capacity than montmorillonite [21]. Moreover, it has been shown that the addition of HNTs as fillers can improve the barrier and mechanical properties of polymers [10].

The purpose of this study is to develop SA antibacterial composite film with controlled release. The lumen of the original HNTs is limited in volume, and the enlargement of the pores will allow more active chemicals to be loaded. Therefore, in this study, HNT were etched with sulfuric acid to further increase its drug loading, and then CIN was loaded on acidified nanoparticles (T-HNTs) to prepare composite film by method of solution casting. The effects of T-HNTs loaded with CIN on the microstructure, barrier performance, transparency, and antibacterial properties of the composite film were studied, and the release behavior of CIN in the composite film was also investigated.

2. Material and Methods

2.1. Materials

Sodium alginate (SA) was obtained from Zhejiang Yinuo Biotechnology Company (Lanxi, China). Halloysite nanotubes (HNTs) were purchased from Xi'an Mingda Biotechnology Co., Ltd. (Xi'an, China), with a purity of 99.96%. Sigma-Aldrich (St. Louis, MO, USA) provided the use of glycerin (MW = 92.09). Cinnamaldehyde (CIN) (Purity 98%) was purchased from Shanghai Macklin Biotech Co., Ltd. (Shanghai, China) The manufacturer of the analytical pure sulfuric acid was the Xilong Chemical Factory in Shantou, Guangdong. The other reagents used in this experiment were all of analytical grade.

2.2. Etching of HNTs

The etching of HNTs was mainly based on the method of Abdullayev et al., and some changes were made in the process of implementation [22]. Five grams of HNTs were added to 500 mL of sulfuric acid (1.0 M) and the dispersion was stirred and kept at 60 °C for 30 h. The treated HNTs (T-HNTs) were then washed with deionized water five times until their pH range was between 6 and 7. Finally, the samples were dried at 50 °C for 24 h and crushed into powder with a mortar.

2.3. Fabrication of T-HNTs-CIN Nanoparticles

The method of T-HNTs loading CIN was mainly modified and implemented by referring to the method of Zou et al. [23]. T-HNTs and CIN were mixed in 30 mL anhydrous ethanol solution, stirred overnight, and then treated with ultrasound for 15 min. The suspension was vacuumed twice for 15 min to remove air to encapsulate CIN into the cavity of HNT. Finally, the suspension was centrifuged to obtain CIN-loaded nanoparticles (T-HNTs-CIN) and dried at room temperature for 24 h.

2.4. Preparation of the SA Composite Film

The nanocomposite film was prepared by the solution casting method. First, 2 g sodium alginate powder was dissolved in 100 mL of distilled water containing glycerol (30 wt% relative to SA) and bathed in water at 50 °C for 30 min to prepare the sodium alginate solution. SA/T-HNTs-CIN dispersion was obtained by dispersing the T-HNTs-CIN nanoparticles in sodium alginate solution by loading a certain amount of CIN on T-HNT (5% by weight relative to SA). Simultaneously, the same amount of T-HNTs (5% by weight relative to SA) or CIN (The amount of CIN supported by 5 wt% T-HNTs) was added to the sodium alginate as the control. The prepared dispersions were poured into the polytetrafluoroethylene plate and dried at room temperature for 36 h to form the film. Then, 1% w/v $CaCl_2$ solution was poured onto the film so that the film was completely covered by the solution. After the reaction for one minute, the $CaCl_2$ solution was poured out and washed twice with distilled water. Finally, after drying for 48 h at room temperature, it was stored in a desiccator at 25 °C and 50% relative humidity. The manufactured films were identified as SA, SA/CIN, SA/T-HNTs, and SA/T-HNTs-CIN, respectively.

2.5. Characterization of Nanoparticles

Field emission transmission electron microscopy (TEM, Tecnai G2 F30 S-Twin) was used to photograph the morphological differences between HNTs and etched HNTs. The nanoparticles were first dispersed in an ethanol solution ultrasonically for 5 min, and then dropped on the surface of the copper grid to be observed after drying.

The change in the porous structure of the nanoparticles was measured by the Micromeritics ASAP2020 system. Before the experiment, the sample was vacuum degassed at 250 °C for 8 h. The specific surface area and pore size distribution of the sample were determined according to the Brunauer Emmett Teller (BET) method and the Barret-Joyner-Halenda (BJH) method, respectively.

2.6. Film Thickness

The thickness of the prepared film was measured by randomly selecting five positions with a micrometer (Ningbo Deli Co., Ltd., Ningbo, China), and the accuracy was 0.001 mm.

2.7. Microscopic Images

The surface of the SA composite film sample was plated with a layer of gold, and then placed in a vacuum environment to observe the surface morphology of the prepared sample using a field emission scanning electron microscope (SEM, NOVA NANOSEM-450, FEI Co., Ltd., Hillsboro, OR, USA)

2.8. Water Vapor Permeability (WVP)

The ability of SA film to block water vapor was measured referring to ASTM E96-95 standard method. Twelve grams of anhydrous silica were put into the weighing cup. The prepared film was cut to a uniform size to seal the rim of the measuring cup. Finally, the weighing cup was placed in a desiccator at 25 °C and 50% relative humidity. The samples were placed in a desiccator to measure the weight of the weighing cup at an interval of 1 h for a total of 12 h, and three samples were set in parallel for each type.

2.9. Mechanical Properties

The mechanical properties of the SA composite film are measured by using a microcomputer electronic tensile testing machine (QLW-5E, Xiamen Qunlong Instrument Co., Ltd., Xiamen, China) to obtain its tensile strength and elongation at break. The stretching speed parameter of the equipment was set at 50 mm/min. The determination of each type of film was repeated five times, and each repetition was from a different sampling unit.

2.10. Light Transmittance and Opacity of Film

The transmittance of SA composite film in ultraviolet (UV) and visible light region was determined by double beam ultraviolet spectrophotometer. The detailed determination method was based on Achachlouei et al. [24]. The film sample was cut into rectangular pieces and installed in a quartz cuvette for spectrum measurement. The wavelength of measurement was selected as 200–800 nm. The opacity of the film can be calculated by Equation (1).

$$\text{Opacity} = (\lg(1/\text{T}))/\text{d} \tag{1}$$

where T is the light transmittance of the film at 600 nm, and d is the thickness of the sample. The measurements were repeated three times for each type of film.

2.11. The Sustained-Release of CIN in Film in Food Simulants

The CIN release from the SA nanocomposite film was assessed following a method adapted from Muller et al. [25]. Sodium alginate is a hydrophilic colloid. When the sodium alginate composite film is in contact with water or aqueous solution, its matrix structure will change. Therefore, isooctane was selected as the food simulant in this experiment. The film, weighing about 2 g, was immersed in 1 L of isooctane solution at 20 °C. A total of 1 mL solution was taken out from the food simulation solution at certain intervals (1 mL of isooctane solution was supplemented after taking it out), and the content of CIN in the solution was measured by a dual-beam UV-Vis spectrophotometer with a wavelength of 279 nm. At the same time, SA/CIN and SA/T-HNTs-CIN corresponding film without CIN were used as controls, and three parallel films were set for each sample. The cumulative release rate of CIN in the film can be calculated by Equation (2).

$$\text{C}\ (\%) = (\text{C} \times \text{V} \times \text{k}/\text{C}_0 \times \text{m}) \times 100 \tag{2}$$

where C is the cumulative release rate of CIN (%), C is the concentration of CIN in isooctane at t (mg/mL), V is the volume of isooctane solution (mL), k is the dilution factor of the solution, C_0 is the initial concentration of CIN (mg/mL), and m is the amount of film (mg).

2.12. In Vitro Antibacterial Activity of the Film in the Release Experiment

In order to assess the antimicrobial properties of the manufactured film, a slight modification of the previous method was used [26]. *Staphylococcus aureus* (*S. aureus*) and *Escherichia coli* (*E. coli*) (both of which were provided by the Microbiology Laboratory of the Faculty of Agriculture and Food Engineering, Kunming University of Science and Technology, Yunnan, China) were stored at −80 °C. The specific operation process of the antibacterial experiment was as follows. First, the long-term stored *S. aureus* and *E. coli* were inoculated with tryptic soy broth (TSB) medium. This step was then repeated to activate the activity of both strains. Then, 100 μL of the activated bacterial solution was added to a 10 mL TSB medium containing 0.2 g of sample, so that the bacterial solution concentration in the medium was about 10^5 CFU/mL. Subsequently, the bacterial solution containing the sample was cultured on a shaker for 12 h, the temperature was set to 37 °C, and the speed was 180 rpm. After the time was up, the bacterial liquid was evenly diluted and spread in tryptic soy agar medium and placed in an incubator at 37 °C. After 18–24 h of colony culture, the number of colonies was calculated. In particular, the sample was regularly taken out of the food simulant and wiped dry for the antibacterial test during the release experiment. Each sample was repeated for three times.

2.13. Statistical Analysis

The data were analyzed by one-way analysis of variance (ANOVA) using SPSS 21.0 (Chicago, IL, USA), followed by Duncan's multiple comparison test at 95% confidence level.

3. Results and Discussion

3.1. Characterization of Nanoparticles

TEM images of the pristine HNTs and the treated HNTs nanoparticles are shown in Figure 1. Both HNTs and T-HNTs are cylindrical, with a central transparent region extending lengthwise along the nanotubes, indicating that the HNTs are hollow and open. No significant changes are observed in the tube length of HNTs and T-HNTs. Interestingly, after HNTs were treated with sulfuric acid, the inner lumen was clearly etched, and their inner diameters were significantly increased by about 10–20 nm. This indicated that sulfuric acid treatment resulted in an increase in the inner diameter of HNTs.

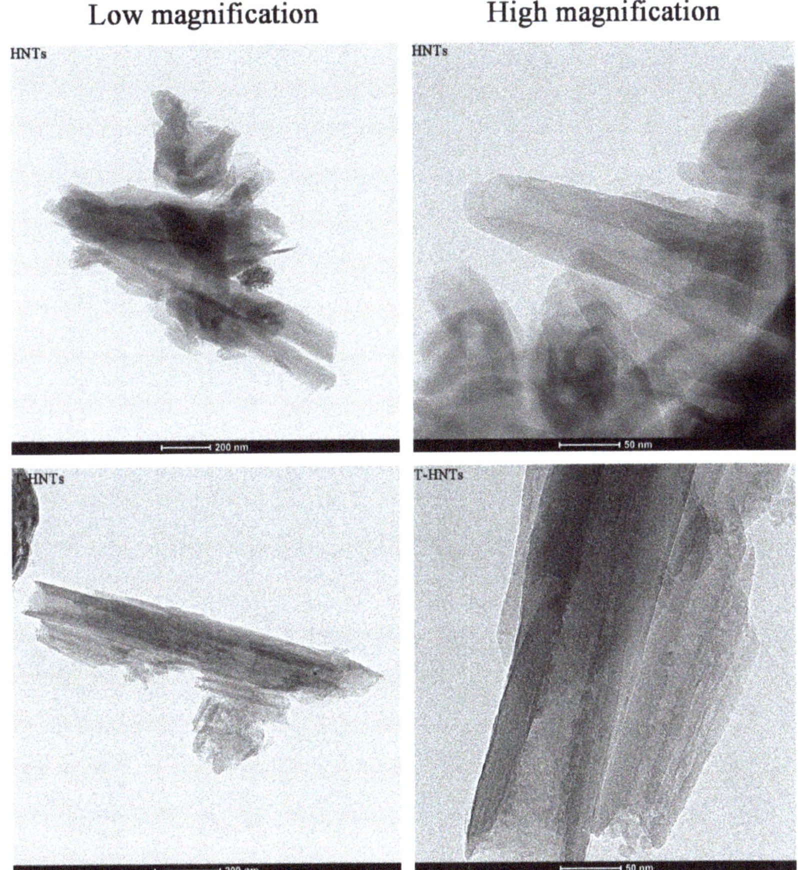

Figure 1. TEM images of HNTs and T-HNTs.

The porous structure parameters of HNTs, T-HNTs, and T-HNTs-CIN are shown in Table 1. After HNT was etched by sulfuric acid, its specific surface area increased from 25.50 to 76.61 m^2/g, and its pore volume increased from 0.30 to 0.39 m^3/g, which were consistent with the TEM observation results. The same trend was found for average pore size of HNTs. The selective dissolution of the AlO_6 octahedral layer in the inner cavity of HNTs and the decomposition of silica would cause the increase of the specific surface area and pore volume of HNTs [27]. Garcia-Garcia et al. found the same trend when treating halloysite nanotubes with acid [28]. CIN itself is volatile, and samples need to be degassed before measurement, which will cause a certain error between the actual measurement result and the theoretical value (the actual measurement result is less than the theoretical value). If the CIN was not successfully loaded onto the T-HNTs, the specific surface area of the T-HNTs-CIN did not change significantly. The actual results showed that the specific surface area and pore volume of the T-HNTs-CIN increased compared with the HNT, but decreased compared with the T-HNTs, indicating that the CIN was successfully loaded onto the T-HNTs.

Table 1. The surface area, pore volume, and average pore size of samples.

Samples	S_{BET} (m^2/g)	Pore Volume (cm^3/g)	Average Pore Size (nm)
HNTs	25.50	0.30	1.88
T-HNTs	76.61	0.39	2.41
T-HNTs-CIN	51.39	0.37	2.13

3.2. *Surface Morphology of Composite Film*

Figure 2 illustrates the surface morphology of four different formulations of SA film. Figure 2a is the SA film, the surface of which is uniform and smooth. The presence of CIN in the SA matrix did not significantly affect the surface of the matrix. However, when T-HNTs were added to the SA matrix, many uniformly distributed white spots appeared on the surface of SA/T-HNTs film. Similarly, SA/T-HNTs-CIN film showed the same phenomenon. This implied that there was no significant difference between T-HNTs and CIN-loaded T-HNTs on the surface morphology of the SA film.

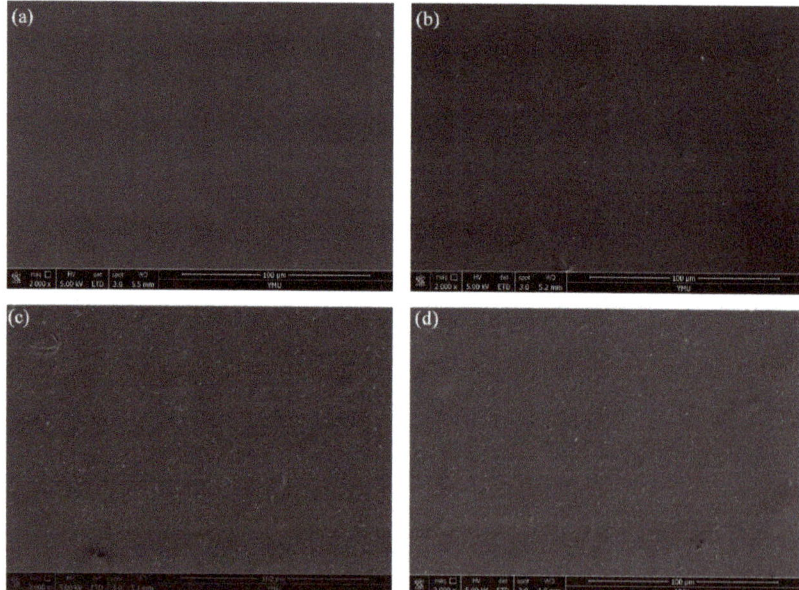

Figure 2. SEM micrographs of the surface of the bio-nanocomposite film: (**a**) SA film, (**b**) SA/CIN film, (**c**) SA/T-HNTs film, and (**d**) SA/T-HNTs-CIN film.

3.3. *WVP*

The key factor in evaluating the feasibility of composite materials in food preservation is to determine the WVP value of composites. During food preservation, edible or biodegradable film can be used to reduce the transfer of moisture from the environment to the inside of the package [29]. The WVP values of neat SA film, as well as the SA nanocomposite film, are shown in Table 2. The WVP value of SA film is 38.2×10^{-2} g mm/h KPa m^2. The barrier of SA composite film to water vapor was significantly enhanced after incorporation of CIN or T-HNTs. When CIN was immobilized on T-HNTs and then incorporated into SA matrix, the WVP value of SA/T-HNTs-CIN film was reduced by 14.7% compared with that of SA film. There are many factors that affect the barrier properties of the film, such as the hydrophobicity, the structure, and the compatibility of the material. CIN is inherently hydrophobic, and its addition might increase the hydrophobic/hydrophilic ratio of the film [30]. As a high aspect ratio aluminosilicate mineral, HNTs have barrier properties to water vapor. HNTs dispersed in the matrix might make the path of water molecules through the matrix become tortuous [10]. All of these might be the reasons for the decrease of WVP value of the film. Yousefi et al. added halloysite nanotubes and *Origanum vulgare* essential oil to the film matrix, which also improved the water resistance of the film [29].

Table 2. Water vapor permeability of SA-based composite film.

Films	Thickness (mm)	WVP ($\times 10^{-2}$ g mm/24 h·KPa·m^2)
SA	0.030 ± 0.001 ^{a}	1.59 ± 0.05 ^{b}
SA/CIN	0.033 ± 0.004 ^{a}	1.31 ± 0.15 ^{a}
SA/T-HNTs	0.035 ± 0.001 ^{a}	1.41 ± 0.02 ^{a}
SA/T-HNTs-CIN	0.033 ± 0.003 ^{a}	1.36 ± 0.11 ^{a}

Data are presented as mean ± standard deviation and different letters (a, b) within the columns shows the significant differences ($p < 0.05$), where ^{a} is the lowest value.

3.4. Mechanical Properties

The effect of T-HNTs nanoparticles loaded with CIN on the mechanical properties of SA composite film was studied. As presented in Table 3, it was evident that with the addition of T-HNTs-CIN nanoparticle, the tensile strength (TS) of the SA/T-HNTs-CIN nanocomposite film improved, and its TS value was increased by 20.8% compared to the SA film. This was similar to the effect of halloysite nanotubes on the mechanical properties of carrageenan/gelatin films previously published by Akrami-Hasan-Kohal et al. [31]. The potential strain-induced arrangement of the clay particle layer in the polymer matrix and the interaction between the polymer and the hydrogen bonds in the clay minerals might contribute to the improvement of the tensile properties of the film [32,33]. The elongation at break (ε) of the film was not changed significantly by the T-HNTs-CIN nanoparticle addition. However, the addition of CIN significantly increased the flexibility of the film. Ahmed et al. also reported that adding clove essential oil to the film matrix increased the flexibility of the film [34]. The presence of the CIN in the SA matrix might hinder the polymer–polymer intermolecular attraction [35]. However, CIN was added to the SA matrix after being loaded by T-HNTs, and most of the CIN was present in the T-HNTs, thereby reducing the plasticizing effect of CIN.

Table 3. Mechanical properties of SA nanocomposite film.

Film	TS (MPa)	ε (%)
SA	66.4 ± 4.28 ^{a}	2.76 ± 0.30 ^{a}
SA/CIN	70.4 ± 2.45 ^{ab}	3.35 ± 0.07 ^{b}
SA/T-HNTs	77.3 ± 7.50 ^{bc}	2.66 ± 0.09 ^{a}
SA/T-HNTs-CIN	80.2 ± 3.81 ^{c}	2.97 ± 0.30 ^{ab}

Data are presented as mean ± standard deviation and different letters (a–c) within the columns shows the significant differences ($p < 0.05$), where ^{a} is the lowest value.

3.5. Light Transmittance and Opacity of Film

The UV-visible spectrum for SA nanocomposite film is displayed in Figure 3. It can be seen that SA film exhibited high light transmittance in both UV and visible regions, especially in the 300–800 nm region with a high light transmittance of more than 80%. The light transmittance of SA/CIN film at 240–300 nm was significantly lower than that of SA film. The anti-UV effect of CIN might be its own aromatic compound, and its chemical bond could absorb UV light [26]. Ahmed et al. had previously found a similar phenomenon in the study of polylactide/cinnamon oil composite films [36]. Moreover, as can be seen from curves c and d in Figure 2, the transmittance of the composite film at all wavelengths was greatly reduced after the addition of T-HNTs in SA film.

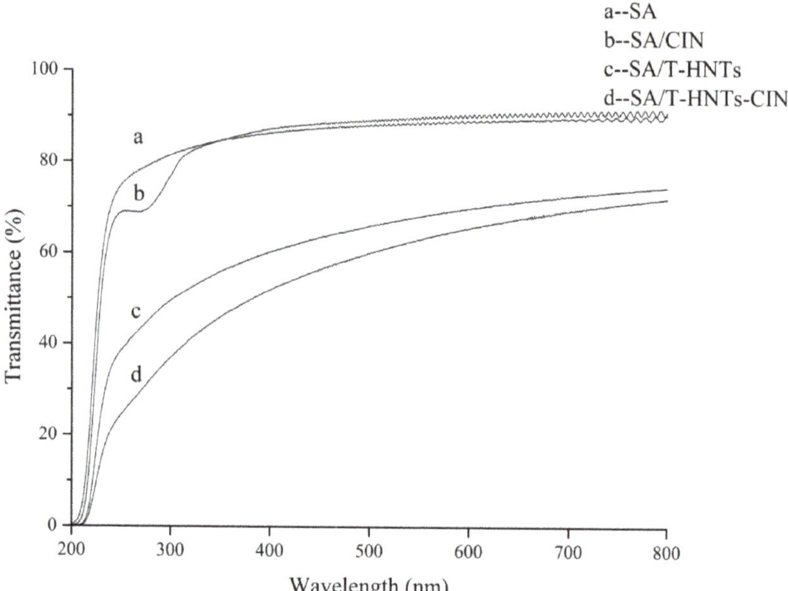

Figure 3. Light transmittance of the nanocomposite film: (**a**) SA film, (**b**) SA/CIN film, (**c**) SA/T-HNTs film, and (**d**) SA/T-HNTs-CIN film.

Table 4 shows the transmittance values of the composite films at 240 (UV-C), 300 (UV-B), 360 (UV-A), and 600 nm (visible light) as well as the opacity of different formulation film. The light transmittances of SA film at 240, 300, 360, and 600 nm are 69.5, 81.8, 85.0, and 87.4%, respectively, and the transmittance value after the addition of T-HNTs-CIN decreased to 22.8, 40.9, 50.1, and 66.1%, respectively. This might be due to the combined action of T-HNTs and CIN. The presence of T-HNTs in the film matrix might block or diffract the light, thus affecting the transmittance of light at all wavelengths [8]. Huang et al. also found a similar phenomenon when halloysite was added during the preparation of agar-based nanocomposite films [37]. The transparency of SA composite film decreased with the addition of HNTs. At the same time, the color of the film was whitened by the addition of the white powder HNTs. It is noteworthy that the addition of T-HNTs-CIN resulted in a greater reduction in the transmittance of UV light than that of visible light. This means that by preparing the composite film with T-HNTs-CIN, the UV barrier properties can be improved without sacrificing the transparency of SA film. The nanocomposite film with high ultraviolet shielding performance has high application potential as a transparent ultraviolet blocking packaging material.

Table 4. Transmittance (%) and opacity and values of nanocomposite film in the visible, UV-A, UV-B, and UV-C regions.

Film Sample	UV-C (240 nm) T (%)	UV-B (300 nm) T (%)	UV-A (360 nm) T (%)	Visible (600 nm) T (%)	Opacity (AU. nm/mm)
SA	69.5 ± 2.29^{d}	81.8 ± 0.67^{d}	85.0 ± 0.32^{c}	87.4 ± 2.45^{c}	2.50 ± 0.21^{a}
SA/CIN	64.4 ± 1.13^{c}	76.1 ± 1.13^{c}	84.8 ± 0.57^{c}	89.9 ± 0.85^{c}	2.00 ± 0.10^{a}
SA/T-HNTs	35.6 ± 1.62^{b}	50.2 ± 1.07^{b}	57.5 ± 0.79^{b}	70.3 ± 0.50^{b}	4.54 ± 0.55^{b}
SA/T-HNTs-CIN	22.8 ± 2.40^{a}	40.9 ± 4.62^{a}	50.1 ± 3.27^{a}	66.1 ± 0.57^{a}	4.78 ± 1.15^{b}

Data are presented as mean ± standard deviation and different letters (^{a}–^{d}) within the columns shows the significant differences ($p < 0.05$), where ^{a} is the lowest value.

3.6. Slow-Release Behavior of the CIN in Food Simulants

Isooctane was used as food simulant to simulate food with hydrophobic fats. The cumulative release of CIN by the SA film without T-HNTs and the SA film with T-HNTs in food simulant are shown in Figure 4.

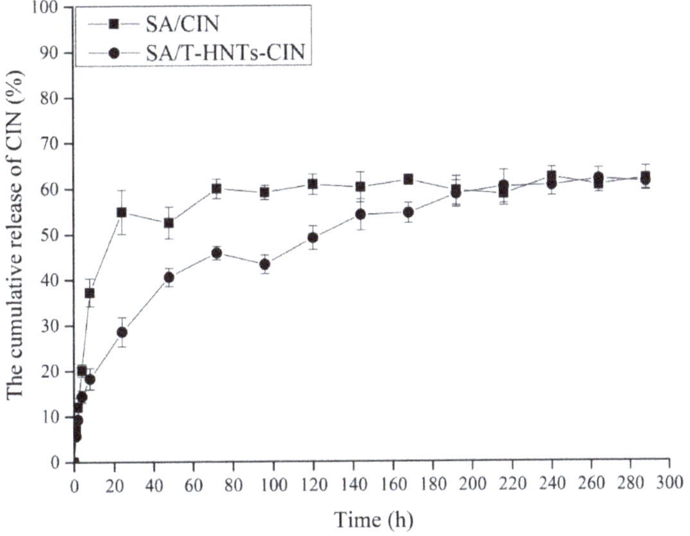

Figure 4. The cumulative release of CIN in the nanocomposite film: SA/CIN film and SA/T-HNTs-CIN film.

The whole process of the release experiment was carried out under stable environmental conditions (temperature of 20 °C, relative humidity 75%). The SA/CIN and SA/T-HNTs-CIN film in food simulant released most of CIN in the first 24 h, and the cumulative release of SA/CIN film was higher (58.95%), which was significantly higher than SA/T-HNTs-CIN film (28.57%). Subsequently, the release rate of CIN in the two films slowed down significantly. Finally, the release of CIN from SA/CIN film reached a peak of 59.97% at 72 h, and the release amount of CIN was about 16.64 mg, while the release of SA/T-HNTs-CIN film reached a stable level at 216 h, and the cumulative release of film stabilized at 60.31% (The release of CIN was about 17.44 mg). The presence of T-HNTs in SA/T-HNTs-CIN film slowed down the release rate of CIN, which in turn delayed the time when CIN reached stability. Shen et al. prepared novel sodium alginate-based double network hydrogel spheres after loading urea on HNTs, which reduced the release rate of urea [38]. The release of CIN in the SA composite film matrix into the food simulant is affected by many factors. First, the liquid molecules in the solvent diffuse from the outer surface of the film into the matrix of the SA composite film. Then, the polymer matrix network relaxes due to the presence of the solvent in the film matrix. Lastly, CIN is released from the relaxed polymer matrix into the food simulating liquid until the thermodynamic equilibrium between the SA composite film and the food simulating liquid is reached. Of course, the last step is influenced not only by mass transfer, but also by the interaction between the volatile compounds and the matrix [25,39].

According to the steps described above for the release of the active compound from the SA composite film matrix, and in combination with the release curves of the two film systems in food simulation solution, it can be seen that the release rate of the SA composite film supported by T-HNTs was significantly slower than that of the SA/CIN film. This observed behavior could be explained by the retarded release of CIN by T-HNTs. The loading of CIN by T-HNTs was mainly the adsorption of intracavity and external surface of T-HNTs to CIN [40]. This increased the mass transfer steps of CIN in the release process, thereby prolonging the release time of CIN. Compared with CIN directly added to SA matrix, CIN was added to SA matrix in a HNTs loaded manner, and CIN was better protected to maintain its activity through the action of HNTs carrier, and the release time of CIN from the controlled release system was prolonged, thus maximizing the function of CIN. Therefore, it can be concluded that the presence of HNTs in SA/T-HNTs-CIN film controlled release system could effectively alleviate the initial burst release of CIN and prolong the action time of CIN. The SA/T-HNTs-CIN film slow-release system might have a good application prospect in the packaging of fatty foods.

3.7. *In Vitro Antibacterial Activity of the Film in the Release Experiment*

The effects of the manufactured film on the antibacterial activity of typical food-borne pathogens (*S. aureus* and *E. coli*) were investigated, and the results are shown in Figure 5. As expected, the presence of T-HNTs in the film matrix has no antibacterial activity against *S. aureus* and *E. coli*. However, on day 0 of the experiment, SA/CIN and SA/T-HNTs-CIN film decreased by 1.34 and 1.35 Log_{10}CFU/mL, respectively, compared to the control group. This might be attributed to the presence of CIN in the polymer matrix. A large number of previous studies have shown that CIN had inhibitory effects on the growth of a variety of bacteria, because CIN might damage the cell membrane of bacteria, leading to changes in cytoplasmic leakage and membrane permeability [41,42]. Similarly, for *E. coli*, the number of colonies in the film containing CIN was significantly reduced by 0.65 and 0.57 Log_{10}CFU/mL, respectively, compared with the control group. Interestingly, we found that *S. aureus* was more susceptible to CIN inhibition than *E. coli*. This phenomenon could be attributed to the fact that the cell wall of *E. coli* (Gram-negative bacterium) has an extra layer of lipopolysaccharide outer membrane than that of *S. aureus* (gram-positive bacterium). The lipopolysaccharide outer membrane has a good barrier effect on hydrophobic substances (CIN). Although it cannot completely block hydrophobic compounds, it limits the penetration of CIN into microbial cells and reduces the inhibitory effect [43].

In addition, with the increase of release time, the bacteriostatic effect of the film containing CIN on the two kinds of bacteria became less and less. There was no significant difference between SA/CIN film and SA in the colony count of *S. aureus* on the second day. However, the SA/T-HNTs-CIN film showed no significant difference in the number of colonies from the SA/T-HNTs film on 7 days. The same trend was observed for *E. coli*, where CIN was fixed by T-HNTs and then added to SA matrix, and the bacteriostatic duration was extended by 4 days compared with SA/CIN films. These results confirmed that the SA/T-HNTs-CIN film had a controlled release effect on CIN, successfully alleviating the interaction between the active compound and the surrounding environment and making it release slowly. This packaging system can be used as a new and promising alternative to renewable food packaging. Of course, future research in this area should focus more on increasing the loading of active substances or using multiple loading systems to improve their applicability.

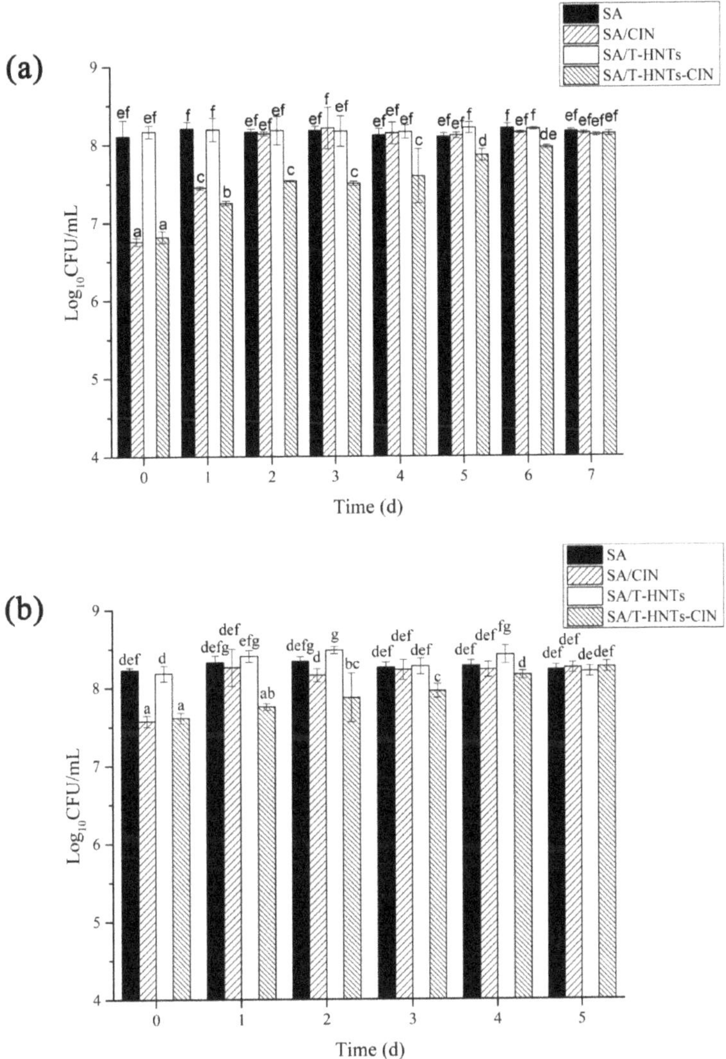

Figure 5. Antimicrobial activity of the composite film: (**a**) *Staphylococcus aureus* and (**b**) *Escherichia coli*. Values followed by different letters (a–f) were significantly different ($p < 0.05$), where a is the lowest value.

4. Conclusions

In this work, the HNTs were etched and then loaded with CIN, and the food functional packaging film with continuous release and antibacterial activity were prepared by solvent volatilization. It was confirmed that the cavities of HNTs were etched and their specific surface area increased significantly after sulfuric acid treatment. CIN was first loaded by T-HNTs and then added to SA matrix, which greatly improved the mechanical properties of the film and the barrier property of water vapor. In addition, the combined effect of T-HNTs and CIN could increase the UV shielding effect while losing less transparency. In the release experiment, the SA/T-HNTs-CIN film could be released continuously in the fatty food simulation solution for 216 h, which was 144 h longer than that of SA/CIN film.

The antibacterial activity of SA/T-HNTs-CIN film against *S. aureus* and *E. coli* was longer by five days and four days than that of SA/CIN film without T-HNTs.

Author Contributions: Conceptualization, R.C. and Y.Q.; methodology, M.Y. (Mingwei Yuan) and M.Y. (Minglong Yuan); formal analysis, R.C. and B.Z.; investigation, R.C. and J.Y.; resources, G.C.; data curation, M.Y. (Mingwei Yuan); writing—original draft preparation, R.C. and Y.Q.; writing—review and editing, R.C. and B.Z. All authors have read and agreed to the published version of the manuscript.

Funding: This research was funded by the National Natural Science Foundation of China (No. 32060569 and 21576126), Yunnan Province "Thousand Talents Program" project training project (YNQR-CYRC-2018-012) and Major Science and Technology Project of Yunnan Province, China (2019ZF010).

Conflicts of Interest: The authors declare no competing financial interest.

References

1. Lisuzzo, L.; Cavallaro, G.; Milioto, S.; Lazzara, G. Layered composite based on halloysite and natural polymers: A carrier for the pH controlled release of drugs. *New J. Chem.* **2019**, *43*, 10887–10893. [CrossRef]
2. Duncan, T.V. Applications of nanotechnology in food packaging and food safety: Barrier materials, antimicrobials and sensors. *J. Colloid Interface Sci.* **2011**, *363*, 1–24. [CrossRef] [PubMed]
3. Rhim, J.-W.; Park, H.-M.; Ha, C.-S. Bio-nanocomposites for food packaging applications. *Progr. Polym. Sci.* **2013**, *38*, 1629–1652. [CrossRef]
4. Norajit, K.; Kim, K.M.; Ryu, G.H. Comparative studies on the characterization and antioxidant properties of biodegradable alginate films containing ginseng extract. *J. Food Eng.* **2010**, *98*, 377–384. [CrossRef]
5. Abdollahi, M.; Alboofetileh, M.; Rezaei, M.; Behrooz, R. Comparing physico-mechanical and thermal properties of alginate nanocomposite films reinforced with organic and/or inorganic nanofillers. *Food Hydrocoll.* **2013**, *32*, 416–424. [CrossRef]
6. King, A.H. Brown seaweed extracts (alginates). *Food Hydrocoll.* **1983**, *2*, 115–188.
7. Rhim, J.W. Physical and mechanical properties of water resistant sodium alginate films. *Lebensm. Wiss. Technol. Food Sci. Technol.* **2004**, *37*, 323–330. [CrossRef]
8. Shankar, S.; Kasapis, S.; Rhim, J.-W. Alginate-based nanocomposite films reinforced with halloysite nanotubes functionalized by alkali treatment and zinc oxide nanoparticles. *Int. J. Biol. Macromol.* **2018**, *118*, 1824–1832. [CrossRef]
9. Kumar, A.; Rao, K.M.; Han, S.S. Development of sodium alginate-xanthan gum based nanocomposite scaffolds reinforced with cellulose nanocrystals and halloysite nanotubes. *Polym. Test.* **2017**, *63*, 214–225. [CrossRef]
10. Lee, M.H.; Kim, S.Y.; Park, H.J. Effect of halloysite nanoclay on the physical, mechanical, and antioxidant properties of chitosan films incorporated with clove essential oil. *Food Hydrocoll.* **2018**, *84*, 58–67. [CrossRef]
11. Krepker, M.; Shemesh, R.; Poleg, Y.D.; Kashi, Y.; Vaxman, A.; Segal, E. Active food packaging films with synergistic antimicrobial activity. *Food Control* **2017**, *76*, 117–126. [CrossRef]
12. Lee, M.H.; Seo, H.-S.; Park, H.J. Thyme Oil Encapsulated in Halloysite Nanotubes for Antimicrobial Packaging System. *J. Food Sci.* **2017**, *82*, 922–932. [CrossRef]
13. Loke, X.-J.; Shemesh, R.; Poleg, Y.D.; Kashi, Y.; Vaxman, A.; Segal, E. Plasma-treated polyethylene coated with polysaccharide and protein containing cinnamaldehyde for active packaging films and applications on tilapia (*Orechromis niloticus*) fillet preservation. *Food Control.* **2021**, *125*, 117–126. [CrossRef]
14. Chuesiang, P.; Sanguandeekul, R.; Siripatrawan, U. Enhancing effect of nanoemulsion on antimicrobial activity of cinnamon essential oil against foodborne pathogens in refrigerated Asian seabass (*Lutes calcarifer*) fillets. *Food Control* **2021**, *122*, 107782. [CrossRef]
15. Otoni, C.G.; de Moura, M.R.; Aouada, F.A.; Camilloto, G.P.; Cruz, R.S.; Lorevice, M.V.; de F F Soares, N.; Mattoso, L.H.C. Antimicrobial and physical-mechanical properties of pectin/papaya puree/cinnamaldehyde nanoemulsion edible composite films. *Food Hydrocoll.* **2014**, *41*, 188–194. [CrossRef]
16. Zhang, R.; Cheng, M.; Wang, X.; Wang, J. Bioactive mesoporous nano-silica/potato starch films against molds commonly found in post-harvest white mushrooms. *Food Hydrocoll.* **2019**, *95*, 517–525. [CrossRef]
17. Zhou, Z.M.; Liu, Y.; Liu, Z.; Fan, L.; Dong, T.; Jin, Y.; Saldana, M.D.A.; Sun, W. Sustained-release antibacterial pads based on nonwovens polyethylene terephthalate modified by beta-cyclodextrin embedded with cinnamaldehyde for cold fresh pork preservation. *Food Packag. Shelf Life* **2020**, *26*, 100554. [CrossRef]
18. Massaro, M.; Barone, G.; Biddeci, G.; Cavallaro, G.; di Blasi, F.; Lazzara, G.; Nicotra, G.; Spinella, C.; Spinelli, G.; Riela, S. Halloysite nanotubes-carbon dots hybrids multifunctional nanocarrier with positive cell target ability as a potential non-viral vector for oral gene therapy. *J. Colloid Interface Sci.* **2019**, *552*, 236–246. [CrossRef]
19. Lee, M.H.; Park, H.J. Preparation of halloysite nanotubes coated with Eudragit for a controlled release of thyme essential oil. *J. Appl. Polym. Sci.* **2015**, *132*. [CrossRef]

20. Abdullayev, E.; Price, R.; Shchukin, D.; Lvov, Y. Halloysite Tubes as Nanocontainers for Anticorrosion Coating with Benzotriazole. *ACS Appl. Mater. Interface* **2009**, *1*, 1437–1443. [CrossRef]
21. Zhong, B.; Wang, S.; Dong, H.; Luo, Y.; Jia, Z.; Zhou, X.; Chen, M.; Xie, D.; Jia, D. Halloysite Tubes as Nanocontainers for Herbicide and Its Controlled Release in Biodegradable Poly(vinyl alcohol)/Starch Film. *J. Agric. Food Chem.* **2017**, *65*, 10445–10451. [CrossRef] [PubMed]
22. Abdullayev, E.; Joshi, A.; Wei, W.; Zhao, Y.; Lvov, Y. Enlargement of Halloysite Clay Nanotube Lumen by Selective Etching of Aluminum Oxide. *ACS Nano* **2012**, *6*, 7216–7226. [CrossRef] [PubMed]
23. Zou, Y.; Zhang, C.; Wang, P.; Zhang, Y.; Zhang, H. Electrospun chitosan/polycaprolactone nanofibers containing chlorogenic acid-loaded halloysite nanotube for active food packaging. *Carbohydr. Polym.* **2020**, *247*, 116711. [CrossRef]
24. Achachlouei, B.F.; Zahedi, Y. Fabrication and characterization of CMC-based nanocomposites reinforced with sodium montmorillonite and TiO2 nanomaterials. *Carbohydr. Polym.* **2018**, *199*, 415–425. [CrossRef]
25. Muller, J.; Quesada, A.C.; Gonzalez-Martinez, C.; Chiralt, A. Antimicrobial properties and release of cinnamaldehyde in bilayer films based on polylactic acid (PLA) and starch. *Eur. Polym. J.* **2017**, *96*, 316–325. [CrossRef]
26. Cui, R.; Jiang, K.; Yuan, M.; Cao, J.; Li, L.; Tang, Z.; Qin, Y. Antimicrobial film based on polylactic acid and carbon nanotube for controlled cinnamaldehyde release. *J. Mater. Res. Technol. Jmr T* **2020**, *9*, 10130–10138. [CrossRef]
27. Zhang, A.-B.; Pan, L.; Zhang, H.; Liu, S.; Ye, Y.; Xia, M.; Chen, X. Effects of acid treatment on the physico-chemical and pore characteristics of halloysite. *Colloids Surf. Physicochem. Eng. Asp.* **2012**, *396*, 182–188. [CrossRef]
28. Garcia-Garcia, D.; Ferri, J.M.; Ripoll, L.; Hidalgo, M.; Lopez-Martinez, J.; Balart, R. Characterization of selectively etched halloysite nanotubes by acid treatment. *Appl. Surf. Sci.* **2017**, *422*, 616–625. [CrossRef]
29. Echeverría, I.; Lopez-Caballero, M.E.; Gomez-Guillen, M.C.; Mauri, A.N.; Montero, M.P. Structure, Functionality, and Active Release of Nanoclay–Soy Protein Films Affected by Clove Essential Oil. *Food Bioprocess. Technol.* **2016**, *9*, 1937–1950. [CrossRef]
30. Yousefi, P.; Hamedi, S.; Garmaroody, E.R.; Koosha, M. Antibacterial nanobiocomposite based on halloysite nanotubes and extracted xylan from bagasse pith. *Int. J. Biol. Macromol.* **2020**, *160*, 276–287. [CrossRef]
31. Akrami-Hasan-Kohal, M.; Ghorbani, M.; Mahmoodzadeh, F.; Nikzad, B. Development of reinforced aldehyde-modified kappa-carrageenan/gelatin film by incorporation of halloysite nanotubes for biomedical applications. *Int. J. Biol. Macromol.* **2020**, *160*, 669–676. [CrossRef] [PubMed]
32. Casariego, A.; Souza, B.W.S.; Cerqueira, M.A.; Teixeira, J.A.; Cruz, L.; Diaz, R.; Vicente, A.A. Chitosan/clay films' properties as affected by biopolymer and clay micro/nanoparticles concentrations. *Food Hydrocoll.* **2009**, *23*, 1895–1902. [CrossRef]
33. Rehan, M.; El-Naggar, M.E.; Mashaly, H.M.; Wilken, R. Nanocomposites based on chitosan/silver/clay for durable multi-functional properties of cotton fabrics. *Carbohydr. Polym.* **2018**, *182*, 29–41. [CrossRef] [PubMed]
34. Ahmed, J.; Mulla, M.; Jacob, H.; Luciano, G.; Bini, T.B.; Almusallam, A. Polylactide/poly(ε-caprolactone)/zinc oxide/clove essential oil composite antimicrobial films for scrambled egg packaging. *Food Packag. Shelf Life* **2019**, *21*, 100355. [CrossRef]
35. Prasetyaningrum, A.; Utomo, D.P.; Raemas, a.A.; Kusworo, T.D.; Jos, B.; Djaeni, M. Alginate/κ-Carrageenan-Based Edible Films Incorporated with Clove Essential Oil: Physico-Chemical Characterization and Antioxidant-Antimicrobial Activity. *Polymers* **2021**, *13*, 354. [CrossRef]
36. Ahmed, J.; Mulla, M.Z.; Arfat, Y.A. Thermo-mechanical, structural characterization and antibacterial performance of solvent casted polylactide/cinnamon oil composite films. *Food Control* **2016**, *69*, 196–204. [CrossRef]
37. Huang, D.J.; Zhang, Z.; Zheng, Y.; Quan, Q.; Wang, W.; Wang, A. Synergistic effect of chitosan and halloysite nanotubes on improving agar film properties. *Food Hydrocoll.* **2020**, *101*, 105471. [CrossRef]
38. Shen, Y.; Wang, H.; Li, W.; Liu, Z.; Liu, Y.; Wei, H.; Li, J. Synthesis and characterization of double-network hydrogels based on sodium alginate and halloysite for slow release fertilizers. *Int. J. Biol. Macromol.* **2020**, *164*, 557–565. [CrossRef]
39. Sanchez-Gonzalez, L.; Chafer, M.; Gonzalez-Martinez, C.; Chiralt, A.; Desobry, S. Study of the release of limonene present in chitosan films enriched with bergamot oil in food simulants. *J. Food Eng.* **2011**, *105*, 138–143. [CrossRef]
40. Lvov, Y.M.; DeVilliers, M.M.; Fakhrullin, R.F. The application of halloysite tubule nanoclay in drug delivery. *Exp. Opin. Drug Deliv.* **2016**, *13*, 977–986. [CrossRef]
41. Zhao, X.; Mu, Y.; Dong, H.; Zhang, H.; Zhang, H.; Chi, Y.; Song, G.; Li, H.; Wang, L. Effect of cinnamaldehyde incorporation on the structural and physical properties, functional activity of soy protein isolate-egg white composite edible films. *J. Food Process. Preserv.* **2021**, *45*, e15143. [CrossRef]
42. Cerqueira, M.A.; Fabra, M.J.; Castro-Mayorga, J.L.; Bourbon, A.I.; Pastrana, L.M.; Vicente, A.A.; Lagaron, J.M. Use of Electrospinning to Develop Antimicrobial Biodegradable Multilayer Systems: Encapsulation of Cinnamaldehyde and Their Physicochemical Characterization. *Food Bioprocess Technol.* **2016**, *9*, 1874–1884. [CrossRef]
43. Fisher, K.; Phillips, C.A. The effect of lemon, orange and bergamot essential oils and their components on the survival of *Campylobacter jejuni*, *Escherichia coli* O157, *Listeria monocytogenes*, *Bacillus cereus* and *Staphylococcus aureus* in vitro and in food systems. *J. Appl. Microbiol.* **2006**, *101*, 1232–1240. [CrossRef] [PubMed]

Article

Fabrication of Caseinate Stabilized Thymol Nanosuspensions via the pH-Driven Method: Enhancement in Water Solubility of Thymol

Wei Zhou [1,2], Yun Zhang [1], Ruyi Li [2], Shengfeng Peng [1,3,*], Roger Ruan [1], Jihua Li [2] and Wei Liu [1]

1 State Key Laboratory of Food Science and Technology, Nanchang University, Nanchang 330047, China; weizhou111@foxmail.com (W.Z.); zhangyun@email.ncu.edu.cn (Y.Z.); ruanx001@umn.edu (R.R.); liuwei@ncu.edu.cn (W.L.)
2 Key Laboratory of Tropical Crop Products Processing of Ministry of Agriculture and Rural Affairs, Agricultural Products Processing Research Institute, Chinese Academy of Tropical Agricultural Sciences, Zhanjiang 524001, China; mwlry19901016@163.com (R.L.); foodpaper@126.com (J.L.)
3 School of Life Sciences, Nanchang University, Nanchang 330031, China
* Correspondence: pengsf@ncu.edu.cn

Abstract: Thymol has been applied as a spice and antibacterial agent in commercial products. However, the utilization of thymol in the food and pharmaceutical field has recently been limited by its poor water solubility and stability. In this work, a caseinate-stabilized thymol nanosuspension was fabricated by pH-driven methods to overcome those limitations. Firstly, the chemical stability of thymol at different pH value conditions was investigated. The physiochemical properties of thymol nanosuspensions were then characterized, such as average particle size, zeta potential, encapsulation efficiency, and loading capacity. Meanwhile, the X-ray diffraction results showed that thymol was present as an amorphous state in the nanosuspensions. The thermal stability of thymol was slightly enhanced by encapsulation through this process, and the thymol nanosuspensions were stable during the long-term storage, and the average particle size of nanosuspensions showed that there was no aggregation of nanosuspensions during storage and high temperature. Finally, the antimicrobial activity of thymol nanosuspensions was evaluated by investigating the minimum inhibitory concentration (MIC) and minimum bactericidal concentration (MBC) against *Salmomella enterca*, *Staphlococcus aureus*, *Escherichia coli*, and *Listeria monocytogenes*. These results could provide useful information and implications for promoting the application of thymol in food, cosmetic, and pharmaceutical commercial products.

Keywords: thymol; nanosuspensions; caseinate; pH-driven method; water solubility

Citation: Zhou, W.; Zhang, Y.; Li, R.; Peng, S.; Ruan, R.; Li, J.; Liu, W. Fabrication of Caseinate Stabilized Thymol Nanosuspensions via the pH-Driven Method: Enhancement in Water Solubility of Thymol. *Foods* **2021**, *10*, 1074. https://doi.org/10.3390/foods10051074

Academic Editors: Hong Wu and Hui Zhang

Received: 8 April 2021
Accepted: 10 May 2021
Published: 12 May 2021

1. Introduction

Thymol (2-isopropyl-5-methylphenol), a monoterpene phenol, is the principal active component of essential oil obtained from thyme, which has been applied as traditional herbs in the cure of disorders affecting the digestive, cardiovascular, and nervous systems [1]. Due to the numerous functions of thymol, such as antioxidant, anti-inflammatory, antimicrobial, and anticarcinogenesis functions, the research on its potential to be utilized as a food preservative and nutraceutical is increasing dramatically [2]. For example, Pan et al. stated that thymol could be utilized in food matrices and depolarize the microbial cytoplasmic membrane to extend the shelf life of productions [3]. Moreover, thymol is already widely applied as a natural feed additive for livestock, which could improve the metabolism and absorption of the nutrients in the animal gut via enhancing digestive enzymes and modulating intestinal microbiota activity to increase the growth indices (including body weight, feed efficiency, and daily growth rate) of them [4–6]. Thymol has been included in the list of 'Generally Recognized As Safe' for application as food additives by the FDA, and it was registered by the European Commission for utilization in the food field due to the lack of intake risk [7,8]. Meanwhile, a comprehensive review has concluded and shown the safety and anticancer activity of thymol [9]. However, its commercial application is limited due to its hydrophobic nature and low water solubility [8,10]. Several types of delivery and encapsulation systems have been investigated and evaluated to enhance the water solubility of thymol; for example, Nasrabadi et al. [11] improved the water solubility and bioaccessibility of thymol via encapsulating it into Pickering emulsions; Cakir et al. [12] fabricated thymol loaded chitosan nanoparticles via ionic gelation method to overcome the limitation of thymol; and nanoliposome was applied by Heckler et al. [13] to encapsulate free thymol as antimicrobials in food products. Normally, thymol as a lipophilic element has been encapsulated in delivery systems via dissolving thymol into organic liquid firstly, which is criticized as leading to high costs and potential damage to the consumer's health. Thus, it is significant to find effective and environmentally friendly methods for handling the key application limitation and challenge of thymol encapsulation.

Recently, an easily scalable and environmentally friendly approach, termed the pH-driven method, was developed [14,15]. It has been utilized to incorporate lipophilic phenols (mainly curcumin) into a variety of delivery systems such as liposomes [16], nanoparticles [17,18], emulsions [19], and oil body [20]. The mechanism of this method is controlling the water solubility of lipophilic phenols via adjusting the pH values. At lower pH values, the lipophilic phenols showed poor water solubility. When the pH value was higher than the pKa of the hydroxyl group of phenols, the deprotonation of the hydroxyl group leads to an increase in the charge and water solubility of phenols. Then, the deprotonated/water-soluble phenols become protonated/water insoluble once the pH value is lower than the pKa. These protonated lipophilic phenols molecules show different behaviors in different systems: (1) In pure water systems, these molecules recrystallize into crystal nucleus, which grows into phenol crystal and precipitates [21]; (2) in the presence of biosurfactant or biopolymers, these amphiphilic molecules are absorbed onto the surface of the crystal nucleus and prevent its growth, and biosurfactant/biopolymer coated phenol nanoparticles are formed [15,17,22]; (3) in the presence of delivery systems with hydrophobic domains (such as liposomes, emulsions, and zein nanoparticles), the protonated lipophilic phenol molecules penetrate into hydrophobic domains of delivery systems [23]. Among these phenol loaded delivery systems fabricated by the pH-driven method, biosurfactant/biopolymer-stabilized phenol nanosuspensions showed great potential due to their easier procedure and higher loading capacity. While the nanosuspensions showed great advantages, they were mainly applied to encapsulate curcumin. Whether they could be applied to encapsulate other lipophilic phenols is still unknown. The influence of phenol properties on the formation and structure of phenol nanosuspensions needs to be explored.

Caseinate as the main source protein obtained from milk has been widely applied in food industries due to its desirable amphiphilic properties and abundant essential amino acids [24]. In previous research, caseinate has been employed concerning hydrophobic nutrients, such as curcumin, beta-carotene [25], and quercetin, which could improve their stability and water solubility. Compared with some other synthetic compounds, for instance, inorganic materials and small molecule surfactants, caseinate as a kind of natural material is more easily accepted by consumers and food factories. Our previous research indicated that caseinate is a wonderful encapsulation material for lipophilic nutrients via the pH-driven method with the highest encapsulation efficiency and loading capacity compared with serval kinds of proteins and polysaccharides [17]. All of this information proved the potential of caseinate as shell and encapsulation compounds to construct nanoparticles with thymol. However, the relevant results and information of nanoparticle fabrication with thymol and caseinate via the pH-driven method still need to be reported.

Therefore, the main objective of this study was to fabricate thymol nanosuspensions stabilized by caseinate via the pH-driven method and to check if the pH-driven method is suitable for thymol nanosuspensions. First, the chemical stability of thymol in different pH value conditions was evaluated. Then, the caseinate-stabilized nanosuspensions were prepared through the pH-driven method, and the stability, encapsulation efficiency, and particle properties were determined. Finally, the antibacterial properties of encapsulated thymol were also investigated. These results will provide important information for the encapsulation of thymol via the pH-driven method, which is useful for application in the fabrication and design of food cosmetics and pharmaceutical products.

2. Materials and Methods

2.1. Materials

Thymol and caseinate powder were purchased from Aladdin Biochemical Technology Co., Ltd. (Shanghai, China). Bacterial strains (CICC 22956, CICC 21600, CICC 10003, CICC 21635) were obtained from China General Microbiological Culture Collection Center (Beijing, China). All other chemicals were of analytical grade.

2.2. Preparation of Nanosuspensions

Thymol nanosuspensions were prepared by a pH-driven method according to our previous study with some modification [15,22]. Thymol powder was weighed and dissolved in 0.1 M NaOH to obtain a thymol alkaline solution (10 mg/mL). Caseinate stock solutions (50 mg/mL) were prepared by dissolving caseinate in PBS (5 mM, pH 6.5) for 4 h and the solution was centrifuged at 8000 rpm for 30 min to remove the undissolved impurities. The solution was then diluted with PBS to different concentrations and stored at 4 °C overnight before use. Thymol alkaline solutions were then added into the caseinate solution (1:1, *v/v*) with constant stirring using a magnetic stir plate. Nanosuspensions with a thymol concentration of 5 mg/mL and different caseinate concentrations (2.5, 5, 10, 15, 20, and 25 mg/mL) were prepared by adjusting the pH value to 6.5 using 6 M HCl solutions.

2.3. The Physical Properties of Nanosuspensions

The average particle size and surface potential of nanosuspensions were determined by Zetasizer Nano ZSP (Malvern Instruments Ltd., Worcestershire, UK) at room temperature. Samples were diluted tenfold with PBS, and each sample was tested three times in parallel.

The microstructure of nanosuspensions was observed using Atomic Force Microscopy (AFM, C300, Nanosurf, Liestal, Switzerland). The nanosuspension solutions were diluted 1000-fold with distilled water and one drop of samples was placed on a freshly cleaved mica substrate. The images of the sample were obtained using the AFM operated with a silicon cantilever force constant of 0.58 N m^{-1} in tapping mode.

The morphology of the nanosuspensions was further confirmed using transmission electron microscopy (TEM). Briefly, nanosuspensions were placed onto a copper mesh grid for 4 min. The sample was then stained with 1% uranyl acetate solution for 1 min and then washed with double distilled water. The sample-loaded grid was then air dried at room temperature and imaged using a TEM (JEM-2000FX, JEOL, Ltd., Tokyo, Japan) operating at a voltage of 200 kV.

The crystalline properties of nanosuspensions were investigated using X-ray diffraction. The blank casein nanoparticles (without thymol) fabricated using the pH-driven method and thymol powder were used as control groups. The emission slit was 1°, the accepted slit was 0.1 mm, and the scanning speed was 2°/min.

The loading capacity (LC) and encapsulation efficiency (EE) were calculated according to the previous method [2]. Samples were extracted with n-hexane (volume ratio 1:9) and the absorbance was then measured at 263 nm using a UV–visible spectrophotometer. The thymol concentration was then calculated using a standard curve prepared. Then, the insoluble part of the thymol was removed by centrifugation (6000 rpm, 10 min), and the thymol in the supernatant fluid was extracted by n-hexane to measure the amount of dissolving thymol in the samples. The free thymol was separated and measured via an ultrafiltration method (3 kDa). The EE and LC of samples were calculated with the following formulas:

$$\text{EE (\%)} = (W_s - W_f)/W_t * 100 \quad (1)$$

$$LC (\%) = (W_s - W_f)/M * 100 \quad (2)$$

where W_t represents the total thymol content, W_s represents the dissolved thymol content, and W_f represents the free thymol content. M is the total mass of the loaded nanosuspensions: Thymol and caseinate.

2.4. The Stability of Thymol Nanosuspensions

Thermal stability: Nanosuspensions were treated at 80 °C for 1 h and sampled every 10 min to measure the properties (retention rate, average diameter, and visual pictures) according to previous methods of nanosuspension after heating.

Storage stability: Samples were stored at room temperature (25 °C) for one month, and their properties (retention rate and average diameter) were investigated every 7 days.

2.5. The Antibacterial Properties of Nanosuspensions

LB liquid medium: 10 g tryptone, 5 g yeast extract, and 5 g sodium chloride were weighed and dissolved in 1000 mL distilled water. The medium was autoclaved at 121 °C for 20 min. The medium was placed at 4 °C for use after cooling.

LB solid medium: 10 g tryptone, 5 g yeast extract, 5 g sodium chloride, and 2 g agar were weighed in 1000 mL distilled water, autoclaved at 121 °C for 20 min, and then poured into 90 mm Petri dishes with about 12–15 mL in each dish. After cooling and solidification, they were placed upside down at 4 °C.

Bacterial resuscitation and rejuvenation: First, 50 μL of glycerin bacteria were vaccinated in 5 mL of LB liquid medium and cultured in a biochemical incubator at 37 °C overnight to turbidity, and then the strains were inoculated from the LB liquid medium to the LB solid medium using inoculation loops to an incubator at 37 °C constant temperature for 10–12 h. Then, the appropriate single colony was inoculated to a liquid medium and cultured to a later growth stage. The bacterial strains were transferred to a solid medium again, sealed by film, and stored at 4 °C with the appropriate single colony observed.

Bacterial culture: The plate was taken out of a refrigerator at 4 °C; a single colony was inoculated in a 5 mL LB liquid medium and cultured in a constant temperature vibration incubator at 37 °C and 220 rpm. For the minimum inhibitory concentration and maximum bactericidal concentration, the bacterial solution was diluted to 10^6 CFU/mL.

MIC measurement: The 96-well microplates method was adopted according to the previous method [13]. The test bactericidal strains were CICC 22956, CICC 21600, CICC 10003, and CICC 21635. First, 100 μL of liquid LB medium was added to each well. Next, 100 μL samples were placed into the first row of the wells and mixed, 100 μL of the mixture was placed into the next well, and so forth. Then, 100 μL bacterial solutions were mixed into each well and incubated (37 °C, 24 h) and put into an incubator at 37 °C for culture for 24 h. The minimum sample concentration with clear culture and no substrate precipitation was the minimum inhibitory concentration (MIC).

MBC measurement: 100 μL was obtained from the MIC well and inoculated in culture, then the cultures were placed in an incubator at 37 °C for 12 h. The sample concentration corresponding to the plate without bacterial colony was the maximum bactericidal concentration (MBC).

2.6. Statistical Analysis

All of the experiments were repeated at least three times, and the results were expressed as mean ± standard deviation. SPSS 18.0 software and the Student–Newman–Keuls (SNK) equation were applied to analyze the significance ($p < 0.05$).

3. Results and Discussion

3.1. Fabrication of Nanosuspensions

During the fabrication of thymol nanosuspensions through the pH-driven method, thymol must be dissolved in the alkaline solutions for at least 5 min. When thymol was dissolved in alkaline solutions, the phenolic hydroxyl group that was thymol ionized and the potential charge of the thymol molecule became negative, which caused an enhancement in its hydrophilic properties and water solubility. It is well known that phenols are unstable under alkaline conditions [26,27]. Therefore, the chemical stability of thymol was firstly investigated via measuring the retention concentration at different pH conditions. As displayed in Figure 1, the retention rate of thymol at a pH range of 7.0 to 10.0 slightly decreased to 93.0% after 24 h, while thymol at pH 11.0 and 12.0 was ultra-stable without any degradation. During the pH-driven process, the thymol was dissolved in alkaline solutions (around pH 12.5), which meant the loss of thymol during the pH-driven process was negligible.

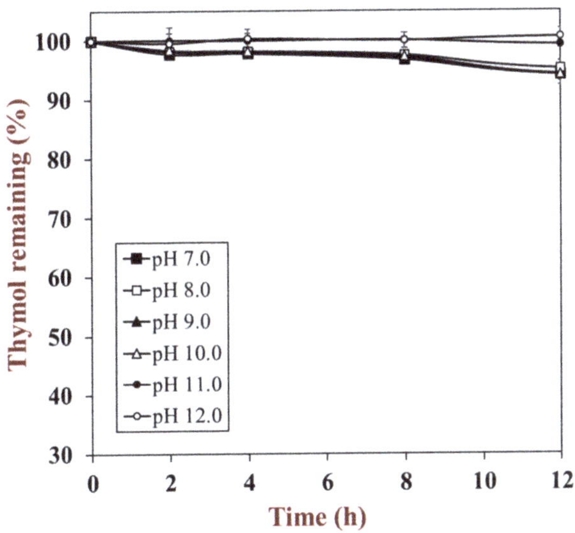

Figure 1. The influence of pH values on the stability of thymol.

Thymol showed relatively high chemical stability compared with other phenols. Our previous study showed that the chemical stability of phenols negatively corresponded to the number of hydroxyl groups [28]. Thymol molecules have only one hydroxyl group, which could explain why they showed extremely high stability. On the other hand, thymol showed higher stability at pH 11.0 and 12.0 compared to lower pH values (7.0 to 10.0), which was similar to curcumin [29]. The chemical stability of quercetin and resveratrol, however, negatively corresponded to the pH values (i.e., the higher the pH, the poorer the chemical stability). The difference could be assigned to the presence of the $-O-CH_3$ group or the $-CH-(CH_3)_2$ group in curcumin and thymol molecules, respectively. There are no other groups in the quercetin and resveratrol except the hydroxyl group.

Initially, the influence of caseinate concentration on the encapsulation efficiency and loading capacity was investigated (Figure 2A). The EE increased with the increase of caseinate concentration from 2.5 to 5 mg/mL and then became steady. The initial low EE of samples could be mainly due to the insufficient caseinate that could not form enough particles for the encapsulation of free thymol, and the free part of thymol was removed by centrifugation. At high and sufficient caseinate concentration, most of the thymol nanosuspensions could be completely coated by caseinate during the change of pH values, and the EE and amount of encapsulated thymol increased. Since part of thymol could be dissolved in water, even the extra protein and nanosuspensions were present with caseinate increasing; no significant increase of EE could be observed. To further investigate the encapsulation level of thymol into nanosuspensions, the influence of caseinate concentration on the LC was also evaluated (Figure 2A). The LC of thymol nanosuspensions decreased with the increase of caseinate concentration, which could again be due to the hypothesis that most of the thymol had been encapsulated. The largest LC of thymol nanosuspensions coated by caseinate was nearly 60%; this phenomenon and result proved that the nanosuspensions coated by caseinate are extremely suitable encapsulation systems for thymol via the pH-driven method.

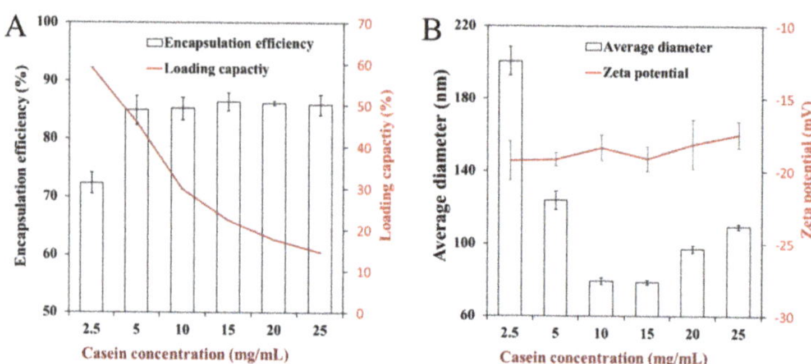

Figure 2. The influence of casein concentration on encapsulation efficiency and loading capacity (**A**), and average diameter and zeta potential (**B**) of thymol nanosuspensions.

The influence of caseinate on the average particle size and zeta potential of nanosuspensions are shown in Figure 2B. It is obvious that the size of nanosuspensions dramatically decreased with the caseinate concentration increasing from 2.5 to 10 mg/mL. The decrease of nanosuspension size could be mainly attributed to the behavior and function of caseinate molecules, which are adsorbed at the surface of thymol particles via their hydrophobic regions to prevent the aggregation of nanosuspensions. When the concentration of caseinate increased from 10 to 25 mg/mL, the size of nanosuspensions increased gently, which could be due to an accumulation of caseinate at the nanosuspension surface. Conversely, there was no significate difference between the zeta potential of nanosuspensions fabricated with different concentration caseinate; all of the nanosuspensions presented a moderately negative charge. The charge of nanosuspensions could be mainly due to the properties of caseinate by which the pH values of water were higher than its isoelectric point, and the surface charge of nanosuspensions was dominated by caseinate, which covered and was present at the surface of nanosuspensions.

The ideal function and behavior of caseinate during the pH-driven method could be related to the emulsifying and binding ability of this kind of protein. It has been reported that amphiphilic biopolymers (such as whey protein isolate) can adsorb to the surface of the lipophilic nanoparticle, prevent contact, and aggregate between them [17]. These data and phenomena showed that the water solubility and dispersibility of thymol could be enhanced by encapsulation via the pH-driven method with caseinate, and this information could promote the development and application of thymol in food and cosmetic productions.

3.2. *Characterization of Nanosuspensions*

A caseinate concentration of 10 mg/mL was used in the following studies since it led to relatively high EE and LC values with a small nanosuspension size distribution. As discussed above, the nanosuspensions with final formal were relatively small (79.4 nm) and negatively charged (−19.4 mV). At the same time, they had high EE (85.2%) and LC (30.0%), and the free thymol was dissolved in water instead of sediment. These results were relatively similar to caseinate-coated curcumin nanoparticles [17]. Previous researchers had also applied some delivery systems to encapsulate thymol and investigated the LC. For example, Shi et al. prepared thymol-loaded solid lipid nanoparticles with a maximum LC of 16% [30], and the LC of thymol/soy protein nano-complexation was 10.36% [2]. All of the data and experiments indicated that the thymol was suitable for the construction of nanosuspensions coated with caseinate via the pH-driven method.

To further prove the fabrication of nanosuspensions, their morphology was observed and detected by AFM and TEM (Figure 3). The AFM results demonstrated that the thymol nanosuspensions are spherical and uniformly distributed in the systems. As expected, the height and size of nanosuspensions were consistent with results provided by dynamic light scattering. In the TEM image, most of the particles were uniformly distributed. The sizes of nanosuspensions were in the range of 50–150 nm, which agrees with the data obtained by dynamic light scattering.

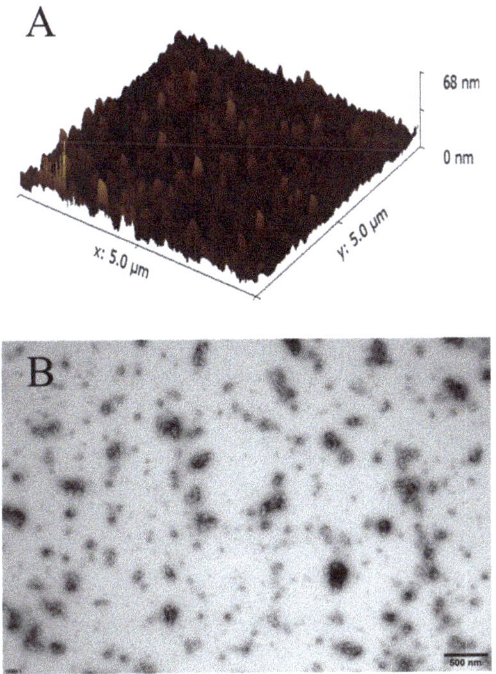

Figure 3. Atom force microscopy (**A**) and transmission electron microscopy (**B**) image of thymol nanosuspension.

The physical states of the thymol in caseinate-coated nanosuspensions were investigated via X-ray diffraction. The diffraction peak of thymol was obtained at 2θ values from 5° to 35° (Figure 4), which indicates the high crystal state of thymol. The relatively smooth curve of caseinate indicated that it was not present in a crystalline structure. As expected, the diffraction peak of the nanosuspensions could not be observed, indicating that all of the thymol are present in an amorphous state in the core of nanosuspensions. These phenomena and data suggest that the incorporation of thymol into the caseinate-coated nanosuspensions could prevent its crystallization, which is consistent with previous results and may be beneficial for the application in commercial products, such as food colloids.

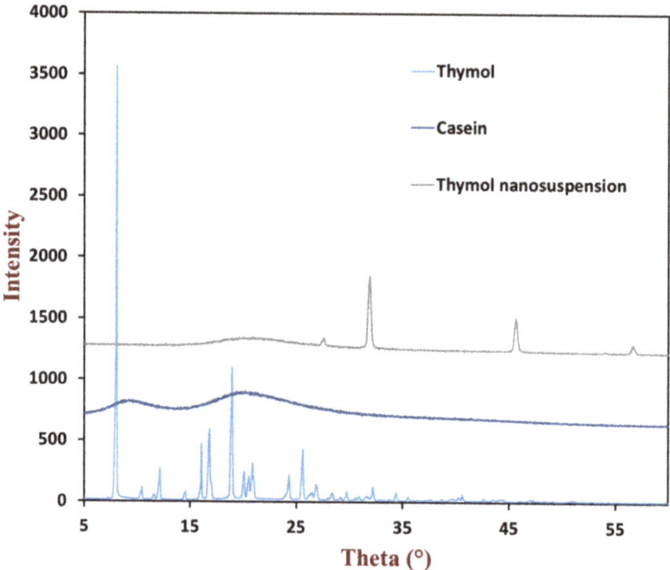

Figure 4. XRD spectra of thymol, casein, and powdered thymol nanosuspension.

3.3. Stability of Nanosuspensions

In food products, most nanosuspensions and food colloids are applied by incorporating them into different products and are exposed to different environmental conditions. Therefore, the thermal and storage stability of nanosuspensions were investigated in this section (Figure 5), which could provide information for the application of thymol in commercial products.

It has been reported that the volatile properties of thymol limit its application in the food industry, and the decrease of thymol mass begins from 50 °C [31]. Thus, the impact of high temperature on the stability of thymol nanosuspensions was investigated at 80 °C. The retention rate of both pure thymol and nanosuspensions decreased with time, and the retention rate of nanosuspensions was slightly higher than pure thymol. These results and phenomena indicated that the coating of caseinate at the surface of nanosuspensions could enhance its thermal stability. Similar results were also reported in nano-complexation of thymol-soy protein isolate, and Chen et al. attributed the enhancement of thermal stability to the shift of thymol evaporation temperature point from 50 to 90 °C and the inclusion complex fabrication [2,32]. Meanwhile, the influence of heat on the average size of nanosuspensions was also investigated, and it is obvious that there are no significate changes. The results of particle size indicate that there are no aggregation and dissociation of caseinate at the surface of the nanosuspensions at elevated temperature, which further proved that caseinate is suitable for the encapsulation of thymol nanosuspensions via the pH-driven method under high temperature. Meanwhile, in the visual figures (Figure 5c), all of the samples remained clear, which indicated there was no significant recrystallization and dissolution of thymol at high temperature. However, in previous research, the curcumin nanoparticles coated by caseinate via a pH-driven shift exhibited a large leakage of curcumin and a slight increase in turbidity [17]. These differences of nanoparticles are mainly due to the difference between thymol and curcumin in their molecular weight or hydrophobicity.

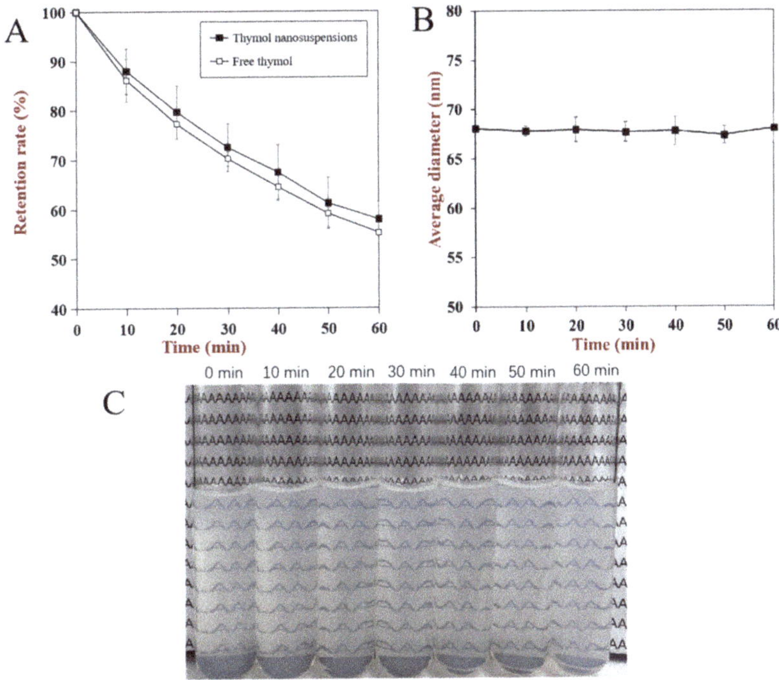

Figure 5. The influence of heating on the thymol retention rate (**A**), average diameter (**B**), and visual appearance (**C**) of nanosuspensions.

To further investigate the stability of thymol nanosuspensions, their storage stability at room temperature was examined via investigating the retention rate and average diameter during 28 days of storage. While the results showed that the retention of thymol nanosuspensions decreased slowly, the value was higher than 90% during the storage period (Figure 6). Meanwhile, there were no significant changes in the average diameter of thymol nanosuspensions, which indicated that the caseinate-coated nanosuspensions via the pH-driven method were stable at long-term storage. However, it has been reported that the retention rate of curcumin nanoparticles coated by caseinate by the pH-driven method decreased dramatically during storage at room temperature [17]. This difference is the same as the thermal stability, which could be mainly due to the difference between the encapsulated lipophilic nutrients.

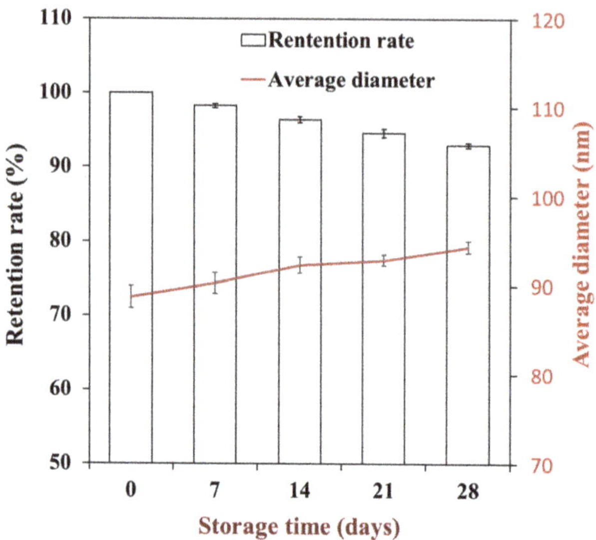

Figure 6. Changes of thymol retention rate and nanosuspension diameter during storage for four weeks.

3.4. MIC and MBC of Thymol

As a natural antibacterial agent, thymol has been applied in many products to extend shelf life, and its antibacterial activity is better than many other plant oil components, such as carvacrol, diacetyl, and eugenol [2,13]. Moreover, it has been reported that the antibacterial activity of thymol could be mainly attributed to the thymol's capacity to change the structure and properties of the outer and cytoplasmic membrane of bacteria [33]. Therefore, to investigate the antibacterial activity of thymol nanosuspensions, the MICs and MBCs of nanosuspensions against four common pathogenic bacteria (*Salmomella enterca*, *Staphlococcus aureus*, *Escherichia coli*, and *Listeria monocytogenes*) were investigated and calculated (Table 1).

Table 1. Minimum inhibitory concentrations (MIC) and minimum bactericidal concentrations (MBC) of nanosuspensions and free thymol for inhibition and inactivation of *L. monocytogenes*, *S. aureusd*, *S. typhimurium*, and *E. coli*.

		Thymol Solutions	Nanosuspensions
L. monocytogenes	MIC (μg/mL)	142	312
	MBC (μg/mL)	142	312
S. aureusd	MIC (μg/mL)	142	156
	MBC (μg/mL)	142	156
S. typhimurium	MIC (μg/mL)	142	156
	MBC (μg/mL)	142	156
E. coli	MIC (μg/mL)	142	312
	MBC (μg/mL)	142	312

The MICs of free thymol against four kinds of bacteria were the same as 142 mg/mL, and the values of all MBCs were the same as MICs. These results and phenomena indicated that thymol could be used as antibacterial and bacteriostatic agents for all of the four kinds of bacteria and there was no significant influence of bacteria kinds on antibacterial activity. However, all of the MICs and MBCs of thymol nanosuspensions were higher than the free thymol, which showed that there was no enhancement in antimicrobial activity of thymol by encapsulating with caseinate via pH-driven methods. These results could be due to the bonds between thymol and caseinate, which prevent the release of thymol from coated nanosuspensions. Thus, the amount of thymol in contact with bacteria decreased at the same concentration, the same antibacterial effect required a higher concentration of thymol. Similar results have also been reported by Wattanasatcha et al., that there was no difference between free thymol and thymol encapsulated in the ethylcellulose/methylcellulose sphere on its antimicrobial activity [34]. Moreover, the MIC/MBC values of thymol encapsulated into liposomes were also lower than free thymol, which has been mainly attributed to the strong binding between thymol/liposomes and the lower release of encapsulated thymol [13]. Therefore, in this work, the decrease in the amount of encapsulated thymol release could be due to the hydrophobic interaction force between thymol and caseinate, which influenced the antibacterial properties of thymol.

4. Conclusions

In this study, caseinate-stabilized thymol nanosuspensions were successfully fabricated via the pH-driven method. Thymol were extremely stable at pH 7.0–12.0 even after incubation for 24 h, which means the loss of thymol during the pH-driven process is negligible. The physchemical properties of thymol nanosuspensions are highly dependent on the caseinate concentration. Caseinate could stabilize thymol nanosuspensions even at a relatively low caseinate concentration, and the loading capacity can be as high as 45.9%. AFM and TEM results showed that the nanosuspensions were uniformly distributed. The thymol nanosuspensions remained stable under high temperature and long-time storage at room temperature. The antimicrobial activity of thymol nanosuspensions was lower than free thymol, which could be due to the bonding between caseinate with thymol and thus retard its release and antimicrobial effects. This work and these results may provide useful information for the design of functional food, cosmetics, and pharmaceutical productions using the pH-driven method.

Author Contributions: W.Z.: Conceptualization, Investigation, Data Curation, Writing—Original Draft. Y.Z.: Investigation. R.L.: Formal analysis, Visualization. S.P.: Validation, Funding acquisition, Methodology, Writing—Review & Editing. R.R.: Writing—Review & Editing, Supervision. J.L.: Supervision, Project administration. W.L.: Supervision, Project administration. All authors have read and agreed to the published version of the manuscript.

Funding: This research was funded by the China Postdoctoral Science Foundation (2020M682111), "Shuangqian Project" of Scientific and Technological Innovation of High-end Talents-Natural Science, Jiangxi Province (S2019GDKX2836), the State Key Laboratory of Food Science and Technology of Nanchang University (No. SKLF-KF-2020015), the Natural Science Foundation of Guangdong Province of China (grant number 2021A1515010901), the Natural Science Foundation of Hainan Province of China (grant number 320QN325), the Central Public-interest Scientific Institution Basal Research Fund for Chinese Academy of Tropical Agricultural Sciences (grant number 1630122017016 and 1630122021006), and the 'Guangdong Special Support Plan' Science and Technology Innovation Young Talents Project (2019TQ05N133).

Acknowledgments: All authors are thankful to their representative universities/institutes for the support and services used in this study.

Conflicts of Interest: The authors declare no conflict of interest.

References

1. Salehi, B.; Mishra, A.P.; Shukla, I.; Sharifi-Rad, M.; Contreras, M.D.M.; Segura-Carretero, A.; Fathi, H.; Nasrabadi, N.N.; Kobarfard, F.; Sharifi-Rad, J. Thymol, thyme, and other plant sources: Health and potential uses. *Phytother. Res.* **2018**, *32*, 1688–1706. [CrossRef] [PubMed]
2. Chen, F.P.; Kong, N.Q.; Wang, L.; Luo, Z.; Yin, J.; Chen, Y. Nanocomplexation between thymol and soy protein isolate and its improvements on stability and antibacterial properties of thymol. *Food Chem.* **2021**, *334*, 127594. [CrossRef] [PubMed]
3. Pan, K.; Chen, H.; Davidson, P.M.; Zhong, Q. Thymol nanoencapsulated by sodium caseinate: Physical and antilisterial properties. *J. Agric. Food Chem.* **2014**, *62*, 1649–1657. [CrossRef]
4. Alagawany, M.; Farag, M.R.; Abdelnour, S.A.; Elnesr, S.S. A review on the beneficial effect of thymol on health and production of fish. *Rev. Aquac.* **2020**, *13*, 632–641. [CrossRef]

5. Khalil, S.R.; Elhakim, Y.A.; El-fattah, A.H.A.; Farag, M.R.; El-Hameed, N.E.A.; L-Murr, A.E.E. Dual immunological and oxidative responses in Oreochromis niloticus fish exposed to lambda cyhalothrin and concurrently fed with Thyme powder (*Thymus vulgaris* L.): Stress and immune encoding gene expression. *Fish Shellfish Immunol.* **2020**, *100*, 208–218. [CrossRef]
6. Ahmadifar, E.; Mansour, M.R.; Amirkolaie, A.K.; Rayeni, M.F. Growth efficiency, survival and haematological changes in great sturgeon (Huso huso Linnaeus, 1758) juveniles fed diets supplemented with different levels of thymol–carvacrol. *Anim. Feed. Technol.* **2014**, *198*, 304–308. [CrossRef]
7. Escobar, A.; Pérez, M.; Romanelli, G.; Blustein, G. Thymol bioactivity: A review focusing on practical applications. *Arab. J. Chem.* **2020**, *13*, 9243–9269. [CrossRef]
8. Marchese, A.; Orhan, I.E.; Daglia, M.; Barbieri, R.; Di Lorenzo, A.; Nabavi, S.F.; Gortzi, O.; Izadi, M.; Nabavi, S.M. Antibacterial and antifungal activities of thymol: A brief review of the literature. *Food Chem.* **2016**, *210*, 402–414. [CrossRef] [PubMed]
9. Islam, M.T.; Khalipha, A.B.R.; Bagchi, R.; Mondal, M.; Smrity, S.Z.; Uddin, S.J.; Shilpi, J.A.; Rouf, R. Anticancer activity of thymol: A literature-based review and docking study with Emphasis on its anticancer mechanisms. *IUBMB Life* **2019**, *71*, 9–19. [CrossRef] [PubMed]
10. Trombetta, D.; Castelli, F.; Sarpietro, M.G.; Venuti, V.; Cristani, M.; Daniele, C.; Saija, A.; Mazzanti, G.; Bisignano, G. Mechanisms of antibacterial action of three monoterpenes. *Antimicrob. Agents Chemother.* **2005**, *49*, 2474–2478. [CrossRef]
11. Nikbakht Nasrabadi, M.; Sedaghat Doost, A.; Goli, S.A.H.; Van der Meeren, P. Effect of thymol and Pickering stabilization on in-vitro digestion fate and oxidation stability of plant-derived flaxseed oil emulsions. *Food Chem.* **2020**, *311*, 125872. [CrossRef]
12. Cakir, M.A.; Icyer, N.C.; Tornuk, F. Optimization of production parameters for fabrication of thymol-loaded chitosan nanoparticles. *Int. J. Biol. Macromol.* **2020**, *151*, 230–238. [CrossRef]
13. Heckler, C.; Marques Maders Silva, C.; Ayres Cacciatore, F.; Daroit, D.J.; da Silva Malheiros, P. Thymol and carvacrol in nanoliposomes: Characterization and a comparison with free counterparts against planktonic and glass-adhered Salmonella. *LWT* **2020**, *127*. [CrossRef]
14. Pan, K.; Luo, Y.; Gan, Y.; Baek, S.J.; Zhong, Q. pH-driven encapsulation of curcumin in self-assembled casein nanoparticles for enhanced dispersibility and bioactivity. *Soft Matter* **2014**, *10*, 6820–6830. [CrossRef]
15. Peng, S.; Li, Z.; Zou, L.; Liu, W.; Liu, C.; McClements, D.J. Improving curcumin solubility and bioavailability by encapsulation in saponin-coated curcumin nanoparticles prepared using a simple pH-driven loading method. *Food Funct.* **2018**, *9*, 1829–1839. [CrossRef] [PubMed]
16. Cheng, C.; Peng, S.; Li, Z.; Zou, L.; Liu, W.; Liu, C. Improved bioavailability of curcumin in liposomes prepared using a pH-driven, organic solvent-free, easily scalable process. *RSC Adv.* **2017**, *7*, 25978–25986. [CrossRef]
17. Peng, S.; Zhou, L.; Cai, Q.; Zou, L.; Liu, C.; Liu, W.; McClements, D.J. Utilization of biopolymers to stabilize curcumin nanoparticles prepared by the pH-shift method: Caseinate, whey protein, soy protein and gum Arabic. *Food Hydrocoll.* **2020**, *107*. [CrossRef]
18. Dai, L.; Zhou, H.; Wei, Y.; Gao, Y.; McClements, D.J. Curcumin encapsulation in zein-rhamnolipid composite nanoparticles using a pH-driven method. *Food Hydrocoll.* **2019**, *93*, 342–350. [CrossRef]
19. Zheng, B.; Peng, S.; Zhang, X.; McClements, D.J. Impact of delivery system type on curcumin bioaccessibility: Comparison of curcumin-loaded nanoemulsions with commercial curcumin supplements. *J. Agric. Food Chem.* **2018**, *66*, 10816–10826. [CrossRef] [PubMed]
20. Zheng, B.; Zhang, X.; Lin, H.; McClements, D.J. Loading natural emulsions with nutraceuticals using the pH-driven method: Formation & stability of curcumin-loaded soybean oil bodies. *Food Funct.* **2019**, *10*, 5473–5484. [CrossRef] [PubMed]
21. Zheng, B.; Zhang, X.; Peng, S.; Julian McClements, D. Impact of curcumin delivery system format on bioaccessibility: Nanocrystals, nanoemulsion droplets, and natural oil bodies. *Food Funct.* **2019**. [CrossRef]
22. Peng, S.; Li, Z.; Zou, L.; Liu, W.; Liu, C.; McClements, D.J. Enhancement of curcumin bioavailability by encapsulation in sophorolipid-coated nanoparticles: An in vitro and in vivo study. *J. Agric. Food Chem.* **2018**, *66*, 1488–1497. [CrossRef] [PubMed]
23. Peng, S.; Zou, L.; Liu, W.; Liu, C.; McClements, D.J. Fabrication and characterization of curcumin-loaded liposomes formed from sunflower lecithin: Impact of composition and environmental stress. *J. Agric. Food Chem.* **2018**, *66*, 12421–12430. [CrossRef]
24. Srinivasan, M.; Singh, H.; Munro, P.A. Formation and stability of sodium caseinate emulsions: Influence of retorting (121 °C for 15 min) before or after emulsification. *Food Hydrocoll.* **2002**, *16*, 153–160. [CrossRef]
25. Chu, B.S.; Ichikawa, S.; Kanafusa, S.; Nakajima, M. Preparation of protein-stabilized β-Carotene nanodispersions by emulsification–evaporation method. *J. Am. Oil Chem. Soc.* **2007**, *84*, 1053–1062. [CrossRef]
26. Leung, M.; Colangelo, H.; Kee, T.W. Encapsulation of curcumin in cationic micelles suppresses alkaline hydrolysis. *Langmuir* **2008**, *24*, 5672–5675. [CrossRef] [PubMed]
27. Zupančič, Š.; Lavrič, Z.; Kristl, J. Stability and solubility of trans-resveratrol are strongly influenced by pH and temperature. *Eur. J. Pharm. Biopharm.* **2015**, *93*, 196–204. [CrossRef] [PubMed]
28. Peng, S.; Zou, L.; Zhou, W.; Liu, W.; Liu, C.; McClements, D.J. Encapsulation of Lipophilic Polyphenols into Nanoliposomes Using pH-Driven Method: Advantages and Disadvantages. *J. Agric. Food Chem.* **2019**, *67*, 7506–7511. [CrossRef] [PubMed]
29. Griesser, M.; Pistis, V.; Suzuki, T.; Tejera, N.; Pratt, D.A.; Schneider, C. Autoxidative and cyclooxygenase-2 catalyzed transformation of the dietary chemopreventive agent curcumin. *J. Biol. Chem.* **2011**, *286*, 1114–1124. [CrossRef]
30. Shi, H.; Huang, S.S.; He, J.B.; Han, L.J.; Zhang, W.N.; Zhong, Q.X. 1-Laurin-3-palmitin as a novel matrix of solid lipid particles: Higher loading capacity of thymol and better stability of dispersions than those of glyceryl monostearate and glyceryl tripalmitate. *Nanomaterials* **2019**, *9*, 489. [CrossRef]

31. Zhu, Z.; Min, T.; Zhang, X.; Wen, Y. Microencapsulation of thymol in Poly(lactide-co-glycolide) (PLGA): Physical and antibacterial properties. *Materials* **2019**, *12*, 1133. [CrossRef]
32. Pires, F.Q.; Pinho, L.; Orsi, D.C.; Silva, I.; Filho, M.C. Thermal analysis used to guide the production of thymol and Lippia origanoides essential oil inclusion complexes with cyclodextrin. *J. Therm. Anal. Calorim.* **2018**, *137*, 543–553. [CrossRef]
33. Hyldgaard, M.; Mygind, T.; Meyer, R.L. Essential oils in food preservation: Mode of action, synergies, and interactions with food matrix components. *Front. Microbiol.* **2012**, *3*, 12. [CrossRef] [PubMed]
34. Wattanasatcha, A.; Rengpipat, S.; Wanichwecharungruang, S. Thymol nanospheres as an effective anti-bacterial agent. *Int. J. Pharm.* **2012**, *434*, 360–365. [CrossRef] [PubMed]

Article

Encapsulation of Hydrophobic and Low-Soluble Polyphenols into Nanoliposomes by pH-Driven Method: Naringenin and Naringin as Model Compounds

Mianhong Chen [1], Ruyi Li [1,*], Yuanyuan Gao [1], Yeyu Zheng [1], Liangkun Liao [1], Yupo Cao [1], Jihua Li [1] and Wei Zhou [1,2,3,*]

[1] Key Laboratory of Tropical Crop Products Processing of Ministry of Agriculture and Rural Affairs, Agricultural Products Processing Research Institute, Chinese Academy of Tropical Agricultural Sciences, Zhanjiang 524001, China; mianhong_chen@163.com (M.C.); fannyloveme@163.com (Y.G.); yeyu_zheng@163.com (Y.Z.); liaoliangkunv@163.com (L.L.); yupo53@163.com (Y.C.); foodpaper@126.com (J.L.)

[2] State Key Laboratory of Food Science and Technology, Nanchang University, Nanchang 330047, China

[3] Hainan Key Laboratory of Storage and Processing of Fruits and Vegetables, Zhanjiang 524001, China

* Correspondence: mwlry19901016@163.com (R.L.); weizhou111@foxmail.com (W.Z.); Tel.: +86-759-222-1090 (R.L.); +86-187-1818-7765 (W.Z.)

Abstract: Naringenin and naringin are a class of hydrophobic polyphenol compounds and both have several biological activities containing antioxidant, anti-inflammatory and anti-tumor properties. Nevertheless, they have low water solubility and bioavailability, which limits their biological activity. In this study, an easy pH-driven method was applied to load naringenin or naringin into nanoliposomes based on the gradual reduction in their water solubility after the pH changed to acidity. Thus, the naringenin or naringin can be embedded into the hydrophobic region within nanoliposomes from the aqueous phase. A series of naringenin/naringin-loaded nanoliposomes with different pH values, lecithin contents and feeding naringenin/naringin concentrations were prepared by microfluidization and a pH-driven method. The naringin-loaded nanoliposome contained some free naringin due to its higher water solubility at lower pH values and had a relatively low encapsulation efficiency. However, the naringenin-loaded nanoliposomes were predominantly nanometric (44.95–104.4 nm), negatively charged (−14.1 to −19.3 mV) and exhibited relatively high encapsulation efficiency (EE = 95.34% for 0.75 mg/mL naringenin within 1% *w/v* lecithin). Additionally, the naringenin-loaded nanoliposomes still maintained good stability during 31 days of storage at 4 °C. This study may help to develop novel food-grade colloidal delivery systems and apply them to introducing naringenin or other lipophilic polyphenols into foods, supplements or drugs.

Keywords: naringenin; naringin; nanoliposomes; pH-driven; stability

Citation: Chen, M.; Li, R.; Gao, Y.; Zheng, Y.; Liao, L.; Cao, Y.; Li, J.; Zhou, W. Encapsulation of Hydrophobic and Low-Soluble Polyphenols into Nanoliposomes by pH-Driven Method: Naringenin and Naringin as Model Compounds. *Foods* **2021**, *10*, 963. https://doi.org/10.3390/foods10050963

Academic Editors: Hong Wu and Hui Zhang

Received: 26 March 2021
Accepted: 27 April 2021
Published: 28 April 2021

1. Introduction

Naringenin (5,7,4′-trihydroxyflavanone) and naringin (5,7,4′-trihydroxyflavone 7-rhamno-glucoside) are flavones abundantly existent in fruits, including grapefruit, tomato skin and oranges [1]. Naringin can be formed by naringenin at the position of the 7th carbon atom with neohesperidose (Figure 1), which promotes its solubility in water. Their health-promoting features have been explored in vivo and in vitro models, such as antioxidant activity [2,3], anti-inflammatory activity [4,5], anticancer activity [6,7], anti-diabetic activity [8] and protective effects on the central nervous system [9] and the cardiovascular system [10]. However, naringenin and naringin are hydrophobic and low-soluble polyphenols with low water solubility of 38 and 500 µg/mL at room temperature, respectively [11,12], and taste bitter, which greatly limits their application in clinical and functional foods. To overcome these limits, the solubility and bioavailability of naringenin and naringin have been improved by encapsulation in various kinds of carefully designed colloidal delivery systems, including cyclodextrins complexation [12], nanovehicles [13,14], liposomal [15] and ternary complex particles [16].

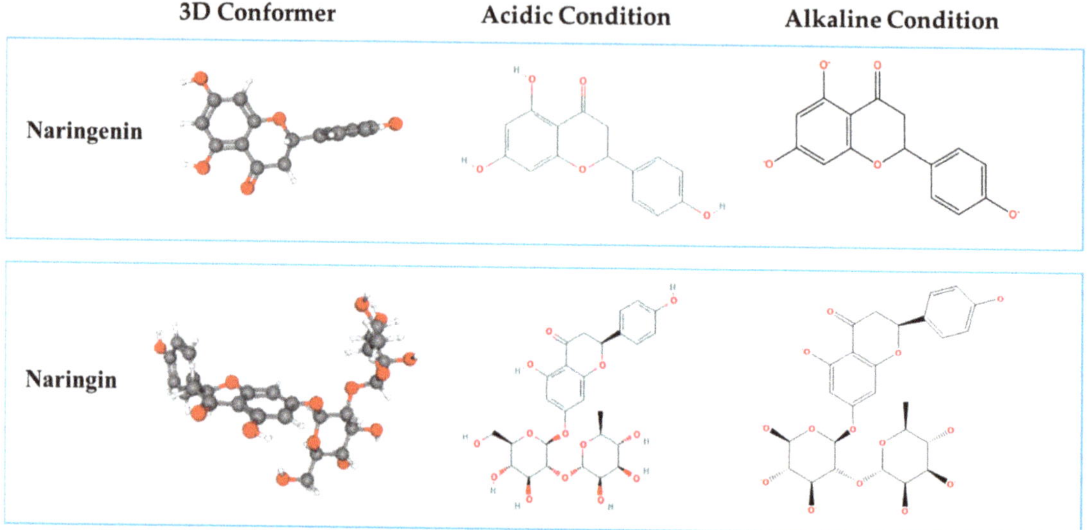

Figure 1. Three dimensions conformer of naringenin and naringin and their hydroxyl groups are non-ionized in acidic condition and negatively charged in alkaline condition.

The pH-driven method is a new method for constructing a delivery system, which has received extensive attention due to its easy-to-handle, cost-saving, energy-saving and organic solvent-free properties during operation. The process of this method includes dissolving the bioactivator under strongly alkaline conditions and then adjusting the pH to neutral or acidic after blending with the carrier system. Hydrophobic polyphenols like curcumin have poor solubility in neutral acidity, while they can be dissolved due to deprotonation at alkaline pH. According to this property of some polyphenols, Pan et al. firstly loaded curcumin into casein nanoparticles by the pH-driven method [17]. Subsequently, a series of studies on the colloidal delivery systems based on the pH-driven method began to emerge, especially for curcumin encapsulation. The successful preparation of a series of colloidal carrier systems loaded with curcumin enhanced its water solubility and bioavailability, which proved the feasibility of preparing the delivery system with the pH-driven method [18–25]. However, the pH-driven method has rarely been applied for the encapsulation of other hydrophobic polyphenols. A potential disadvantage of this method is that the encapsulation of polyphenols needs to be carried out under a highly alkaline conditions, which could facilitate the chemical degradation of polyphenols [26]. Peng and co-workers have demonstrated that the pH-driven method can be used to load certain kinds of lipophilic polyphenols such as curcumin and resveratrol but cannot encapsulate quercetin because the degradation rate of quercetin was extremely fast at alkaline pH values [26].

Liposomes are microcosmic phospholipid vesicles with a bilayered membrane structure. The hydrophilic segments of the phospholipids are located on the inside and outside surface of liposomes, while the hydrophobic tail forming lipid bilayers are separated from aqueous phase [27]. Nanoliposome is a nanometric versions of liposome, which can load bioactive substances with different lipophilicities into the various parts of liposomes, such as phospholipid bilayers, hydrophilic core or bilayer interface [28]. Therefore, nanoliposomes are an extremely promising encapsulation technology in the field of nutraceutical products. The superiority of nanoliposome technology is embodied in the protection of sensitive bioactivators, storage stability, excellent loading capacity, improved bioavailability and sustained-release ability [29]. The utilization of high-pressure homogenizer to prepare liposomes can overcome the traditional liposome preparation method that uses detergents or organic solvents that are either undesirable or not allowed ingredients in foods [25].

Naringenin and naringin, with hydrogen donor count as 3 and 8, respectively, whose hydroxyl groups are non-ionized under acidic conditions (Figure 1). When the pH changes to alkaline, they are easy to be negatively charged due to the deprotonation of the hydroxyl group, which extremely enhances their hydrophilic ability. We hypothesized that naringenin and naringin lose their charges due to the protonation of the same group under neutral and acidic conditions, which would reduce their water solubility. Therefore, when the naringenin/naringin alkaline solution and nanoliposome solutions are mixed, the pH is adjusted to an acidic condition, which causes the naringenin or naringin to be embedded into the hydrophobic region within nanoliposomes. In this study, we first determined the stability of naringenin and naringin under alkaline conditions and the change in solubility of naringenin and naringin in the transition from pH 12.0 to neutral or acidic were also measured, which ensured that they can be encapsulated into liposomes by the pH-driven method. Subsequently, nanoliposomes with different concentrations of naringenin or naringin and different contents of lecithin were prepared by microfluidazition and pH-driven methods, and their encapsulation efficiency and maximum loading capacity were determined. The structures of bioactivator-loaded nanoliposomes, particle size and ζ-potentials of nanoliposomes were determined to understand the bioactivator-loaded procedure using the pH-driven method. Finally, the stability of the naringenin-loaded nanoliposomes was studied. This study may lead to develop novel food-grade colloidal delivery systems and apply to introducing naringenin and naringin into foods, supplements, or drugs.

2. Materials and Methods

2.1. Chemicals

Naringenin (97%) and naringin (95%) powder were purchased from Macklin (Shanghai, China); lecithin from soybean (>70%) and dimethyl sulfoxide (DMSO) were obtained from Aladdin (Shanghai, China). Hydrochloric acid, sodium hydroxide, sodium dihydrogen phosphate, ethanol and other chemicals were all of analytical purity.

2.2. Stability of Naringenin and Naringin in Alkaline Condition

Sodium hydroxide solution (6.0 M) was added to 50 mM sodium dihydrogen phosphate solution to prepare a range of buffers with a pH value ranging from 7.0 to 12.0. Then 0.25 mL of naringenin or naringin ethanol solution (2.0 mg/mL) was mixed with buffer solutions (9.75 mL) of different pH values, respectively. These samples with different pH values were stirred using a magnetic stirrer at ambient temperature. At predetermined every 10 min (ended of 60 min), 0.3 mL of mixture was taken out and diluted with phosphate buffer solution (50 mM, pH 7.0) for the determination of naringenin and naringin content with a UV-vis spectrometer (U-T6A, Yipu Instrument Manufacturing Co., Ltd., Shanghai, China) at 288 and 282 nm, respectively.

2.3. Effect of pH Shift on the Solubility of Naringenin and Naringin

Naringenin was dissolved in NaH_2PO_4 buffer (50 mM, pH 12.0) at 5.0 mg/mL, then the pH values of the solutions were adjusted to 7.0–2.0 with 6.0 M hydrochloric acid. After storage for 24 h at ambient temperature, the suspension was centrifuged, and the supernatant was diluted with DMSO to an appropriate concentration for UV-vis measurement. The determination process for naringin was similar to that described above.

2.4. Narigenin-Loaded and Naringin-Loaded Nanoliposome Preparation

Nanoliposomes were produced by the high-pressure homogenization using a microfluidizer. Soybean lecithin (1.2% and 2.5%, w/v) was suspended in sodium dihydrogen phosphate solution (5 mM, pH 6.0) and stirred for 4 h to ensure complete hydration. Then the soybean lecithin suspension was passed through the microfluidizer 3 times under a working pressure of 121.0 MPa to fabricate nanoliposomes. The nanoliposome solutions were placed in 4 °C before using.

Naringenin was encapsulated into the nanoliposomes by the pH-driven method. Briefly, a range of alkaline naringenin solutions with various naringenin concentration were fabricated by dissolving naringenin in 0.08 M NaOH solution. The alkaline naringenin solutions and nanoliposome solutions were blended in various proportions, and then the mixed systems were rapidly adjusted to pH 7.0, 6.0 and 5.0 by 1.0 M HCl. The final lecithin content within nanoliposome solutions has two levels of 1.0% and 2.0% (w/v). Meanwhile, the final naringenin concentrations were 0.75, 1.00, 1.25, or 1.50 mg/mL with 1.0% (w/v) lecithin and the naringenin concentrations were 2.00 or 3.00 mg/mL with 2.0% (w/v) lecithin, respectively. All the naringenin-loaded nanoliposome samples were equilibrated at room temperature for 24 h.

A series of naringin-loaded nanoliposome (the final naringin concentrations were 1.00, 1.50, or 2.00 mg/mL with 1.0% (w/v) lecithin and 3.00 mg/mL with 2.0% (w/v) lecithin, respectively) were prepared only at the final pH of 6.0 according to the above method.

2.5. Encapsulation Efficiency and Loading Capacity of Naringenin and Naringin

The encapsulation efficiency (EE) and loading capacity (LC) were measured by a centrifugation method. Briefly, the bioactivator-loaded nanoliposome suspension was centrifuged at 11,000× g for 15 min at 4 °C to remove any unencapsulated bioactivator precipitate. Then, the bioactivator-loaded nanoliposome suspensions were centrifuged with ultrafiltration centrifuge tubes (MWCO: 10 kDa) at 11,000× g for 30 min and the subnatant was used to determine the concentration of free bioactivator. The concentration of bioactivator in the supernatant for first centrifugation and the subnatant for second centrifugation were measured using a UV-vis spectrophotometer. The EE and LC of the bioactivator-loaded nanoliposomes without free bioactivator were calculated according to the following expressions:

$$\text{EE }(\%) = \frac{m_N - m_F}{m_I} \times 100 \quad (1)$$

$$\text{LC }(\%) = \frac{m_N - m_F}{m_M} \times 100 \quad (2)$$

where m_N is the mass of bioactivator in the nanoliposome solution, m_F is the mass of free bioactivator in the nanoliposome solution, m_I is the initial mass of bioactivator in the system, and m_M is the total mass of the bioactivator-loaded nanoliposomes (bioactivator + lecithin).

2.6. Naringenin-Loaded and Naringin-Loaded Nanoliposomes Characterization

A combined dynamic light scattering (DLS)-electrophoresis instrument (ZETASIZER PRO, Malvern Instruments, Worcestershire, U.K.) was used to determine the mean particle size, polydispersity index (PDI) and ζ-potentials of naringenin-loaded and naringin-loaded nanoliposomes at 25 °C. The refractive indexes of the nanoliposomes and the dispersion phase water were set to 1.45 and 1.33, respectively. The nanoliposome suspension was diluted 20 times with buffer solution before measuring to avoid multiple scattering interferences.

2.7. Storage Stability

The effect of storage conditions (e.g., temperature and time) on the stability of nanoliposome suspensions were determined to assess their long-term physicochemical stability. The samples (1.0% w/v lecithin) with naringenin concentrations of 1.00 and 1.25 mg/mL were placed in 4 °C, 25 °C and 37 °C within 31 days. During storage, their encapsulation efficiency, mean particle size and ζ-potentials were recorded according to the same procedure described above at different time intervals.

2.8. Atomic Force Microscopy

Atomic force microscopy (AFM) was used for obtaining the microstructure of the blank nanoliposomes and naringenin-loaded nanoliposomes with a concentration of 1.0 and 1.25 mg/mL naringenin before and after 31 days of storage at 4 °C. A droplet of nanoliposome suspension was placed on to a newly cleaved silicon substrate and air-dried for 30 min, and then picture of the naringenin-loaded nanoliposomes were recorded through AFM (C300, Nanosurf, Liestal, Switzerland). AFM images were recorded in tapping mode with a scan time of 3 s and scan points of 512 to obtain sufficient data for statistical analysis.

2.9. Statistical Analysis

At least three freshly prepared samples were measured, and the results were expressed as mean ± standard deviation. Analysis of variance (ANOVA) was used to determine significance at the $p < 0.05$ level using Tukey's HSD test.

3. Results and Discussion

3.1. Stability of Naringenin and Naringin in Alkaline Conditions

Naringenin molecule contains three hydroxyl groups that can be used as hydrogen donors and their pKa values are 7.86, 9.20 and 9.79 (obtained from chemicalize.com), respectively. The hydroxyl groups could be gradually ionized around and above their pKa value, leading to an increase in the negative charge of the naringenin and its hydrophilicity. Since naringenin was dissolved under alkaline conditions, it is necessary to investigate its stability under alkaline conditions. The chemical stability of naringenin at various pH conditions (from 12.0 to 7.0) was evaluated through measuring the change in the concentration remaining versus time (Figure 2a). The degradation rate of naringenin was about 5% when stored for 10 min at pH 12.0 and the degradation did not increase as the time was extended to 60 min. Ionized hydroxyl groups (phenate ion) of the polyphenols at high pH values are particularly susceptible to oxidation causing the degradation of polyphenols [26]. With the decrease of pH from 11.0 to 7.0, naringenin was relatively stable at this pH range and rarely degraded within 60 min.

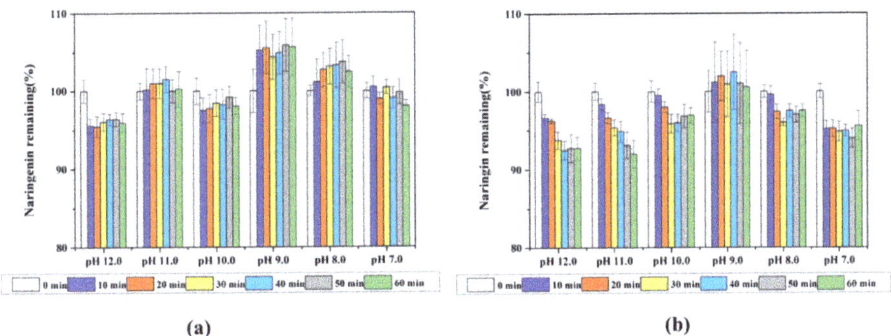

Figure 2. The change of naringenin (**a**) and naringin (**b**) remaining in different alkaline solutions over time.

Compared with naringenin, naringin molecule contains eight hydroxyl groups that can be used as hydrogen donors, and its solubility in alkaline solutions is higher than naringenin. The change of naringin remaining in alkaline solutions over time is shown in Figure 2b. The degradation rate of naringin was about 8% when stored for 60 min at pH 12.0 and 11.0, suggesting that naringin was relatively stable in alkaline solutions within a short time.

3.2. *Solubility of Naringenin and Naringin with pH-Shift*

In order that naringenin can be moved from the aqueous phase into the hydrophobic domain of the nanoliposome instead of simply blending in the nanoliposome solution, we determined the solubility of naringenin after the pH-shift. Naringenin with a concentration of 5.00 mg/mL could quickly dissolved in a pH 12.0 phosphate buffer solution with a color of orange-yellow (Figure 3a). After adjusting the pH to neutral and acidic, naringenin could quickly restore electrical neutrality, resulting in reduced solubility and precipitated from solution. The solubility of naringenin decreased as the solution of pH-shift range increased. Similar to other hydrophobic polyphenols, such as curcumin, quercetin and resveratrol, the hydroxyl groups on their molecules became electronegative in a strongly alkaline solution which increased their water solubility, but then their water solubility decreased due to the molecules regaining neutrality when the pH was adjusted to neutral [26]. When the naringenin solution changed the pH from 12.0 to 7.0, the solubility of naringenin decreased from 5.00 mg/mL to 94.59 μg/mL, indicating that about 98% of naringenin had been precipitated.

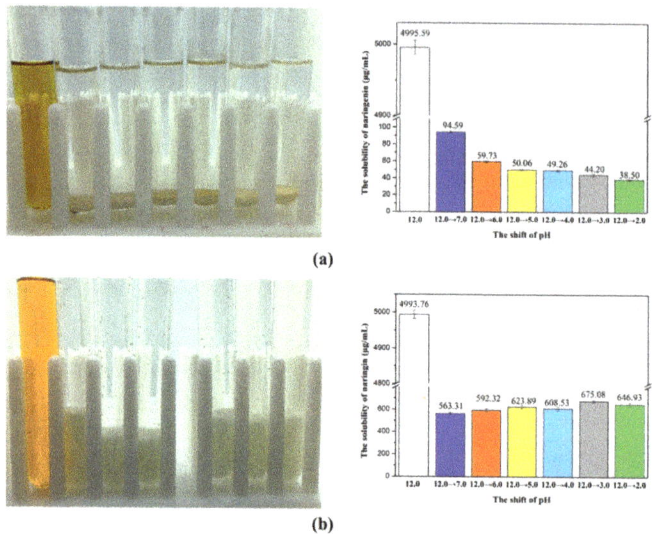

Figure 3. The image and solubility of naringenin (**a**) and naringin (**b**) dissolved at pH 12.0 solution (5.00 mg/mL) and then these samples were adjusted from pH 12.0 to 7.0, 6.0, 5.0, 4.0, 3.0 and 2.0 (left to right), which caused the bioactivators to restore electrical neutrality and reduce solubility.

Unlike naringenin, when naringin was dissolved in a NaH_2PO_4 buffer of pH 12.0 and then the pH was adjusted to acidity, naringin did not immediately precipitate out but some naringin slowly precipitated over time. However, even after standing for 24 h, the solubility of naringin in the pH 6.0 buffer still reached 592.32 μg/mL, which was about ten times that of naringenin (Figure 3b). Additionally, the solubility of naringin did not decrease as the transition pH decreases (Figure 3b). This result means that when using the pH-driven method to prepare naringin-loaded nanoliposomes, some free naringin may be blended in the nanoliposomes solution.

3.3. *Encapsulation Efficiency (EE) and Loading Capacity (LC) of Naringenin and Naringin in Nanoliposomes*

The EE and LC of nanoliposomes fabricated with different naringenin and lecithin concentrations at pH 7.0, 6.0 and 5.0 were determined, respectively (Figure 4). When the concentration of naringenin was 0.75 mg/mL within 1.0% *w/v* lecithin, the EE of naringenin in the nanoliposomes increased as the shift of pH went down during preparation, being 70.12%, 90.30% and 95.34% for pH 7.0, 6.0 and 5.0, respectively. The same trend appeared at naringenin concentration of 1.00 and 1.25 mg/mL, and the EE of 1.00 mg/mL nanoliposome were 63.89%, 86.49% and 91.47%, while 1.25 mg/mL nanoliposome were 63.47%, 81.39% and 85.20% for pH 7.0, 6.0 and 5.0, respectively. This is may be because more naringenin emerged into the hydrophobic cavity of the liposome as the decreased solubility of naringenin from alkaline solution (pH 12.0) to weakly acidic solution, so the EE of naringenin-loaded nanoliposomes increased [26]. However, when the naringenin concentration reached 1.50 mg/mL, the EE decreased as the pH decreased. When the feeding level of naringenin raised, the EE of nanoliposome gradually decreased, and the nanoliposome solution gradually changed from transparent to cloudy (Figure 4c). The maximum load of naringenin in the nanoliposome solutions (1.0% *w/v* lecithin + 1.25 mg/L naringenin) was 1.02 mg/mL at pH 6.0. When the concentration of naringenin attained 1.50 mg/mL, the maximum LC and EE began to decrease significantly, which means that nanoliposome may began to aggregate and easily formed naringenin crystals to initiate precipitation. Therefore, we tried to increase the amount of lecithin to load more naringenin. When the lecithin content reached 2.0% (*w*/*v*) and the concentration of naringenin was 2.00 mg/mL, the amount of naringenin loaded in nanoliposomes was 1.77 mg at pH 6.0, which was about twice than 1.00 mg/mL with 1.0% (*w*/*v*) lecithin, and their EE were close as 88.56% and 86.49%, respectively. Wang and co-workers prepared naringenin-loaded liposomes by thin-film hydration method, and the EE of the naringenin-loaded liposomes (4.0 mg/mL, 6% *w/v* lecithin) was 72.2% [13]. Compared with the thin-film hydration, the pH-driven is a simple and effective method without the use of high temperature or organic solvents, and also has a higher EE [24–26].

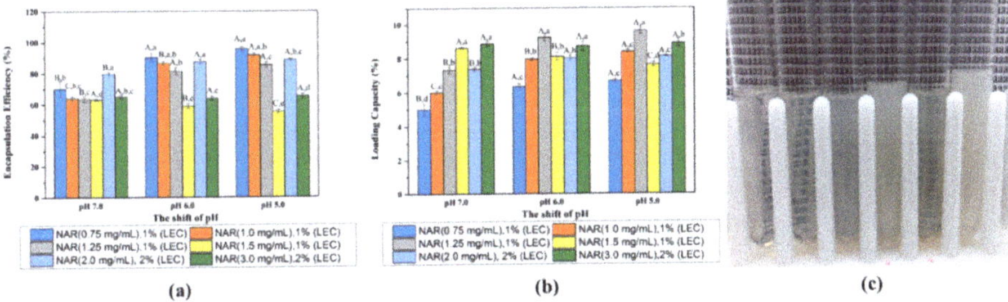

Figure 4. Encapsulation efficiency (**a**) and loading capacity (**b**) of naringenin-loaded nanoliposomes with pH shift to 7.0, 6.0 and 5.0; The image (**c**) of naringenin-loaded nanoliposome solutions with naringenin concentrations of 0.75, 1.00, 1.25 and 1.50 mg/mL with 1% (*w*/*v*) lecithin, while naringenin concentrations of 2.00 and 3.00 mg/mL with 2% (*w*/*v*) lecithin from left to right at pH 6.0. Samples denoted with different letters (A–C) and (a–d) were significantly different ($p < 0.05$) when compared between different pH regions (same naringenin level) and different naringenin levels (same pH region), respectively.

LC was also used to evaluate the efficiency of loading naringenin with lecithin. The maximum loading capacity appeared as a naringenin concentration of 1.25 mg/mL were 9.23% and 9.63% for pH 6.0 and 5.0, while it was 8.63% at 1.50 mg/mL naringenin concentration for pH 7.0. The maximum loading capacity of nanoliposomes at pH 6.0 and 5.0 was close (without significant difference) when enough feeding naringenin was provided, which suggesting that the LC of the nanoliposome itself would not be affected in these pHs. When the nanoliposomes were prepared at pH 6.0, the loading capacity increased from 6.34% to 9.23% with the increase of naringenin concentration from 0.75 to 1.25 mg/mL and then dropped to 8.11% for 1.50 mg/mL naringenin.

The EE and LC of nanoliposomes fabricated with different naringin and lecithin concentrations at pH 6.0 were determined, respectively (Figure 5). After removing the free naringin by centrifugation and ultrafiltration, the encapsulation efficiency of naringin-loaded nanoliposomes was between 45.67%–64.54%. Moreover, the free naringin in the naringin-loaded nanoliposome solutions was between 17.78%–39.60%, which was contribute to the low encapsulation efficiency. The loading capacity of naringin-loaded nanoliposomes were 6.06%, 8.01%, 8.37% and 8.24% at naringin concentration of 1.0, 1.5, 2.0 and 3.0 mg/mL, respectively. Compared with naringenin-loaded nanoliposomes, the naringin-loaded nanoliposome solution did not become cloudy as the concentration of naringin increased (Figure 5b). Additionally, the maximum encapsulation efficiency and loading capacity of the naringin-loaded nanoliposomes were lower, which may attribute to its higher molecular weight and solubility in pH 6.0.

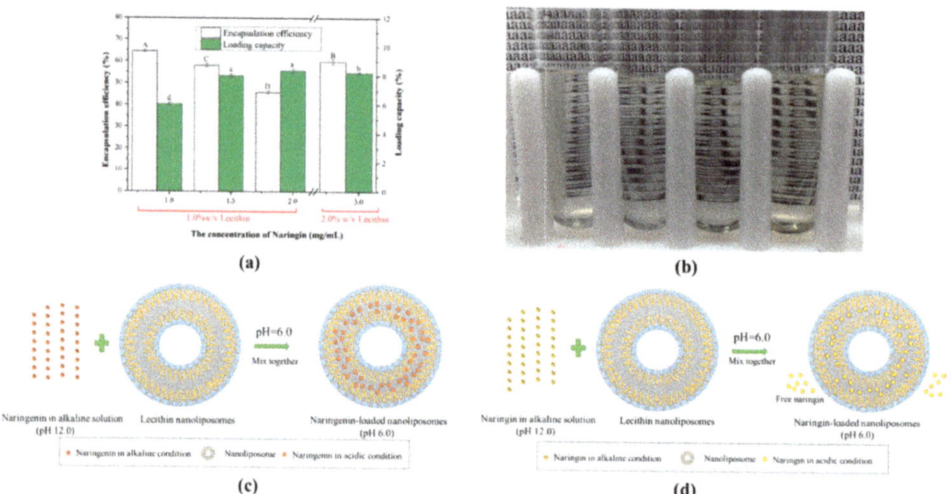

Figure 5. Encapsulation efficiency and loading capacity (**a**) of naringin-loaded nanoliposomes with pH shift to 6.0; The image (**b**) of naringin-loaded nanoliposome solutions with naringin concentrations of 1.0, 1.5, and 2.0 mg/mL with 1% (*w*/*v*) lecithin, while naringin concentrations of 3.0 mg/mL with 2% (*w*/*v*) lecithin from left to right at pH 6.0. Schematic mechanism of the formation of naringenin-loaded (**c**) and naringin-loaded (**d**) nanoliposomes based on pH-driven method. Samples denoted with different letters (A–D) and (a–d) were significantly different ($p < 0.05$) in EE and LC when compared between different naringin levels, respectively.

Above all, these results suggested that naringenin can be encapsulated into nanoliposomes by an easy pH-driven method according to the reduction in its water-solubility after the pH changed to acidity, which causes the naringenin can be moved from the aqueous solution into the lipid bilayers within nanoliposomes (Figure 5c). However, the naringin-loaded nanoliposomes contained some free naringin in solutions due to its higher water-solubility at acidic conditions (Figure 5d).

3.4. *Characterization of Naringenin-Loaded and Naringin-Loaded Nanoliposomes*

The mean particle size, PDI and ζ-potential changes of blank nanoliposomes and naringenin-loaded nanoliposomes were determined using DLS, as showed in Figure 6. The particle size of the blank nanoliposomes after passing through the microfluidizer consists of 1.0% or 2.0% *w/v* lecithin were about 33 and 35 nm, respectively. The results showed that the influence of different pH values and lecithin concentration on mean particle size changes of blank nanoliposome seems to be negligible (Figure 6a). When the naringenin was loaded into nanoliposomes, the particle size of nanoliposomes increased significantly. For the shift of pH, all the mean particle size of the naringenin-loaded nanoliposomes increased as the pH decreased. When the lecithin content in the system was 1.0% (*w*/*v*), the average diameters of nanoliposomes increased as the concentration of naringenin increased, which was related to the increased encapsulation efficiency and suggested that the larger particle diameter while the more naringenin loaded into nanoliposomes. The average diameters of the nanoliposomes with naringenin concentration of 1.00 mg/mL were 53.47, 58.85 and 62.63 nm for pH 7.0, 6.0 and 5.0, respectively. While the naringenin concentration was 1.25 mg/mL, the average diameters of the nanoliposomes were 61.56, 75.71 and 83.48 nm for pH 7.0, 6.0 and 5.0, respectively. For the 2.0% (w/v) lecithin concentration, the average diameters of nanoliposomes increased similarly as the shift of pH decreased and the concentration of naringenin increased. Particularly, the average diameters of nanoliposomes with a naringenin concentration of 2.00 mg/mL were 52.20, 57.68 and 63.32 nm for pH 7.0, 6.0 and 5.0, and with a naringenin concentration of 3.00 mg/mL were 75.44, 99.40 and 104.40 nm for pH 7.0, 6.0 and 5.0, respectively. The PDI of all the nanoliposome samples was relatively small (PDI < 0.3) apart from the 0.75 mg/mL naringenin-loaded nanoliposomes, suggesting that their particle size distributions were relatively narrow (Figure 6b,c). Compared with the blank nanoliposome, the PDI of the naringenin-loaded nanoliposomes with a naringenin concentration of 0.75 mg/mL increased, indicating that part of the unloaded and loaded nanoliposomes form a fairly wide particle size distribution (Figure 6c). When the feeding concentration of naringenin increased, the PDI of nanoliposomes decreased slightly. In particular, for the concentrations of 1.50 and 3.00 mg/mL naringenin, the PDI of nanoliposomes dropped significantly, from 0.262 (blank nanoliposome) to 0.164 and 0.149, respectively. The reason may be that the hydrophobicity of nanoliposomes increased after loading naringenin, which inhibited the aggregation of blank nanoliposomes, and the number of blank liposomes should be very small under this condition [22,30]. The ζ-potentials of those samples were negative and ranged from −14.01 to −19.30 mV (Figure 6d), and the pH and concentration of naringenin didn't seem to cause significant changes in the ζ-potentials of nanoliposomes.

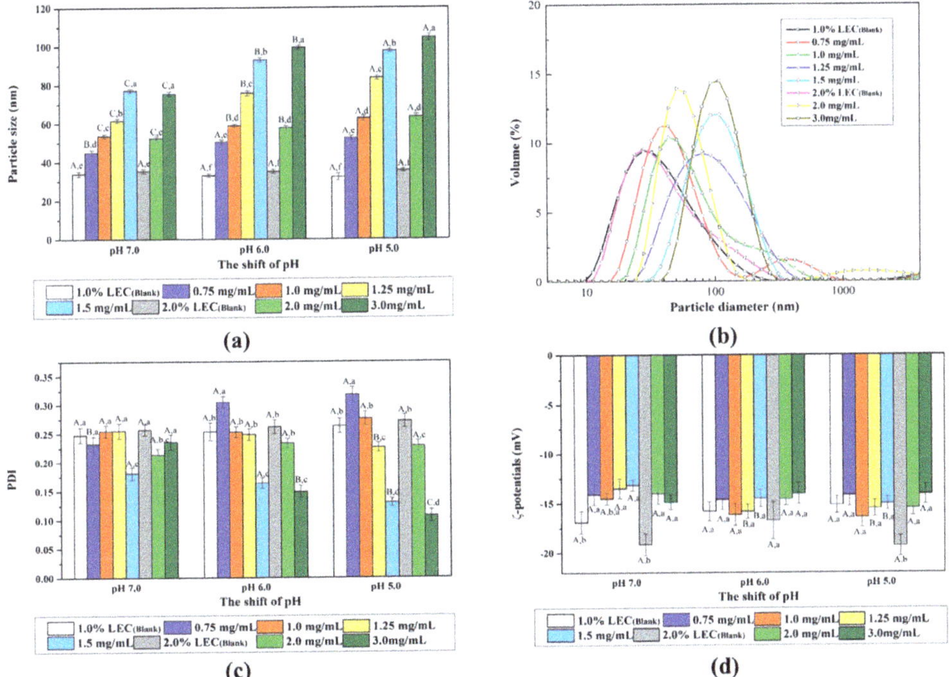

Figure 6. Particle size (**a**), particle size distribution (**b**) at pH 6.0, polydispersity index (**c**, PDI) and ζ-potentials (**d**) of naringenin-loaded nanoliposomes with pH shift to 7.0, 6.0 and 5.0. Samples denoted with different letters (A–C) and (a–f) were significantly different ($p < 0.05$) when compared between different pH regions (same naringenin level) and different naringenin levels (same pH region), respectively.

Different from the nanoliposomes loaded with naringenin, the particle size, PDI and ζ-potentials of the naringin-loaded nanoliposomes did not change significantly after loading with naringin (Table 1). Unlike naringenin, naringin may has weak hydrophobicity and electrostatic interaction with nanoliposomes without forming a larger particle size [30].

Table 1. The mean particle size, PDI and ζ-potentials of naringin-loaded nanoliposomes.

Naringin Concentration (mg/mL)	Mean Particle Size (nm)	PDI	ζ-Potentials (mV)
Blank nanoliposomes	34.93 ± 0.98 [a]	0.262 ± 0.012 [b]	−16.74 ± 1.87 [a]
1.0 (1% *w/v* lecithin)	34.03 ± 0.69 [a]	0.230 ± 0.010 [a]	−12.94 ± 1.31 [a]
1.5 (1% *w/v* lecithin)	34.25 ± 0.72 [a]	0.237 ± 0.010 [a,b]	−16.16 ± 1.60 [a]
2.0 (1% *w/v* lecithin)	35.09 ± 0.86 [a]	0.240 ± 0.012 [a,b]	−13.90 ± 1.07 [a]
3.0 (2% *w/v* lecithin)	35.71 ± 0.49 [a]	0.260 ± 0.011 [b]	−16.97 ± 1.48 [a]

Samples denoted with letters (a,b) were significantly different ($p < 0.05$) when compared between different naringin levels.

3.5. Stability of Naringenin-Loaded Nanoliposomes

Liposomes need to have good storage stability in order to become a commercial product, which means that they must remain integrity during the entire life cycle of the product. We first investigated the change of the encapsulation efficiency of 1.00 and 1.25 mg/mL naringenin loaded nanoliposomes at 4 °C, 25 °C and 37 °C (Figure 7b–d). At 4 °C, the encapsulation efficiency of 1.00 mg/mL naringenin-loaded nanoliposomes solution remained constant at about 86% after one month of storage, while the nanoliposome solution with a concentration of 1.25 mg/mL dropped from 83.58% to 70.48%. At 25 °C, the encapsulation efficiency of all the nanoliposome solutions decreased, to 74.67% and 50.24%, respectively. At 37 °C, the encapsulation efficiency of 1.00 mg/mL naringenin loaded nanoliposomes solution decreased from 86.23% to 69.69%, and the 1.25 mg/mL naringenin loaded nanoliposomes solution decreased from 83.58% to 54.26%. With the increase of storage temperature, the nanoliposome solution finally changed from milky white to yellow at higher temperature (Figure 7a), indicating that the reduction of encapsulation efficiency is related to the precipitation of naringenin. These results suggest that at high temperatures, nanoliposomes loaded with more naringenin are more hydrophobicity and likely to accelerate aggregation, oxidation, leakage and other chemical reactions during storage, which cause the precipitation of naringenin crystals and the reduction in encapsulation efficiency [22]. Similar to previous research results, conventional liposomes without modification are thermodynamically unstable systems that tend to aggregate, fuse, degrade, or hydrolyze, causing leakage of loaded compounds [31–33]. However, the combination of synthetic polymers or biopolymer with liposomes can easily improve the stability and site-specific targeting of liposomes through surface modification [33–35].

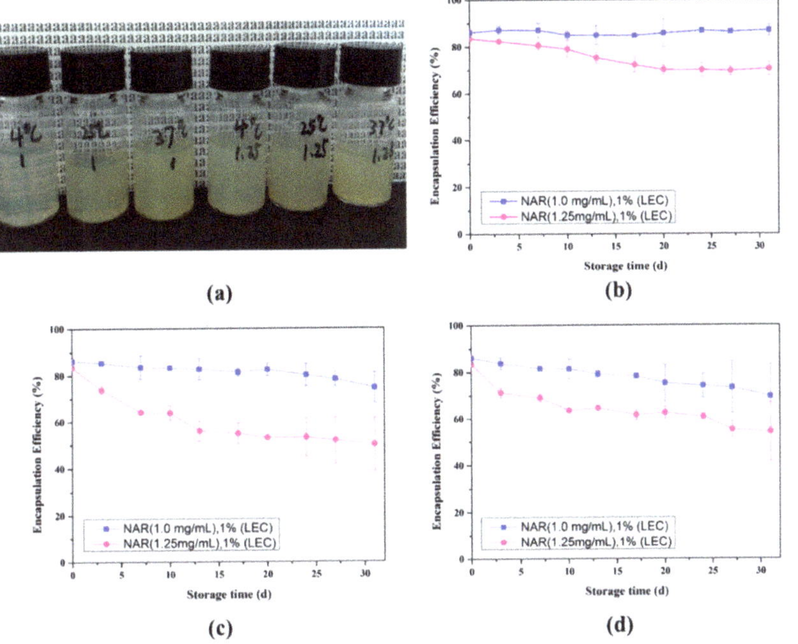

Figure 7. The image (**a**) of naringenin loaded nanoliposome solutions with naringenin concentrations of 1.00 and 1.25 mg/mL with 1% (*w*/*v*) lecithin after 31 days storage at different temperatures; Changes of encapsulation efficiency at 4 °C (**b**), 25 °C (**c**) and 37 °C (**d**).

The storage stability of the nanoliposomes were also determined by recording changes in their mean particle size, PDI and ζ-potentials (Figure 8). There was little change in the mean particle size, PDI and ζ-potentials of the 1.00 and 1.25 mg/mL naringenin loaded nanoliposomes solutions when they were placed in 4 °C within 31 days. At 25 °C, the mean particle size of 1.00 mg/mL naringenin loaded nanoliposomes increased from 65.88 nm to 87.14 nm at 17-day, then slightly decreased to 83.83 nm at day 31. Similarly, the 1.25 mg/mL naringenin loaded nanoliposomes increased from 91.15 nm to 95.80 nm at 10-day, then slightly decreased to 93.55 nm at day 28. The change of the particle size of nanoliposomes with different naringenin concentration at 37 °C also followed the trend of first increasing and then decreasing. These results indicate that storing naringenin-loaded nanoliposomes at high temperature is prone to aggregation in the early stage, and then the particle size decreased in the later stage due to oxidation and leakage of the nanoliposomes [19]. In the early stage of storage, the aggregation of nanoliposomes may cause little increase the PDI, while the PDI of the later supersaturated nanoliposomes may decrease due to the precipitation of part of naringenin. Similar to Homayoonfal et al.'s study of the stability of anthocyanin compound loaded nanoliposomes at 37 °C, there were no significant difference in the parameters of PDI and ζ-potential after one month of storage, but the particle diameter and encapsulation efficiency significantly increased and decreased, respectively [36]. Although the instability mechanism is not yet clear, it may be because of the aggregation and leakage of nanoliposomes and chemical degradation of naringenin in nanoliposomes solutions, which results in color changes or precipitation at high store temperature [20,37,38].

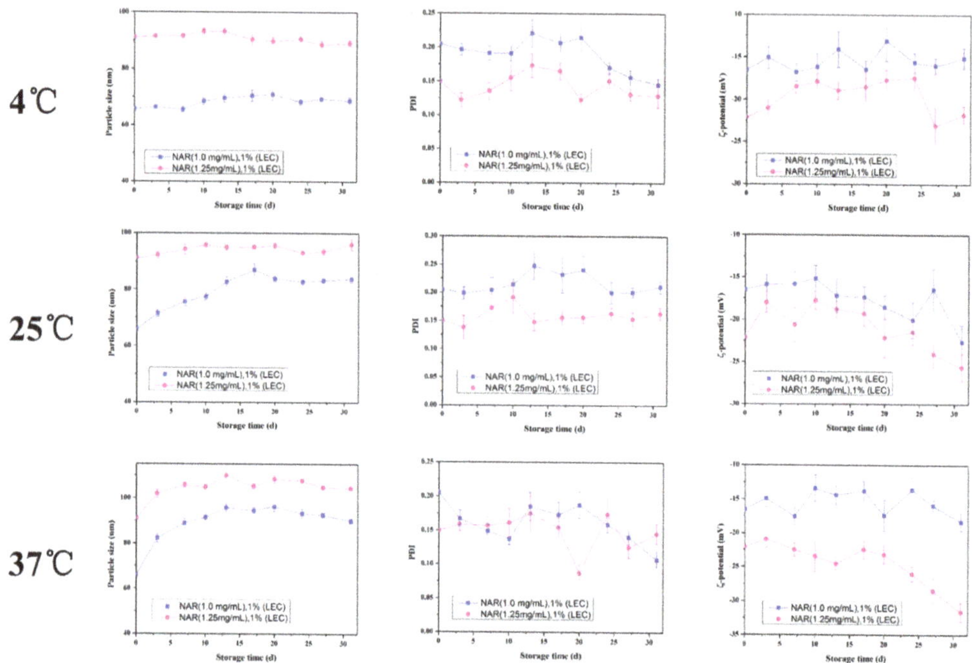

Figure 8. Changes of the naringenin-loaded nanoliposomes in mean particle size, PDI and ζ-potential during storage at 4, 25, and 37 °C for 31 days.

3.6. Micosturcture of Naringenin-Loaded Nanoliposomes

The microstructure of the blank nanoliposomes and naringenin loaded nanoliposomes were obtained using AFM. Figure 9 illustrates representative 3D pictures of the blank and nanoliposomes with different naringenin concentration obtained before and after storage at 4 °C for 31 days. Even the nanoliposomes after storage for 31 days, all the nanoliposomes displayed as smooth sphericities that were uniformly distributed throughout the whole pictures, and the particle sizes were similar with those measured by laser granulometry method.

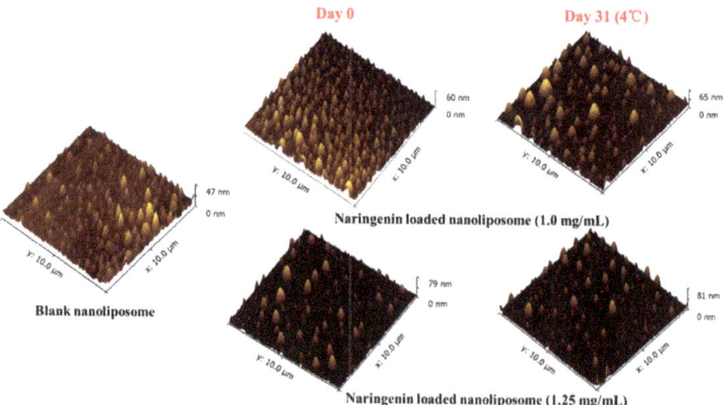

Figure 9. Atomic forces microscopy 3D images of blank nanoliposomes and naringenin-loaded nanoliposomes with a concentration of 1.00 and 1.25 mg/mL after 0 and 31 days of storage at 4 °C.

4. Conclusions

In conclusion, we have proved that naringenin could be encapsulated into nanoliposomes by an easy pH-driven method according to the reduction in its water-solubility after the pH changed to acidity, which causes the naringenin can be moved from the aqueous solution into the lipid bilayers within nanoliposomes, while the naringin-loaded nanoliposomes contained some free naringin due to its higher water-solubility at lower pH values. Experiments showed that the naringenin-loaded nanoliposomes were predominantly nanometric, negatively charged and exhibited relatively high encapsulation efficiency. Moreover, the naringenin-loaded nanoliposomes still maintain good stability during storage at 4 °C for 31 days, while they were unstable at high store temperatures. In general, an easy method for manufacturing naringenin-loaded nanoliposomes has been constructed that may help to develop novel food-grade colloidal delivery systems and apply to introducing naringenin or other lipophilic polyphenols into foods, supplements, or drugs.

Author Contributions: Conceptualization, M.C. and R.L.; methodology, R.L. and W.Z.; software, Y.G. and Y.Z.; validation, M.C., W.Z. and R.L.; formal analysis, R.L. and L.L.; investigation, M.C. and R.L.; resources, Y.G. and L.L.; data curation, Y.C. and W.Z.; writing—M.C. and R.L.; writing—review and editing, R.L. and W.Z.; visualization, Y.G., Y.C. and Y.Z.; supervision, W.Z. and J.L.; project administration, R.L. and J.L.; funding acquisition, R.L. and W.Z. All authors have read and agreed to the published version of the manuscript.

Funding: This research was funded by the National Natural Science Foundation of China (grant number 32001689), the Natural Science Foundation of Guangdong Province of China (grant number 2021A1515010901), the Natural Science Foundation of Hainan Province of China (grant number 320QN325), the Central Public-interest Scientific Institution Basal Research Fund for Chinese Academy of Tropical Agricultural Sciences (grant number 1630122017016 and 1630122021006), Innovative Research Team Construction Project for Modern Agricultural Industry Common Key Technologies of Guangdong Province (grant number 2020KJ116 and 2020KJ117), and the 'Guangdong Special Support Plan' Science and Technology Innovation Young Talents Project (2019TQ05N133).

Institutional Review Board Statement: Not applicable.

Informed Consent Statement: Not applicable.

Data Availability Statement: Not applicable.

Acknowledgments: All authors are thankful to their representative universities/institutes for the support and services used in this study.

Conflicts of Interest: The authors declare no conflict of interest.

References

1. Raeisib, S.; Chavoshia, H.; Mohammadic, M.; Ghorbanid, M.; Sabzichie, M.; Ramezani, F. Naringenin-loaded nano-structured lipid carrier fortifies oxaliplatin-dependent apoptosis in HT-29 cell line. *Process Biochem.* **2019**, *83*, 168–175. [CrossRef]
2. Rao, K.; Imran, M.; Jabri, T.; Ali, I.; Perveen, S.; Shafiullah; Ahmed, S.; Shah, M.R. Gum tragacanth stabilized green gold nanoparticles as cargos for naringin loading: A morphological investigation through afm. *Carbohyd. Polym.* **2017**, *174*, 243. [CrossRef] [PubMed]
3. Hermenean, A.; Ardelean, A.; Stan, M.; Herman, H.; Mihali, C.V.; Costache, M.; Dinischiotu, A. Protective effects of naringenin on carbon tetrachloride-induced acute nephrotoxicity in mouse kidney. *Chem. Biol. Interact.* **2013**, *205*, 138–147. [CrossRef]
4. Bodet, C.; La, V.D.; Epifano, F.; Grenier, D. Naringenin has anti-inflammatory properties in macrophage and ex vivo human whole-blood models. *J. Periodontal Res.* **2008**, *43*, 400–407. [CrossRef] [PubMed]
5. Fouad, A.A.; Albuali, W.H.; Jresat, I. Protective effect of naringenin against lipopolysaccharide-induced acute lung injury in rats. *Pharmacology* **2016**, *97*, 224–232. [CrossRef]
6. Yen, H.R.; Liu, C.J.; Yeh, C.C. Naringenin suppresses TPA-induced tumor invasion by suppressing multiple signal transduction pathways in human hepatocellular carcinoma cells. *Chem. Biol. Interact.* **2015**, *235*, 1–9. [CrossRef] [PubMed]
7. Leonardi, T.; Vanamala, J.; Taddeo, S.S.; Davidson, L.A.; Murphy, M.E.; Patil, B.S.; Wang, N.; Carroll, R.J.; Chapkin, R.S.; Lupton, J.R.; et al. Apigenin and naringenin suppress colon carcinogenesis through the aberrant crypt stage in azoxymethane-treated rats. *Exp. Biol. Med.* **2010**, *235*, 710–717. [CrossRef]

8. Lin, C.Y.; Ni, C.C.; Yin, M.C.; Lii, C.K. Flavonoids protect pancreatic beta-cells from cytokines mediated apoptosis through the activation of PI3-kinase pathway. *Cytokine* **2012**, *59*, 65–71. [CrossRef]
9. Al-Dosari, D.I.; Ahmed, M.M.; Al-Rejaie, S.S.; Alhomida, A.S.; Ola, M.S. Flavonoid naringenin attenuates oxidative stress, apoptosis and improves neurotrophic effects in the diabetic rat retina. *Nutrients* **2017**, *9*, 1161. [CrossRef]
10. Zbarsky, V.; Datla, K.P.; Parkar, S.; Rai, D.K.; Aruoma, O.I.; Dexter, D.T. Neuroprotective properties of the natural phenolic antioxidants curcumin and naringenin but not quercetin and fisetin in a 6-OHDA model of Parkinson's disease. *Free Radic. Res.* **2005**, *39*, 1119–1125. [CrossRef]
11. Lucas-Abellán, C.; Pérez-Abril, M.; Castillo, J.; Serrano, A.; Mercader, M.T.; Fortea, M.I.; Gabaldón, J.A.; Núñez-Delicado, E. Effect of temperature, pH, β- and HP-β-cds on the solubility and stability of flavanones: Naringenin and hesperetin. *LWT Food Sci.Technol.* **2019**, *108*, 233–239. [CrossRef]
12. Uchiyama, H.; Kadota, K.; Nakanishi, A.; Tandia, M.; Tozuka, Y. A simple blending with α-glycosylated naringin produces enhanced solubility and absorption of pranlukast hemihydrate. *Int. J. Pharmaceut.* **2019**, *567*, 118490. [CrossRef]
13. Shpigelman, A.; Shoham, Y.; Israeli-Lev, G.; Livney, Y.D. β-Lactoglobulin–naringenin complexes: Nano-vehicles for the delivery of a hydrophobic nutraceutical. *Food Hydrocoll.* **2014**, *40*, 214–224. [CrossRef]
14. Budel, R.G.; Silva, D.A.D.; Moreira, M.P.; Dalcin, A.J.F.; Boeck, C.R. Toxicological evaluation of naringin-loaded nanocapsules in vitro and in vivo. *Colloids Surf. B Biointerfaces* **2020**, *188*, 110754. [CrossRef] [PubMed]
15. Wang, Y.W.; Wang, S.C.; Firempong, C.K.; Zhang, H.Y.; Wang, M.M.; Zhang, Y.; Zhu, Y.; Yu, J.N.; Xu, X.M. Enhanced Solubility and Bioavailability of Naringenin via Liposomal Nanoformulation: Preparation and In Vitro and In Vivo Evaluations. *AAPS PharmSciTech* **2017**, *18*, 586–594. [CrossRef] [PubMed]
16. Yin, X.; Fu, X.; Cheng, H.; Wusigale; Liang, L. α-Tocopherol and naringenin in whey protein isolate particles: Partition, antioxidant activity, stability and bioaccessibility. *Food Hydrocoll.* **2020**, *106*, 105895. [CrossRef]
17. Pan, K.; Luo, Y.; Gan, Y.; Baek, S.J.; Zhong, Q. pH-driven encapsulation of curcumin in self-assembled casein nanoparticles for enhanced dispersibility and bioactivity. *Soft Matter.* **2014**, *10*, 6820–6830. [CrossRef] [PubMed]
18. Guo, Q.; Bayram, I.; Zhang, W.; Su, J.; Shu, X.; Yuan, F.; Mao, L.; Gao, Y. Fabrication and characterization of curcumin-loaded pea protein isolate-surfactant complexes at neutral pH. *Food Hydrocoll.* **2021**, *111*, 106214. [CrossRef]
19. Mcclements, D.J.; Peng, S.; Li, Z.; Zou, L.; Liu, W.; Liu, C. Enhancement of curcumin bioavailability by encapsulation in sophorolipid-coated nanoparticles: An in Vitro and in Vivo Study. *J. Agric. Food Chem.* **2018**, *66*, 1488–1497.
20. Peng, S.; Li, Z.; Zou, L.; Liu, W.; Liu, C.; Mcclements, D.J. Improving curcumin solubility and bioavailability by encapsulation in saponin-coated curcumin nanoparticles prepared using a simple pH-driven loading method. *Food Funct.* **2018**, *9*, 1829–1839. [CrossRef]
21. Ma, Y.; Chen, S.; Liao, W.; Zhang, L.; Liu, J.; Gao, Y. Formation, physicochemical stability, and redispersibility of curcumin-loaded rhamnolipid nanoparticles using the pH-driven method. *J. Agric. Food Chem.* **2020**, *68*, 7103–7111. [CrossRef] [PubMed]
22. Zhan, X.; Dai, L.; Zhang, L.; Gao, Y. Entrapment of curcumin in whey protein isolate and zein composite nanoparticles using pH-driven method. *Food Hydrocoll.* **2020**, *106*, 105839. [CrossRef]
23. Zheng, B.; Peng, S.; Zhang, X.; Mcclements, D.J. Impact of delivery system type on curcumin bioaccessibility: Comparison of curcumin-loaded nanoemulsions with commercial curcumin supplements. *J. Agric. Food Chem.* **2018**, *66*, 10816–10826. [CrossRef] [PubMed]
24. Cheng, C.; Peng, S.; Li, Z.; Zou, L.; Liu, W.; Liu, C. Improved bioavailability of curcumin in liposomes prepared using a pH-driven, organic solvent-free, easily scalable process. *RSC Adv.* **2017**, *7*, 25978–25986. [CrossRef]
25. Peng, S.; Zou, L.; Liu, C.; Mcclements, D.J. Fabrication and characterization of curcumin-loaded liposomes formed from sunflower lecithin: Impact of composition and environmental stress. *J. Agric. Food Chem.* **2018**, *66*, 12421–12430. [CrossRef]
26. Peng, S.; Zou, L.; Zhou, W.; Liu, W.; Mcclements, D.J. Encapsulation of lipophilic polyphenols into nanoliposomes using pH-driven method: Advantages and disadvantages. *J. Agric. Food Chem.* **2019**, *67*, 7506–7511. [CrossRef]
27. Su, C.; Liu, Y.; He, Y.; Gu, J. Analytical methods for investigating in vivo fate of nanoliposomes: A review. *J. Pharm. Anal.* **2018**, *8*, 219–225. [CrossRef]
28. Mozafari, M.R. Nanoliposomes: Preparation and analysis. *Methods Mol. Biol.* **2010**, *60*, 29–50.
29. Khorasani, S.; Danaei, M.; Mozafari, M.R. Nanoliposome technology for the food and nutraceutical industries. *Trends Food Sci. Technol.* **2018**, *79*, 106–115. [CrossRef]
30. Chen, Y.; Zhao, Z.; Xia, G.; Xue, F.; Zhang, Y. Fabrication and characterization of zein/lactoferrin composite nanoparticles for encapsulating 7, 8-dihydroxyflavone: Enhancement of stability, water solubility and bioaccessibility. *Int. J. Biol. Macromol.* **2019**, *146*, 179–192. [CrossRef]
31. Peng, S.; Zou, L.; Liu, W.; Li, Z.; Hu, X.; Chen, X.; Liu, C. Hybrid liposomes composed of amphiphilic chitosan and phospholipid: Preparation, stability and bioavailability as a carrier for curcumin. *Carbohydr. Polym.* **2017**, *156*, 322–332. [CrossRef] [PubMed]
32. Li, J.; Chang, C.; Zhai, J.; Yang, Y.; Yu, H. Ascorbyl palmitate effects on the stability of curcumin-loaded soybean phosphatidylcholine liposomes. *Food Biosci.* **2021**, *41*, 100923. [CrossRef]
33. Tan, C.; Wang, J.; Sun, B. Biopolymer-liposome hybrid systems for controlled delivery of bioactive compounds: Recent advances. *Biotechnol. Adv.* **2021**, *48*, 107727. [CrossRef]

34. Liu, Y.; Xie, X.; Chen, H.; Hou, X.; He, Y.; Shen, J.; Shi, J.; Feng, N. Advances in next-generation lipid-polymer hybrid nanocarriers with emphasis on polymer-modified functional liposomes and cell-based-biomimetic nanocarriers for active ingredients and fractions from Chinese medicine delivery. *Nanomed. Nanotechnol. Biol. Med.* **2020**, *29*, 102237. [CrossRef] [PubMed]
35. Mukherjee, A.; Waters, A.K.; Kalyan, P.; Achrol, A.S.; Kesari, S.; Yenugonda, V.M. Lipid-polymer hybrid nanoparticles as a nextgeneration drug delivery platform: State of the art, emerging technologies, and perspectives. *Int. J. Nanomed.* **2019**, *14*, 1937–1952. [CrossRef] [PubMed]
36. Homayoonfal, M.; Mousavi, S.M.; Kiani, H.; Askari, G.; Desobry, S.; Arab-Tehrany, E. Encapsulation of berberis vulgaris anthocyanins into nanoliposome composed of rapeseed lecithin: A comprehensive study on physicochemical characteristics and biocompatibility. *Foods* **2021**, *10*, 492. [CrossRef] [PubMed]
37. Li, R.; Dai, T.; Tan, Y.; Fu, G.; Wan, Y.; Liu, C.; Mcclements, D.J. Fabrication of pea protein-tannic acid complexes: Impact on formation, stability, and digestion of flaxseed oil emulsions. *Food Chem.* **2020**, *310*, 125828. [CrossRef] [PubMed]
38. Li, R.; Dai, T.; Zhou, W.; Fu, G.; Wan, Y.; McClements, D.J.; Li, J. Impact of pH, ferrous ions, and tannic acid on lipid oxidation in plant-based emulsions containing saponin-coated flaxseed oil droplets. *Food Res. Int.* **2020**, *136*, 109618. [CrossRef]

 foods

MDPI

Article

Fabrication of Nano/Micro-Structured Electrospun Detection Card for the Detection of Pesticide Residues

Kun Feng [1], Meng-Yu Zhai [1], Yun-Shan Wei [1], Min-Hua Zong [1], Hong Wu [1,*] and Shuang-Yan Han [2,*]

[1] Guangdong Province Key Laboratory for Green Processing of Natural Products and Product Safety, School of Food Science and Engineering, South China University of Technology, Guangzhou 510640, China; fengkun_89@163.com (K.F.); mengyu.zhai@outlook.com (M.-Y.Z.); 13650501669@163.com (Y.-S.W.); btmhzong@scut.edu.cn (M.-H.Z.)

[2] College of Biosciences and Bioengineering, South China University of Technology, Guangzhou 510640, China

* Correspondence: bbhwu@scut.edu.cn (H.W.); syhan@scut.edu.cn (S.-Y.H.)

Abstract: A novel nano/micro-structured pesticide detection card was developed by combining electrospinning and hydrophilic modification, and its feasibility for detecting different pesticides was investigated. Here, the plain and hydrophilic-modified poly(ε-caprolactone) (PCL) fiber mats were used for the absorption of indolyl acetate and acetylcholinesterase (AChE), respectively. By pre-treating the fiber mat with ethanol, its surface wettability was improved, thus, promoting the hydrolysis of the PCL fiber mat. Furthermore, the absorption efficiency of AChE was improved by almost two times due to the increased hydrophilicity of the modified fiber mat. Noteworthily, this self-made detection card showed a 5-fold, 2-fold, and 1.5-fold reduction of the minimum detectable concentration for carbofuran, malathion, and trichlorfon, respectively, compared to the national standard values. Additionally, it also exhibited good stability when stored at 4 °C and room temperature. The food detection test showed that this nano/micro-based detection card had better detectability than the commercial detection card. Therefore, this study offers new insights into the design of pesticide detection cards, which also broadens the application of electrospinning technique.

Keywords: nano/micro-structure; pesticide detection card; electrospinning; hydrophilic modification; sensitivity

Citation: Feng, K.; Zhai, M.-Y.; Wei, Y.-S.; Zong, M.-H.; Wu, H.; Han, S.-Y. Fabrication of Nano/Micro-Structured Electrospun Detection Card for the Detection of Pesticide Residues. *Foods* **2021**, *10*, 889. https://doi.org/10.3390/foods10040889

Academic Editor: Roberto Romero-González

Received: 17 March 2021
Accepted: 13 April 2021
Published: 19 April 2021

1. Introduction

Pesticides are effective ingredients that have been extensively used to control agricultural diseases. However, their excessive use poses a challenge to the sustainable development of agriculture, and the administration of pesticide residues on the surface of food is harmful to human health. For this reason, considerable progress has been made recently in the determination of pesticides. Current approaches are mainly carried out by laboratory-scale analytical methods, such as HPLC, MC, GC, etc. [1,2]. Nevertheless, their inherent disadvantages, such as complicated pre-treatment, high costs, and time-consuming procedures, generally limit their applications [3]. Therefore, research into other effective, convenient, and reliable detection strategies has gained increasing attention in recent years [4–6]. Specifically, rapid detection based on enzyme acetylcholinesterase (AChE) inhibition has become widely accepted for pesticide residue analysis, ascribing to its simple principle and visual evaluation [7]. In this regard, the rapid detection card, generally made of glass fiber, qualitative filter paper, or absorbent paper, has been identified as the preferred and direct method for determining pesticides. Nevertheless, the challenge associated with these cards is the discrepancy between the sensitivity of the analysis and the increased awareness of food safety, making it imperative to seek new types of carriers for the production of detection cards.

Recently, extensive evidence has demonstrated the great potential of nano/micromaterials in the field of catalysis, owing to their unique structure and surface properties [8,9]. Herein, we try to further improve the conventional detection card's performance by supplying the nano/micromaterials as the immobilization carrier for the enzyme and substrate. Electrospinning—a mild, convenient, versatile, and cost-effective technique for developing micro-/nano-structured vehicles—is associated with a broad range of applications in pharmacy, tissue engineering, food packaging, etc. [10]. Specifically, the superiority of electrospun fibers as the carrier for chemical biosensors and immobilization has been highlighted in recent years due to their desirable properties, such as a high surface area to volume ratio, porous structure, and tunable porosity [11,12]. However, electrospun fibers used as the detection card matrix for the pesticide detection card have been less explored, except for in our previous research study [13]. It is known that the electrospun fiber mat could not only be used as the encapsulation vehicle but also the absorption carrier [14]. Our published work has already designed a novel detection card by encapsulating the enzyme (AChE) and substrate (indolyl acetate, IA) into the electrospun nanofibers and we also demonstrated their proper role for determining pesticides more sensitively. Unfortunately, the deformation problem of this card, stemming from its hydrophilic feature, calls for the exploration of hydrophobic materials. Additionally, it is also interesting to determine the pesticide detection performance of the card by using the electrospun fiber mat as the absorption matrix for the enzyme and substrate.

Herein, a biocompatible and biodegradable polymer, Poly(ε-caprolactone) (PCL), was applied to prepare the detection card. To improve enzyme absorption on the fiber mat, hydrophilic modification of the hydrophobic PCL fiber mat was performed, and the related properties were investigated systematically. Subsequently, the plain and modified PCL fiber mat were employed to absorb IA and AChE, respectively. Furthermore, the key factors including the concentration of the enzyme and substrate, inhibition time, and color development time were optimized to improve the performance of the detection card. The detection ability of this card for different pesticides was examined and compared with the corresponding values specified by the Standardization Administration of China (SAC). Lastly, the feasibility of this card for detecting pesticide residues in real foods was studied by comparison with the commercial card. This study offers new access for the sensitive and convenient detection of pesticide residues, which is expected to inspire the further application of electrospinning in the field of food and agriculture.

2. Materials and Methods

2.1. Materials

Poly(ε-caprolactone) (PCL) was purchased from Sinopharm Chemical Reagent Co., Ltd. (Beijing, China). AChE (200 U/g, from head of fly) and IA were purchased from ShanghaiYuanye Bio-Technology Co., Ltd. (Shanghai, China). Trichlorfon and carbaryl were purchased from Aladdin (Shanghai, China). Malathion and carbofuran were purchased from the National Information Center for Certified Reference Materials (Beijing, China). The commercial rapid detection card was obtained from Dayuan Oasis Food Safety Technology Co., Ltd. (Guangzhou, China). Analytical pesticide standards (100 μg/mL) were stored at 4 °C and spiked to the desired concentrations. Samples of cabbage and apples were purchased from a local market (Guangzhou, China).

2.2. Preparation of the Detection Card Matrix

PCL solutions in the concentration range of 100–150 g/L were prepared by dissolving certain amounts of PCL into a mixed solvent of methanol and chloroform at different volume ratios (0, 1:9, 3:7, 5:5). Then, the sealed solution was placed on a magnetic stirrer and stirred at room temperature in the dark for more than one hour to obtain the stable solution. The conductivity and viscosity of different spinning solutions were measured by Brookfield digital viscometer (Model DV-II t Pro, Brookfield Engineering, Inc., Middleboro, MA, USA) and conductivity meter (DDS-11A, Shanghai Leici Chuangyi Instrument Co., Ltd., Jiading, Shanghai, China), respectively. Then, electrospinning was conducted by putting the solution into a plastic syringe (5 mL) with a 20-gauge steel needle which was connected to a high voltage power supply. The voltage and the distance between the needle tip and collector were set in the range of 11–17 kV and 11–15 cm, respectively. The injection rate was controlled by a syringe pump (NE-300, New Era Pump Systems Inc., Farmingdale, NY, USA) in the range of 1.5 to 3.5 mL/h. The fiber mat was fabricated at 24 ± 2 °C and under 55 ± 5% relative humidity for a period of time.

2.3. Hydrophilic Modification of the Matrix

Here, the PCL fiber mat was modified by one step or two steps of treatment. For the one-step modification, the PCL fiber mat was immersed in the NaOH solution with different concentrations (1 M, 2 M, 4 M, and 6 M) for a certain period of time, while the two-step modification was performed by initially infiltrating in the 70% ethanol for 15 min before the same NaOH immersion process. Then, the fiber mat was washed with deionized water and dried in a vacuum drying oven.

2.4. Characterization of the Matrix

The morphologies of the original and modified PCL fiber were observed by scanning electron microscopy (SEM, S-3700N, Hitachi High-Tech Ltd., Chiyoda-ku, Japan). Before the observation, the electrospun fiber mat was sputter-coated under vacuum conditions and observed at 15 kV. Then, the obtained SEM image was processed by the Nano Measure 1.2 software, and the distribution of the fibers was further calculated by analyzing around 50 fibers. The thickness of the fiber mat was measured by a digital thickness gauge (Syntek, Deqing Shengtai Electronic Technology Co., Ltd., Deqing, China).

The changes that occurred in the polymer molecules were examined by employing attenuated total reflection-Fourier transform infrared spectroscopy (ATR-FTIR) (VERTEX 70, Bruker Co., Ettlingen, Germany). All the spectra were recorded in the frequency region of 500–4000 cm^{-1} at a spectral resolution of 4 cm^{-1}.

To characterize the surface wettability of the fiber mat, the surface contact angle tester (OCA40, DataPhysics Instruments GmbH, Filderstadt, Germany) was used to determine the contact angles of different samples. Briefly, the measurement was conducted by dropping 2 μL of deionized water onto the surface of the fiber mat with a drop rate of 2 μL/s. After 10 s, the contact angle of the fiber mat was measured by taking a screenshot. Each sample was tested 3 times, and the average value was calculated.

2.5. Optimization of the Immobilization and Determination Conditions

In this study, two fiber mats, namely the enzyme (AChE) fiber mat (AFM) and substrate (IA) fiber mat (IFM), were used for the preparation of the nano/micro-structured detection card. For optimizing the suitable AChE concentration for the preparation of AFM, the IFM was prepared by immersing the fiber mat in 5 mg/mL IA solution, and the cards were then dried in a vacuum oven. Similarly, the fiber mat was treated with 8 mg/mL of AChE solution when the optimal IA solution was investigated. Here, PBS (pH 7.4) and 1 mg/L of malathion were selected as control and positive sample, respectively. On the other hand, the absorption efficiencies of the fiber mat and other commonly used absorption materials (qualitative and quantitative filter paper) were examined by determining the amount of released enzyme from the carrier materials according to the Bradford method. Before doing this, the carrier materials were immersed into 5 mg/mL of AChE solution for 24 h and dried in a vacuum drying oven.

The principle of this detection method is that AChE can catalyze the hydrolysis of colorless IA to produce indoxyl, as depicted in Figure 1, which then becomes blue due to the rapid oxidation by air. The blue color of indigo can be well distinguished with the naked eye. Hence, the existence of pesticides can be analyzed according to the color change as a consequence of the inhibition of the pesticide on AChE activity. On the basis of understanding the principle of this determination method, inhibition and color development time were two critical factors that significantly influenced the result. Herein, the inhibition time was optimized in the range of 5–15 min, where the color development time was set at 15 min. Similarly, the inhibition time was set at 10 min for exploring the appropriate color development time (5–25 min). PBS (pH 7.4) and 0.5 mg/L malathion were applied as control and positive samples, respectively.

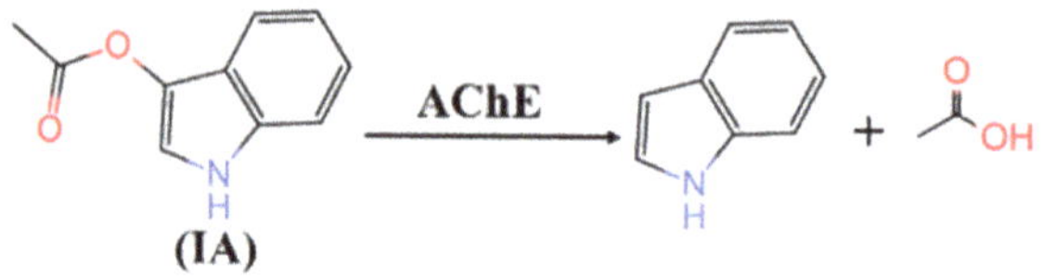

(IA)
AChE
+

Figure 1. The reaction of indoxyl acetate hydrolysis, catalyzed by AChE.

2.6. Performance of the Detection Card and Its Real-Life Application

To evaluate the efficacy of this detection card, a series of concentrations of these two classical pesticides, including organophosphorus (OP) (omethoate and malathion) and carbamate (CM) (carbaryl and carbofuran), were determined and compared with their corresponding low limit of detection values reported in SAC. In brief, the pesticides were diluted to a series of concentrations by PBS. An aliquot (50 μL) of sample solution was dropped on the AFM for analysis, and PBS served as the control group. All the detection procedures were performed according to the conditions optimized in the above section.

The storage stabilities of the detection card under 4 °C and room temperature (RT) were evaluated for 60 days. Malathion and PBS were served as the positive and control group to examine its detection efficacy periodically.

To further verify the determination performance of this card, organic cabbage and carrots were tested as samples with different concentrations of malathion (0, 5, 10, 20 μg/mL). Different volumes of malathion were sprayed on sample surfaces based on the sample weight (1 mL/g) and stored at room temperature for 24 h. Subsequently, 5 g of the samples were immersed in 10 mL PBS and then shaken by hand. After the stabilization for 2 min, the supernatant was analyzed using self-made and commercial rapid detection cards.

2.7. Statistical Analysis

Statistical analysis was performed using one-way analysis of variance (ANOVA). A value of $p \leq 0.05$ was considered statistically significant.

3. Results and Discussion

3.1. Preparation of the Nano/Micro-Structured Immobilization Matrix

Recently, the improved attention on food safety and pursuit of the sustainable development of the agricultural industry has promoted the exploration of efficient and convenient pesticide detection approaches. Specifically, our group has creatively investigated the potent application of electrospinning in designing rapid detection cards to determine different pesticides. The detection card was composed of two electrospun PVA fiber mats, in which the enzyme and substrate were encapsulated. Furthermore, it was found to be able to efficiently and sensitively detect different pesticides. However, the high hydrophilicity of PVA made the fiber mat easier to deform, which would impede its application to some extent. Given this situation, it is meaningful to investigate the performance of the detection card, taking a hydrophobic material as the card matrix. Additionally, electrospinning was also supposed as a proper absorption matrix in the field of analysis [15]. However, knowledge of its capability for use as a detection card is lacking. Therefore, as shown in Figure 2, this study offered another novel route for the preparation of pesticide detection card by electrospinning, where the electrospun fiber mats were employed as the immobilization matrix for the absorption of the enzyme (AChE) and substrate (AI) and the corresponding detection principle was depicted in Figure 2.

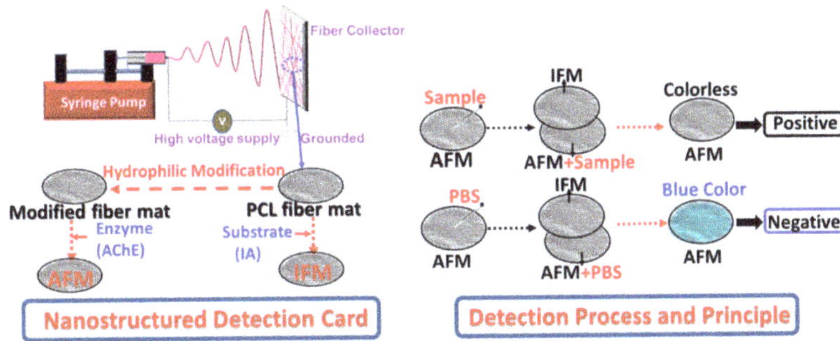

Figure 2. Preparation process and pesticide determination principle of the nano/micro-structured detection card.

Here, an environmentally friendly material, PCL, was utilized to fabricate the detection card matrix, owing to its proper biodegradable and hydrophobic properties [16]. To successfully obtain the electrospun fiber mat, the appropriate solvent for PCL electrospinning should be initially investigated. It is known that solvents with low boiling points are more suitable for electrospinning [17]. Hence, a commonly used solvent, chloroform ($CHCl_3$), was selected for hydrophobic PCL electrospinning, and methanol (CH_3OH) was further blended with it to decrease the fast evaporation of the solvent during the electrospinning process because of the lower boiling point of the $CHCl_3$ [18]. Furthermore, as shown in Table S1, the conductivity and viscosity of the solution were increased with the increase in the ratio of the CH_3OH, which was similar to a previous study [19]. The fiber with a good morphology was obtained at the volume ratio of 3:7 (CH_3OH: $CHCl_3$) (Figure S1). Then, the suitable concentration of PCL for electrospinning was examined, and it was found that smooth fibers were obtained when the concentration was 125 g/L (Figure S2). The balance between the conductivity and viscosity, which was one key factor that would significantly influence the electrospinnability, was disrupted when the PCL concentration was 100 or 150 g/L, resulting in the beaded fiber morphology [20]. Thereby, 125 g/L of PCL concentration prepared in a blended solvent (CH_3OH: $CHCl_3$ = 3:7, v/v) was adopted for the preparation of the card matrix.

Additionally, the influence of different electrospinning parameters, including spinning voltage, distance, and injection rate, on the fiber morphology was studied. The SEM images of the fibers prepared under different parameters and their relevant fiber diameter distributions are shown in Figures S3–S5. It can be seen that smooth and uniform fibers were produced when the voltage was 13 kV. Lower voltages result in the accumulation of the electrospinning solution at the tip of the needle due to the weak electrical force [21]. While a high applied voltage increases the rate at which a polymer filament is drawn out of the Taylor cone, resulting in greater fiber elongation and a reduced fiber diameter [22,23]. In addition, fiber morphology was improved with the increase in the flow rate, but further increases in the flow rate could obtain thicker fibers. The reason behind this phenomenon was that the flow rate generally determined the amount of electrospinning solution at the tip of the needle and a higher flow rate would produce larger droplets and result in a lower charge density required for electrostatic repulsive forces to overcome the surface tension of the solution and the extra solution cannot be stretched during the electrospinning process [24]. Similarly, good fiber morphology was achieved when the distance was 13 cm. Higher or lower than this value would change the electric field distribution, disturbing the balance between the electrical force and surface tension [25]. Overall, the appropriate parameters for preparing of the PCL fiber mat were a spinning voltage of 13 kV, distance of 13 cm, and injection rate of 2.5 mL/h. The average diameter of the fibers fabricated under this condition was 1.07 μm.

3.2. Hydrophilic Modification of the Matrix and Its Characterization

In this study, the detection card consisted of an enzyme (AChE) loaded fiber mat (AFM) and a substrate (IA) loaded fiber mat (IFM). The AFM was fabricated through the absorption of the AChE onto the fiber mat. Unfortunately, the absorption efficacy was low because of the hydrophobic property of the electrospun PCL fiber mat. A previous study indicated that the absorption efficiency of the enzyme could be remarkably enhanced with an increase in hydrophilicity [26,27]. Therefore, hydrophilic modification of the PCL fiber mat might be a proper way to improve the detection performance of the designed detection card. To our knowledge, there were different approaches for the hydrophilic modification of PCL materials [28–30]. Nevertheless, alkaline hydrolysis was considered here on account of its simple and efficient features and two corresponding hydrophilic modification models, shown in Figure 3, were systemically investigated in this part. It was found that the contact angle changed from 133.1° to 129.0°, demonstrating that the hydrophilicity of the fiber mat was improved by direct treatment with the NaOH solution. This result could be attributed to the breaking of the easter linkages along the polymer backbone when the PCL fiber mat was immersed in the NaOH solution, leading to hydroxyl and carboxylic acid groups being exposed at the polymer surface [31,32]. Similar hydrophilicity modifications through the hydrolysis by alkaline have been conducted for other hydrophobic polymer membranes [33–35]. Noteworthily, as depicted in Model 2, the contact angle of the PCL fiber mat changed dramatically from 131.1° to 0° when it was presoaked with ethanol, which indicated that the surface of the fiber mat suffered a transformation from hydrophobicity to hydrophilicity. This phenomenon can be ascribed to the fact that the ethanol pretreatment can improve the surface wettability of the fiber mat, as described in a previous study [36]. If the NaOH solution was directly applied for the modification, the interaction between the NaOH solution and the PCL fiber would be limited because of its hydrophobic surface. However, the fiber mat can be pre-infiltrated when it was treated with ethanol, which was more conducive to the interaction between the PCL molecule and NaOH. Additionally, it is noteworthy that the size of the fiber also decreased from the microfiber to nanofiber scale after modification, while maintaining overall fiber structure. This structure change is favorable in the immobilization of the enzyme. Hence, Model 2 was utilized to modify the hydrophobic fiber mat. The influence of the NaOH concentration and the related hydrolysis time on the properties of the fiber mat were subsequently investigated.

To further illustrate the hydrolysis behaviors of the PCL fiber mat through two different modification models, ATR-FTIR was employed to characterize the interactions that occurred in the PCL molecule by observing the changes of the specific peaks. The spectra of different fiber mats are displayed in Figure 4. The intense peaks at 1723 cm^{-1} that appeared in different samples represented the presence of the ester carbonyl group (-CO stretching) in PCL polymer. Additionally, peaks at 2865 and 2942 cm^{-1} were related to the asymmetric and symmetric CH_2 stretching. Apart from these characteristic peaks, there were two extra bands in the spectrum of modified fiber mat (model 2), one was at the wavenumber range of 1500–1700 cm^{-1} (C=O group) and the other was at the wavenumber range of 3250–3750 cm^{-1} (OH stretching vibrations). This can be explained that the presence of -OH (hydroxyl) functional group and -COOH (carboxyl) group in the PCL membrane after the hydrolysis of the ester carbonyl group [37]. Furthermore, it was concluded that Model 2 was more beneficial for the modification of the PCL fiber mat.

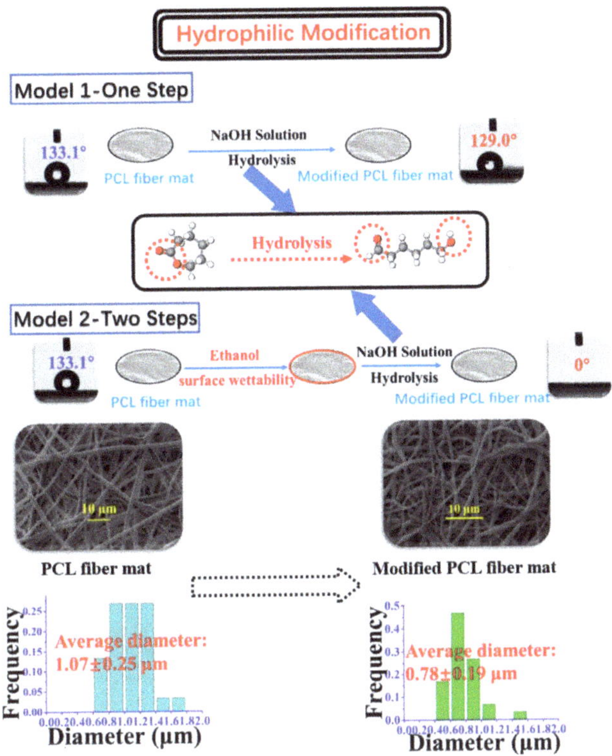

Figure 3. Two models used for the hydrophilic modification of PCL fiber mat and the properties of plain and modified fiber mat (n = 50).

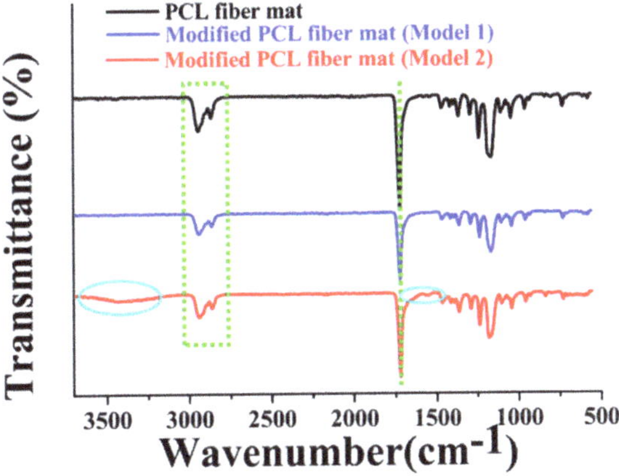

Figure 4. ATR–FTIR spectra of different samples.

From the hydrolysis results displayed in Figure 5, it is clear that the thickness of the fiber mat decreased after treatment with NaOH, and the surface of the fiber mat could successfully be changed from hydrophobic to hydrophilic when the concentration of the NaOH solution was above 2 M. During the modification process, the fiber surface suffered different degrees of hydrolysis. The higher the NaOH concentration, the higher the degree of fiber hydrolysis. During the hydrolysis procession, the contact angle was decreased, probably due to the increase in surface free energy [38]. When the specific surface free energy of the fiber mat was increased to a critical value, the droplets dripping onto the surface of the fiber mat could easily penetrate into the fiber mat. However, further improving the concentration of the NaOH solution, the thickness of the fiber mat was seriously decreased and became fragile. Herein, 2 M and 4 M were determined as the suitable modification concentration to obtain the hydrophilic matrix.

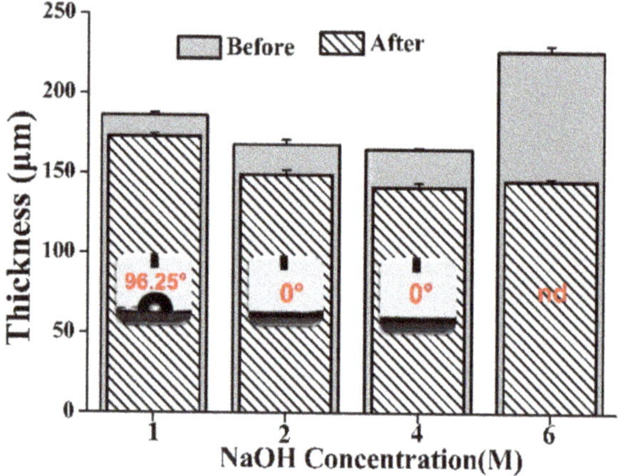

Figure 5. Effect of NaOH concentration on the thickness and contact angle of fiber mat before and after modification (nd, not determined, $n = 3$).

Apart from the concentration, the effect of the treatment time on the properties of the modification fiber mat was examined. As displayed in Figure 6, 39.4° and 73.2° of reduction in the contact angle of the fiber mats occurred when the fiber mat was treated with 2 M and 4 M NaOH solution for 0.5 h, respectively, and the obtained fiber mats were still hydrophobic. Meanwhile, a considerable change in the wettability occurred when the hydrolysis time was longer than 0.5 h and the contact angles of these two modified fiber mats were both 0°, which led to an understanding that the improvement of the hydrophilicity of the modified fiber mat could be achieved by postponing the hydrolysis time. As the degree of fiber hydrolysis increased, the polar group hydroxyl and carboxyl content on the surface gradually increased. Additionally, many dips appeared on the fiber surface, which was in accordance with another similar study [39]. The increase in surface roughness may contribute to the hydrophilic property of the fiber mat [40]. Hence, appropriate modification was performed with the 2 M NaOH solution for 1 h.

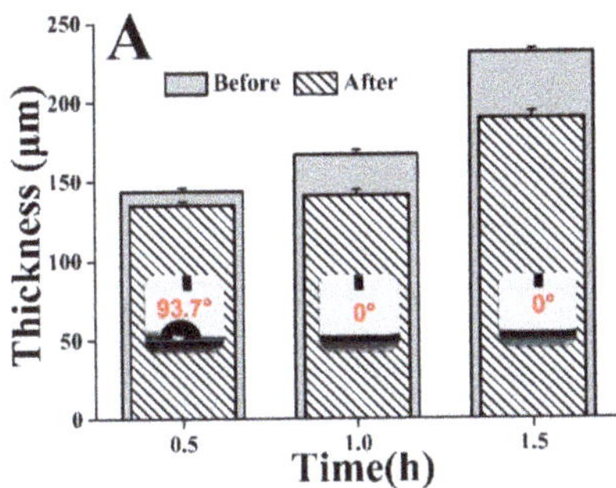

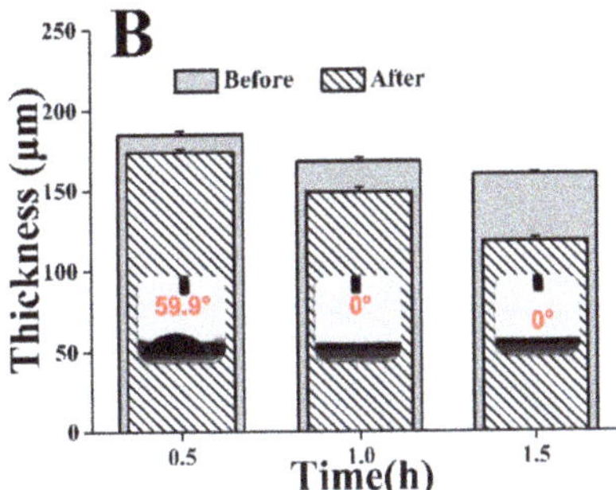

Figure 6. Effect of NaOH concentration and hydrolysis time on the thickness and contact angle of fiber mat before and after modification ((**A**), 2 M NaOH; (**B**), 4 M NaOH; $n = 3$).

3.3. Fabrication and Measurement Conditions of the Detection Card

The rationale for this detection card is based on the hydrolysis of the substrate (IA) by the acetylcholinesterase (AChE) to produce indole phenol, which then becomes blue indigo due to the oxidization. In contrast, the card will be colorless if the AFM is treated with a positive sample (pesticide) as the enzyme activity can be inhibited by the pesticide. Here, to obtain a reliable detection approach, it is essential to investigate the influence of the two vital factors (absorption time and enzyme/substrate concentration) on the detection performance of this card and eventually to figure out the appropriate synthesis conditions. Here, the detection result could be obtained by simply distinguishing the color of the detection card with the naked eye. As shown in Figure 7, it appeared that the suitable immobilization time and concentration for the AChE were 24 h and 5 mg/mL, respectively. Less immobilization time or a lower concentration of AChE would result in a false-positive result. Likewise, 12 h and 5 mg/mL were found to be the proper time and concentration for the immobilization of IA to obtain a reliable detection result. On the other hand, considering the absorption efficiency of the AFM, the absorption amount of AChE for the commercial filter papers, plain and modified PCL fiber mats were determined and compared. As shown in Figure 8, the modified fiber mats showed a significantly higher absorption efficiency than the other two commercial applied filter papers ($n = 3$, $p < 0.05$), which may be attributable to the fact that the special nano-structure was more beneficial for the immobilization of the enzymes. More importantly, the fiber mat exhibited an improved absorption efficiency after the hydrophilicity modification, and a nearly 2-fold enhancement of the absorption amount was achieved. The probable reason was that the hydrophobicity property could hamper the interaction between the enzyme and the matrix. Moreover, the fiber morphology changing from microfiber to nanofiber after the modification might be another vital contributor to the improved loading efficiency of the enzyme. Furthermore, the pits on the surface of the modified fiber would naturally increase the specific area, thus, promoting the immobilization of the enzyme.

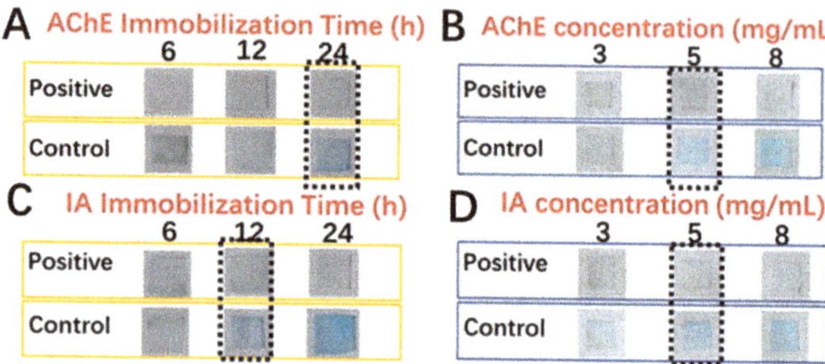

Figure 7. Color development of the detection card prepared under different conditions (**A**), AChE concentration 5 mg/mL, IA concentration 8 mg/mL, IA immobilization time 24 h; (**B**), AChE immobilization time 24 h, IA concentration 8 mg/mL, IA immobilization time 24 h; (**C**), AChE concentration 5 mg/mL, AChE immobilization time 24 h, IA concentration 8 mg/mL; AChE immobilization time 24 h; (**D**), AChE concentration 5 mg/mL, AChE immobilization time 24 h, IA immobilization time 12 h).

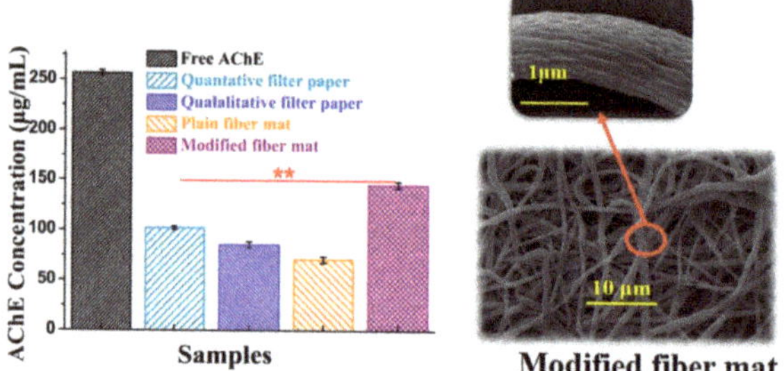

Figure 8. Absorption amounts of AChE for different matrixes and the surface morphology of modified fibers ($n = 3$, **, $p < 0.01$).

In addition to the card matrix preparation, the measurement parameters, mainly including enzyme inhibition time and color development time, are particularly important for the precise determination of the pesticides. It was known that the substrate IA could be decomposed into indole phenol by the catalysis of AChE, and then blue indigo would be formed after oxidization. In contrast, the inhibition of AChE activity by the pesticide would result in the colorless appearance of the detection card. The inhibition time, referred to as the reaction time between the sample and AFM and the color development time, indicated the time that AFM with IFM. Generally, the inhibition time has a considerable influence on color development. If the inhibition time was too short, the enzyme could not react with the sample sufficiently, leading to a negative result. Nevertheless, if the inhibition time was too long, it would be time-consuming and decrease productivity. On the other hand, time was required for AFM to react with IFM in order to develop the distinct blue color. Similarly, a shorter or longer color development time would limit the real-life application of the detection card. For this reason, the exploration work of these two factors was conducted, and the related results were displayed in Figure 9. The suiFireaction condition was found to be an inhibition time 10 min and a color development time 10 min, which was superior to the commercial detection card.

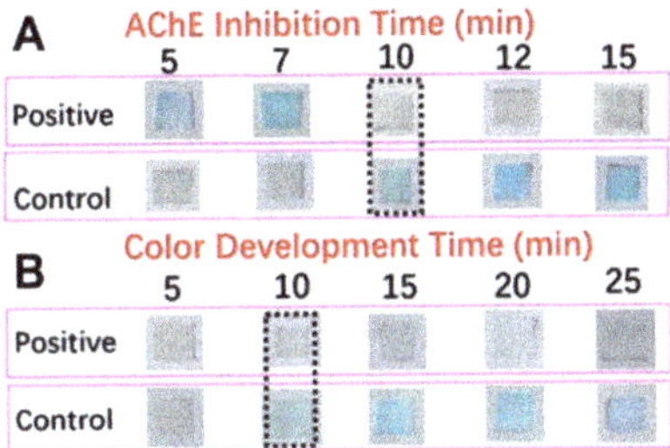

Figure 9. Color development of the detection card prepared under different conditions (**A**), color development time 20 min; (**B**), AChE inhibition time 10 min).

3.4. Detection Performance of the Detection Card and Its Real-Life Application

Nowadays, consumers would like to select foods with fewer pesticides. Pesticide detection with high sensitivity seems to be the preferred strategy and would have great application potential in the food and agricultural industry. To check the detectability of this card, four pesticides representing two main types of pesticide, organophosphorus and carbamate, were selected to be tested. Additionally, the minimum detectable concentration of this card for each pesticide was compared with its corresponding limit of detection (LOD) value specified by SAC [41]. Results depicted in Table 1 elucidated that this nano/micro-structured detection card exhibited a comparative value with the corresponding LOD values. It is noteworthy that 5-fold, 2-fold, and 1.5-fold reductions in the detectable concentrations of the rapid detection card for carbofuran, malathion, and trichlorfon were obtained compared to the national standard value, suggesting that this developed card could be able to detect these types of pesticides.

Table 1. The minimum detectable concentrations (MDC) of the card for different pesticides and the corresponding limit of detection (LOD) values specified by Standardization Administration of China (SAC).

Pesticide		LOD (mg/L)	MDC (mg/L)	Color Development			
Organophosphorus	Trichlorfon	0.3	0.2	0	0.1	0.2	0.3
	Malathion	2.0	1.0	0	0.5	1.0	2.0
Carbamate	Carbaryl	2.5	2.5	0	1.0	2.0	2.5
	Carbofura	0.5	0.1	0	0.05	0.1	0.5

The shelf life of the as-prepared detection card is another important factor that influences its real-life applications. In this study, the storage stability of this detection card under 4 °C and RT were investigated. As shown in Figure 10, all the detection cards in PBS groups had a blue color, while other detection cards were colorless, demonstrating that this detection card had good stability. It could effectively detect the pesticide even though they were stored for 60 days. Hence, this desirable property of the self-made card also contributes to its real-life application.

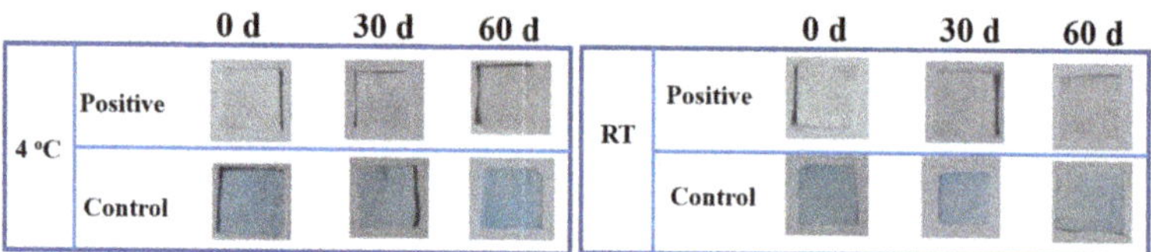

Figure 10. Storage stability of the detection card under 4 °C and room temperature (RT).

To verify the real-life performance and practicability of this self-made detection card, pesticide residues on cabbage and carrot were measured according to the above-mentioned conditions, and comparisons were also performed with the commercial detection card. As shown in Table 2, the commercial card cannot detect the malathion even at its concentration of 20 µg/mL, since the control group still showed a blue color. However, the obtained detection card could accurately detect 5 µg/mL malathion, indicating the existence of pesticide on the cabbage. Similarly, the detection results for the carrot also revealed that this self-made detection card had a better detectability than the commercial card. The card with a nano/micro-structure could determine pesticides on the real vegetables accurately and efficiently. Therefore, this study opens up a new way to develop the pesticide rapid detection vehicle, promoting the sustainable development of the agricultural and food industry.

Table 2. Color development results for real-life food detection.

Sample	Detection Card	Spraying Concentration of Malathion (µg/mL)			
		0	5	10	20
Cabbage	Commercial				
	NMF				
Carrot	Commercial				
	NMF				

Note: NMF, nano/micro-structured detection card.

4. Conclusions

To satisfy the improved requirement of safety food, a novel nano/micro-structured pesticide detection card was creatively fabricated by taking the electrospun fiber mat as the card matrix. This detection card has good storage stability and a low minimum detectable concentration, the application of which involves a short reaction time, simple operation, and minimum use of human and material resources as compared to the traditional detection approaches. In addition, experiments with real samples also demonstrated its feasibility and superiority on pesticide detection over the commercial card. More importantly, this detection method can even be performed by nonprofessional individuals, making it more in line with the market requirement. Therefore, the present study offers a new route for designing a rapid detection card for pesticides, which could promote the application of the electrospinning technique and nano/micro material in the agricultural and food industry.

Supplementary Materials: The following are available online at https://www.mdpi.com/article/10.3390/foods10040889/s1, Figure S1: SEM images of electrospun fibers at different ratios of CH_3OH to $CHCl_3$ (the electrospinning conditions were: voltage of 13 kV, flow rate of 2.5 mL/h and distance of 13 cm); Figure S2: SEM images of electrospun fibers at different concentrations of PCL (the electrospinning conditions were: voltage of 13 kV, flow rate of 2.5 mL/h and distance of 13 cm); Figure S3: SEM images and fiber diameter distributions of PCL fibers under different electrospinning voltages (the electrospinning conditions were: flow rate of 2.5 mL/h and distance of 13cm); Figure S4: SEM images and fiber diameter distributions of electrospun PCL fibers under different flow rates (the electrospinning conditions were voltage 13 kV and distance 13 cm); Figure S5: SEM images and fiber diameter distributions of electrospun PCL fibers under different distances (the electrospinning conditions were: voltage of 13 kV and flow rate of 2.5 mL/h); Table S1: Effect of PCL concentration and volume ratio of $CHCl_3$ to CH_3OH on the properties of electrospinning solution.

Author Contributions: Conceptualization, K.F., M.-Y.Z.; formal analysis, K.F., M.-Y.Z. and Y.-S.W.; investigation, K.F., M.-Y.Z.; visualization, K.F.; writing—original draft, K.F.; writing—review and editing, M.-H.Z., H.W. and S.-Y.H. All authors have read and agreed to the published version of the manuscript.

Funding: This research was funded by the Key Area R&D Program of Guangdong Province (No. 2019B020211002), the Collaborative Innovation Center for Sports of Guangdong Province (No. 2019B110210004), the Science and Technology Project of Guangzhou City (No. 201804010151), and the Natural Science Foundation of Guangdong Province (No. 2017A030313148).

Data Availability Statement: Data are contained within the article or supplementary material.

Acknowledgments: We acknowledge the Key Area R&D Program of Guangdong Province (No. 2019B020211002), the Collaborative Innovation Center for Sports of Guangdong Province (No. 2019B110210004), the Science and Technology Project of Guangzhou City (No. 201804010151), and the Natural Science Foundation of Guangdong Province (No. 2017A030313148) for financial support.

Conflicts of Interest: The authors declare no conflict of interest.

References

1. Huang, Y.S.; Shi, T.; Luo, X.; Xiong, H.L.; Min, F.F.; Chen, Y.; Nie, S.; Xie, M.Y. Determination of multi-pesticide residues in green tea with a modified QuEChERS protocol coupled to HPLC-MS/MS. *Food Chem.* **2019**, *275*, 255–264. [CrossRef] [PubMed]
2. Rimayi, C.; Odusanya, D.; Mtunzi, F.; Tsoka, S. Alternative calibration techniques for counteracting the matrix effects in GC–MS-SPE pesticide residue analysis—A statistical approach. *Chemosphere* **2015**, *118*, 35–43. [CrossRef] [PubMed]
3. Wang, X.; Mu, Z.D.; Shangguan, F.Q.; Liu, R.; Pu, Y.P.; Yin, L.H. Rapid and sensitive suspension array for multiplex detection of organophosphorus pesticides and carbamate pesticides based on silica–hydrogel hybrid microbeads. *J. Hazard. Mater.* **2014**, *273*, 287–292. [CrossRef] [PubMed]
4. Samsidar, A.; Siddiquee, S.; Shaarani, S.M. A review of extraction, analytical and advanced methods for determination of pesticides in environment and foodstuffs. *Trends Food Sci. Technol.* **2018**, *71*, 188–201. [CrossRef]
5. Nowicka, A.B.; Czaplicka, M.; Kowalska, A.A.; Szymborski, T.; Kami´nska, A. Flexible PET/ITO/Ag SERS Platform for Label-Free Detection of Pesticides. *Biosensors* **2019**, *9*, 111. [CrossRef]
6. Song, Y.G.; Shan, B.X.; Feng, B.W.; Xu, P.F.; Zeng, Q.; Su, D. A novel biosensor based on ball-flower-like Cu-hemin MOF grown on elastic carbon foam for trichlorfon detection. *RSC Adv.* **2018**, *8*, 27008–27015. [CrossRef]
7. Xu, B.B.; Guo, J.C.; Fu, Y.S.; Chen, X.Y.; Guo, J.H. A review on microfluidics in the detection of food pesticide residues. *Electrophoresis* **2020**, *41*, 821–832. [CrossRef]
8. Arduini, F.; Cinti, S.; Scognamiglio, V.; Moscone, D. Nanomaterials in electrochemical biosensors for pesticide detection: Advances and challenges in food analysis. *Microchim. Acta* **2016**, *183*, 2063–2083. [CrossRef]
9. Ayi, A.A.; Ashishie, P.B.; Khansi, E.E.; Ogar, J.O.; Anyama, C.A.; Inah, B.E. Nano/micro-structured materials: Synthesis, morphology and applications. In *Nanostructures*; IntechOpen: Rijeka, Croatia, 2020.
10. Wu, J.; Wang, N.; Zhao, Y.; Jiang, L. Electrospinning of multilevel structured functional micro-/nanofibers and their applications. *J. Mater. Chem. A* **2013**, *1*, 7290–7305. [CrossRef]

11. Ahmed, J. Electrospinning for the manufacture of biosensor components: A mini-review. *Med. Devices Sens.* **2020**, *4*, e10136.
12. Zhu, S.L.; Nie, L.H. Progress in fabrication of one-dimensional catalytic materials by electrospinning technology. *J. Ind. Eng. Chem.* **2020**, *93*, 28–56. [CrossRef]
13. Zhai, M.Y.; Feng, K.; Hu, T.G.; Zong, M.H.; Wu, H. Development of a novel nano-based detection card by electrospinning for rapid and sensitive analysis of pesticide residues. *J. Sci. Food Agric.* **2020**, *100*, 4400–4408. [CrossRef]
14. Li, D.W.; Wang, Q.Q.; Huang, F.L.; Wei, Q.F. Electrospun nanofibers for enzyme immobilization. In *Electrospinning: Nanofabrication and Applications*, 1st ed.; Ding, B., Wang, X.F., Yu, J.Y., Eds.; William Andrew: New York, NY, USA, 2019; pp. 756–781.
15. Lu, P.; Murray, S.; Zhu, M. Electrospun nanofibers for catalysts. In *Electrospinning: Nanofabrication and Applications*, 1st ed.; Ding, B., Wang, X.F., Yu, J.Y., Eds.; William Andrew: New York, NY, USA, 2019; pp. 695–717.
16. Sunandar, M.; Yulianti, E.; Deswita, D.; Sudaryanto, S. Study of solid polymer electrolyte based on biodegradable polymer polycaprolactone. *Malays. J. Fundam. Appl. Sci.* **2019**, *15*, 467–471. [CrossRef]
17. Chen, J.Y.; Su, C.Y.; Hsu, C.H.; Zhang, Y.H.; Zhang, Q.C.; Chang, C.L.; Hua, C.C.; Chen, W.C. Solvent effects on morphology and electrical properties of poly (3-Hexylthiophene) electrospun nanofibers. *Polymers* **2019**, *11*, 1501. [CrossRef]
18. Semiromi, F.B.; Nejaei, A.; Shojaee, M. Effect of methanol concentration on the morphology and wettability of electrospun nanofibrous membranes based on polycaprolactone for oil-water separation. *Fiber. Polym.* **2019**, *20*, 2453–2460. [CrossRef]
19. Tungprapa, S.; Puangparn, T.; Weerasombut, M.; Jangchud, I.; Fakum, P.; Semongkhol, S.; Meechaisue, C.; Supaphol, P. Electrospun cellulose acetate fibers: Effect of solvent system on morphology and fiber diameter. *Cellulose* **2007**, *14*, 563–575. [CrossRef]
20. Jun, Z.; Hou, H.Q.; Schaper, A.; Wendorff, J.H.; Greiner, A. Poly-L-lactide nanofibers by electrospinning–Influence of solution viscosity and electrical conductivity on fiber diameter and fiber morphology. *e-Polymers* **2003**, *3*, 009.
21. Fallahi, D.; Rafizadeh, M.; Mohammadi, N.; Vahidi, B. Effect of applied voltage on jet electric current and flow rate in electrospinning of polyacrylonitrile solutions. *Polym. Int.* **2008**, *57*, 1363–1368. [CrossRef]
22. Zhang, C.W.; Liu, D.D.; Xu, C.; Liu, D.Z.; Wang, B. Tensile properties of ag-EVOH electrospinning nanofiber mats for large muscle scaffolds. *Mech. Adv. Mater. Struc.* **2020**, *27*, 1312–1321. [CrossRef]
23. Ballengee, J.B.; Pintauro, P.N. Morphological control of electrospun Nafion nanofiber mats. *J. Electrochem. Soc.* **2011**, *158*, B568–B572. [CrossRef]
24. Eslamian, M.; Khorrami, M.; Yi, N.; Majd, S.; Abidian, M.R. Electrospinning of highly aligned fibers for drug delivery applications. *J. Mater. Chem. B* **2019**, *7*, 224–232. [CrossRef]
25. Chowdhury, M.; Stylios, G.K. Analysis of the effect of experimental parameters on the morphology of electrospun polyethylene oxide nanofibres and on their thermal properties. *J. Text. Inst.* **2012**, *103*, 124–138. [CrossRef]
26. Liu, X.J.; Chen, W.W.; Lian, M.L.; Chen, X.; Lu, Y.L.; Yang, W.S. Enzyme immobilization on ZIF-67/MWCNT composite engenders high sensitivity electrochemical sensing. *J. Electroanal. Chem.* **2019**, *833*, 505–511. [CrossRef]
27. Wang, Q.Q.; Peng, L.; Du, Y.Z.; Xu, J.; Cai, Y.B.; Feng, Q.; Huang, F.L.; Wei, Q.F. Fabrication of hydrophilic nanoporous PMMA/O-MMT composite microfibrous membrane and its use in enzyme immobilization. *J. Porous Mater.* **2013**, *20*, 457–464. [CrossRef]
28. Zhao, L.Q.; Ma, S.Y.; Pan, Y.W.; Zhang, Q.Y.; Wang, K.; Song, D.M.; Wang, X.X.; Feng, G.W.; Liu, R.M.; Xu, H.J.; et al. Functional modification of fibrous PCL scaffolds with fusion protein VEGF-HGFI enhanced cellularization and vascularization. *Adv. Healthc. Mater.* **2016**, *5*, 2376–2385. [CrossRef]
29. Ghaee, A.; Bagheri-Khoulenjani, S.; Afshar, H.A.; Bogheiri, H. Biomimetic nanocomposite scaffolds based on surface modified PCL-nanofibers containing curcumin embedded in chitosan/gelatin for skin regeneration. *Compos. Part B Eng.* **2019**, *177*, 107339. [CrossRef]
30. Razmshoar, P.; Bahrami, S.H.; Akbari, S. Functional hydrophilic highly biodegradable PCL nanofibers through direct aminolysis of PAMAM dendrimer. *Int. Polym. Mater. Polym. Biomater.* **2020**, *69*, 1069–1080. [CrossRef]
31. Bosworth, L.A.; Hu, W.X.; Shi, Y.N.; Cartmell, S.H. Enhancing biocompatibility without compromising material properties: An optimised NaOH treatment for electrospun polycaprolactone fibres. *J. Nanomater.* **2019**, *2019*, 4605092. [CrossRef]
32. Wang, W.G.; Caetano, G.; Ambler, W.S.; Blaker, J.J.; Frade, M.A.; Mandal, P.; Diver, C.; Bártolo, P. Enhancing the hydrophilicity and cell attachment of 3D printed PCL/Graphene scaffolds for bone tissue engineering. *Materials* **2016**, *9*, 992. [CrossRef]
33. Sadeghi, A.R.; Nokhasteh, S.; Molavi, A.M.; Khorsand-Ghayeni, M.; Naderi-Meshkin, H.; Mahdizadeh, A. Surface modification of electrospun PLGA scaffold with collagen for bioengineered skin substitutes. *Mater. Sci. Eng. C Mater.* **2016**, *66*, 130–137. [CrossRef]
34. Liu, Q.F.; Wang, Q.; Yan, S.H. Modification of PVDF membranes by alkaline treatment. *Appl. Mech. Mater.* **2013**, *316–317*, 1033–1036. [CrossRef]
35. Abedi, M.; Sadeghi, M.; Chenar, M.P. Improving antifouling performance of PAN hollow fiber membrane using surface modification method. *J. Taiwan Inst. Chem. E* **2015**, *55*, 42–48. [CrossRef]
36. Boo, C.; Lee, J.; Elimelech, M. Omniphobic polyvinylidene fluoride (PVDF) membrane for desalination of shale gas produced water by membrane distillation. *Environ. Sci. Technol.* **2016**, *50*, 12275–12282. [CrossRef] [PubMed]
37. Augustine, R.; Kalarikkal, N.; Thomas, S. Effect of zinc oxide nanoparticles on the in vitro degradation of electrospun polycaprolactone membranes in simulated body fluid. *Int. J. Polym. Mater. Polym. Biomater.* **2016**, *65*, 28–37. [CrossRef]

38. Prakash, C.J.; Raj, C.C.; Prasanth, R. Fabrication of zero contact angle ultra-super hydrophilic surfaces. *J. Colloid Interface Sci.* **2017**, *496*, 300–310. [CrossRef] [PubMed]
39. Yeo, A.; Wong, W.J.; Teoh, S.H. Surface modification of PCL-TCP scaffolds in rabbit calvaria defects: Evaluation of scaffold degradation profile, biomechanical properties and bone healing patterns. *J. Biomed. Mater. Res. A* **2010**, *93*, 1358–1367. [CrossRef] [PubMed]
40. Li, H.B.; Shi, W.Y.; Zeng, X.H.; Huang, S.F.; Zhang, H.X.; Qin, X.H. Improved desalination properties of hydrophobic GO-incorporated PVDF electrospun nanofibrous composites for vacuum membrane distillation. *Sep. Purif. Technol.* **2020**, *230*, 115889. [CrossRef]
41. SAC. Rapid determination for organophosphate and carbamate pesticide residues in vegetables. In *GB/T 5009.199–2003*; Standardization Administration of China: Beijing, China, 2003.

Review

Current Progress in the Utilization of Soy-Based Emulsifiers in Food Applications—A Review

Lingli Deng

College of Biological Science and Technology, Hubei Key Laboratory of Biological Resources Protection and Utilization, Key Laboratory of Green Manufacturing of Super-Light Elastomer Materials of State Ethnic Affairs Commission, Hubei Minzu University, Enshi 445000, China; lingli0312@gmail.com; Tel.: +86-0718-8438-945

Abstract: Soy-based emulsifiers are currently extensively studied and applied in the food industry. They are employed for food emulsion stabilization due to their ability to absorb at the oil–water interface. In this review, the emulsifying properties and the destabilization mechanisms of food emulsions were briefly introduced. Herein, the effect of the modification process on the emulsifying characteristics of soy protein and the formation of soy protein–polysaccharides for improved stability of emulsions were discussed. Furthermore, the relationship between the structural and emulsifying properties of soy polysaccharides and soy lecithin and their combined effect on the protein stabilized emulsion were reviewed. Due to the unique emulsifying properties, soy-based emulsifiers have found several applications in bioactive and nutrient delivery, fat replacer, and plant-based creamer in the food industry. Finally, the future trends of the research on soy-based emulsifiers were proposed.

Keywords: emulsions; soy protein; soy polysaccharide; soy lecithin; delivery

Citation: Deng, L. Current Progress in the Utilization of Soy-Based Emulsifiers in Food Applications—A Review. *Foods* **2021**, *10*, 1354. https://doi.org/10.3390/foods10061354

Academic Editors: Hong Wu and Hui Zhang

Received: 20 May 2021
Accepted: 11 June 2021
Published: 13 June 2021

1. Introduction

The colloidal mixture of two immiscible liquids gives rise to an emulsion where one liquid is dispersed in another liquid. The solvent or substance that tends to form droplets in the emulsion is defined as the dispersed, while the liquid that surrounds the droplet is defined as the external or continuous phase. An emulsifying agent is a surface-active molecule with the ability to be adsorbed at the oil–water interface of the newly formed emulsion during the emulsification and protects the droplets from immediate coalescence. A surface-active molecule can be considered as a proper emulsifier if it possesses the ability towards frequent adsorption at the oil-water or air–water interface and causes a decrease in interfacial tension. There are low-molecular weight emulsifiers (e.g., soy lecithin) and high-molecular weight emulsifiers (e.g., soy protein and polysaccharides) in soy components. Low molecular weight emulsifiers decrease the surface or interfacial tension to a greater extent than the proteins or polysaccharides due to the partitioning of the entire molecule between the two phases. However, the proteins and polysaccharides would undergo a conformational change at the interface and there might be no clear definition of hydrophilic or hydrophobic groups, resulting in relatively high surface tension compared to that of the low molecular emulsifiers. Although low molecular weight emulsifiers are more effective in emulsion formation, the high molecular weight emulsifiers are more effective in the formation of the viscoelastic film around the oil droplets, which favors the stabilization of the food emulsions. In 2019, the world soybean production was 336.5 million metric tons, which accounted for approximately 58.9% of the world's oilseed production. Soybeans contain about 40% protein and 20% oil and are considered to be a major source of proteins and oils, accompanied by the production of natural surfactants: soy protein, soy lecithin, and soy polysaccharides. Soy protein is the most popular plant protein source to serve as an ingredient in food formulation. It should be noticed that the functional properties of soy-based emulsifiers might be different among the ingredients from different suppliers.

However, the functional properties (e.g., solubility) of commercial soy-based emulsifiers depend on the extraction and processing method utilized for its preparation.

2. Emulsifying Properties

Two aspects of emulsifiers are commonly studied in literature during the consideration of their emulsifying potential. These aspects include emulsifying activity and stability. Basic information regarding their emulsifying potential can be acquired by the determination of their efficacy to reduce the interfacial tension of emulsion [1]. Typically, interfacial tension is measured as a function of the increasing level of emulsifier and then calculated the surface tension versus concentration profile of emulsifier. However, the electrical properties of emulsifiers may have an impact on the stability and performance of an emulsion. Although various methods are available to measure the electrical properties of emulsifiers, micro-electrophoresis is considered the simplest and most widely used method. The important aspect of emulsifying properties includes their emulsifying activity, which can be assessed by two comparatively indexes: emulsifying capacities and emulsifying activity index. The emulsifying capacity of an emulsifier is described as the maximum oil content that can be emulsified in an aqueous medium containing a specific amount of the emulsifier without reverting or breaking down the emulsion, while the emulsifying activity index is described as the amount of oil that can be emulsified per unit emulsifier.

Various analytical techniques and protocols are present for the investigation of emulsion stability in different aspects [2]. Among all, the emulsion stability index (ESI) established by Pearce and Kinsella [3], is a widely used and simplest method based on the analysis of the turbidity of the diluted emulsion at a specific wavelength. The phenomena affecting the emulsion stability mainly include coalescence, flocculation, creaming, sedimentation, Ostwald ripening, and phase inversion (Figure 1). Density-driven gravitation separation is the most easily observed emulsion instability phenomena in food emulsion storage, including the upward movement induced creaming and the downward movement induced sedimentation. Coalescence is the merging process of small emulsion droplets into larger emulsion droplets, resulting in the formation of a distinct layer of oil on the surface (oiling-off). Flocculation phenomena are also derived from the aggregation of emulsion droplets, while the droplets would not merge but remain as individual droplets. Phase inversion is a process that the dispersed phase becomes the continuous phase and vice versa, which has also been applied as an emulsification method to make fine emulsions. Ostwald ripening is the growth of one emulsion droplet at the expense of the smaller one due to the difference in chemical potential of the emulsion droplets. The destabilization of the soy-based stabilized emulsions is usually affected by the nature of the molecules (three-dimensional structure, hydrophobicity, charge, solubility, molecular flexibility) and conditions of the environment (e.g., pH and ionic strength) [4].

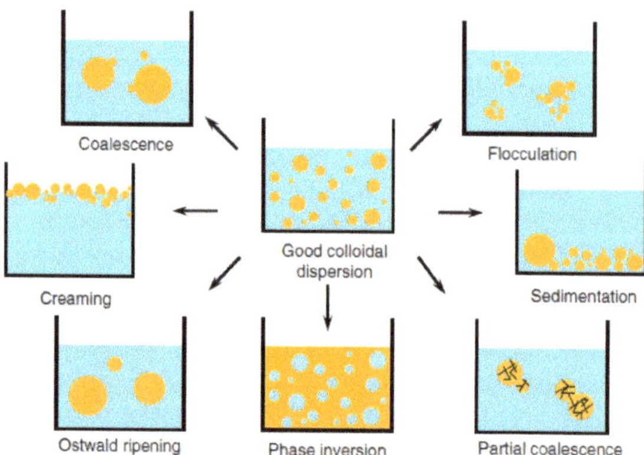

Figure 1. Mechanism for the destabilization of emulsion system [5].

3. Soy Proteins

Soy proteins are mainly composed of 40% 7S (β-conglycinin) and 30% 11S (glycinin) fractions. The 7S globulin has been shown to have better emulsifying properties than the 11S globulin, which is due to the 11S globulin's more stable oligomeric form. Rivas and Sherman [6] found that 7S formed a stronger viscoelastic film at the oil–water interface than 11S at all the tested pH and NaCl concentrations. They concluded that the 7S globulin molecules have a higher degree of intra- and intermolecular cohesion, resulting in more organized films at the interface. Glycinin cannot adsorb quickly to the air–water interface due to its low surface hydrophobicity, large molecular size, and low molecular flexibility. Liu et al. [7] observed that the acidic subunits of glycinin adsorb more quickly to the air–water interface than raw glycinin. Soy protein acts as a macromolecular surfactant to stabilize oil-in-water emulsion systems in food products such as sausages, ice cream, yogurt, and coffee whitener [8]. It has been proposed that the surface hydrophobicity and solubility are the major factors determining emulsifying activity, while the molecular flexibility of the proteins is important for emulsion stability [9]. However, the solubility of soy proteins was limited in the food matrix due to the isoelectric point being around 5.0. Moreover, the globular structure of the soy proteins prevents the exposure of the hydrophobic residues and retards the conformational change of protein when adsorbing at the interface, preventing them from becoming more efficient at reducing interfacial tension [10]. In recent years, soy proteins are considered effective starting materials to produce functional emulsifiers with modified physico-chemical properties by using different enzymatic, physical, chemical, and biological methods [11].

3.1. Enzyme Treatment

Enzyme treatment has been an effective method to modify the functional properties of proteins due to the advantages of high specificity and tunable properties of the hydrolyzed peptides. Enzymatic hydrolysis could reduce the molecular weight and expose the groups (hydrophobic or hydrophilic) buried in the globular structure of the soy protein, which might elevate the emulsifying properties. The tunable functional properties make the hydrolyzed soy proteins important ingredients in the food industry [10]. In a recent report, three different enzymes were reported namely flavorzyme, alcalase, and protamex to hydrolyze SPI to obtain hydrolysate with degrees of hydrolysis of 3%, 7%, and 11% [12]. Flavorzyme was found to induce soy protein nanoparticles (SPNPs) at all DH, whereas protamex showed a limited effect on the formation of SPNPs. The SPNPs attached more rapidly at the oil–water interface as compared to native SPI, indicating the good emulsifying potential of the emulsifier. The peptides with decreased molecules show the advantage of higher mobility when absorbing at the interface, which might also be a disadvantage in maintaining the emulsion stability. Therefore, even though most of the previous studies indicated that that enzymatic hydrolysis improved some of the functional properties of soy proteins, the degree of hydrolysis is the critical parameter for enzyme modification [13]. Enzyme treatment could not only hydrolyze the protein, but also crosslink protein to modify the emulsifying properties. Zang et al. [14] hydrolyzed soy protein by papain with a hydrolysis degree of 6% and then reacted with transglutaminase (TGase) to partial crosslink soy protein hydrolysate. Compared with the raw SPI, transglutaminase treated SPI and SPI hydrolysate, emulsion stabilized by transglutaminase treated SPI hydrolysate exhibited the lowest particle size, creaming index, flocculation degree, coalescence degree, and highest freeze–thaw stability.

3.2. Thermal Treatment

The conformation and aggregation behavior of proteins are greatly influenced by heat treatment, resulting in modified functional properties [15]. The protein molecules tend to unfold the globular structure and expose the hydrophobic groups. It has been reported that the emulsions stabilized by 75 °C; heated soy protein showed lower emulsion droplet size compare to the unheated soy protein. The heated soy protein showed a higher extent of adsorption at the interface, with a higher ratio of β-conglycinin among the absorbed protein. The acidic and basic subunits of glycinin remained in the serum phase in similar amounts in these emulsions as in the unheated samples. Heat-induced dissociation of β-conglycinin has previously been shown to result in the formation of soluble complexes [16,17]. However, high-temperature treatment (95 °C for 15 min) might cause dissociation and denaturation of both β-conglycinin and glycinin, resulting in the formation of soluble aggregates between the basic subunit of glycinin and the β-subunits of β-conglycinin. The majority of the aggregates formed during heating adsorb at the interface during homogenization, which could explain the decreased emulsion stability [15].

3.3. Non-Thermal Processing

Non-heat processing (such as ultrasonication and high pressures) has long been researched to modify the functional property of soy proteins. Ultrasound is well-known energy and time-saving technique. The use of ultrasound in various industries is becoming more common. Ultrasound destroys noncovalent interactions and disulfide bonds through thermal, mechanical, and chemical effects [18], thus causing protein subunits to dissociate and aggregate, resulting in the modification of solubility, emulsifying, foaming, and gelation properties [19]. Jambrak et al. [20] found that ultrasound treatment (20 kHz) increased the emulsifying and foaming ability of soy proteins, which might be ascribed to the higher adsorption at the emulsion droplet interface of the denatured soy protein [10]. It has been revealed that ultrasound treatment increased the surface hydrophobicity and zeta potential value of soy-protein isolate–citrus-pectin electrostatic complexes, and significantly decreased emulsion droplet size was observed. The results showed that ultrasonic cavitation effects changed the structure of both biomacromolecules and increased the electrostatic interactions between soy-protein isolate and citrus pectin, both of which led to the complex's improved emulsifying properties [21]. Albano et al. [22] have also observed reduced particle size in the soy protein and pectin complex stabilized emulsion after ultrasonication. Ren et al. [23] compared the effects of ultrasonic cavitation treatment and hydrodynamic cavitation on the functional properties of soy protein isolate. The particle size and viscosity of SPI were reduced and the surface hydrophobicity was increased after ultrasonic cavitation or hydrodynamic cavitation treatment, resulting in improved solubility, emulsifying activity index, emulsion stability index, and foaming capacity. However, significantly decreased foam stability was observed after ultrasonic cavitation and hydrodynamic cavitation treatment, which might be ascribed to the weaker protein-protein interaction as reflected by the decreased viscosity.

When proteins are subjected to high pressures (HP), it is known that protein molecules would undergo conformational changes, which may lead to modification of the emulsifying properties. Several studies have been conducted to see whether high-pressure treatment can be applied to change the emulsifying properties of soy proteins. Molina, Papadopoulou [9] reported that the emulsifying activity was increased by high-pressure treatment at neutral, while no improvement was observed for the emulsifying stabilities and solubilities. They observed the highest emulsifying activity indexes were 400 MPa and 200 MPa for β-conglycinin and glycinin, respectively. Puppo et al. [24] compared the effect of high pressure on emulsifying properties of soy protein isolate at acidic (pH 3.0) and alkaline (pH 8.0) conditions. High-pressure treatment (200 MPa) of SPI at alkaline condition induced a reduction of droplet size and an increase of depletion flocculation. High-pressure treatment induced a significant increase of adsorbed proteins at the oil–water interface at both pH conditions. Torrezan et al. [25] found in both the low pH range (2.66–4.34) and the near-neutral range (5.16–6.84), increasing the soy protein concentration (0.32–3.68%) caused a reduction in emulsifying activity index. In acidic conditions, the emulsifying activity index was higher at low-pressure treatments whereas, in the near-neutral pH range, the best emulsifying activity was at the middle range of pressure treatment. The critical effect of the initial protein concentration was confirmed by Wang et al. [26], who observed higher EAI value when lower soy protein concentration (1%) was applied for high-pressure treatment.

Apart from some common non-heating processes that have been applied for modification of the emulsifying properties of soy protein, pulsed electric fields (PEF) [27], extrusion process [28], and radiation [29] have also been studied.

3.4. Glycation Modification

Traditional chemical modifications of proteins, such as acylation, phosphorylation, and alkylation, are less researched in food applications due to safety and environmental concerns [30]. Protein non-enzyme glycation, commonly regarded as the initial stage of the Maillard reaction, has been widely researched in the food industry to modify the functional properties of protein due to the relatively mild and safe reaction conditions, and no extraneous chemicals were needed [31]. Thus, this makes glycation a promising method for protein modification in the food industry [32]. There have been emerging studies that focused on improving the stability of glycated soy protein hydrolysates stabilized emulsions. Recent studies that focused on the glycation of polysaccharides with soy proteins for improved emulsifying properties are listed in Table 1. Covalent attachment between proteins and polysaccharides may enhance the protein functionality to act as both emulsifier and stabilizer. Most of the investigations (Table 1) conducted on Maillard conjugates followed a similar trend that glycated conjugates increased the emulsification ability and emulsion stability. The main advantages of the soy protein-polysaccharide conjugates synthesized by the Maillard reaction include the increased functional characteristic and solubility over a wide range of environments, such as very low pH, and very high ionic strength [33]. In the case of larger molecular weight polysaccharides, the conjugate-stabilized emulsion may have a thicker stabilizing layer than the protein-stabilized ones (Figure 2). It has been reported that the well-prepared soy protein hydrolysate–polysaccharide conjugates substantially improve emulsifying and stabilizing properties as compared with soy protein hydrolysates and its native proteins [34]. Apart from polysaccharides, some low molecular weight carbohydrates such as glucose and maltose have also been reported to be conjugated with soy protein hydrolysates. Yang et al. [35] studied the effect of the chain length of the carbohydrate on the interfacial and structural characteristics of the conjugates of soy protein hydrolysates (Mw > 30 kDa) produced by the Maillard reaction. The results of the study revealed that increasing carbohydrate chain length increases the emulsion stability of the conjugates. Zhang et al. [36] reported that the soy protein hydrolysate–dextran conjugate-based emulsions exhibited better freeze–thaw stability compared with the SPI-dextran conjugates, especially in the case of 3% DH. For the lowest creaming index and best freeze–thaw stability, the optimum wet Maillard reaction conditions included a soy protein hydrolysate/dextran ratio of 2:3 in which the dispersion of 40 g/L of soy protein hydrolysate was prepared in phosphate buffer (pH 8), which was then incubated for 1 h at 85 °C. The surface activity measurements indicated the closely packed soy peptide–dextran conjugates which formed a thick adsorbed layer at the oil–water interface. Even though the Maillard reaction has long been regarded as an effective method to improve the emulsifying properties of soy protein [31], there are still concerns about the conformational change, surface polarity, specific interactions among the components. Due to the complexity of the Maillard reaction, the productivity, stability, and repeatability of the conjugates are concerns for commercial use.

Table 1. Lists of recently reported studies focusing on polysaccharide glycation with soy proteins for improved emulsifying properties.

Emulsifier	Glycation Conditions	Main Conclusions	References
Soy protein isolate–glucose	Wet heating, 50, 60, 70, 80, and 90 °C, 5 h	The EAI and ESI of the soy protein–glucose isolate were markedly improved under different reaction temperature conditions in comparison to that of untreated SPI.	[37]
Soy protein–maltose	Wet heating, 100 °C, 2 h	The 1-butyl-3-methylimidazolium chloride was proved to be a proper medium for protein glycation to increase glycation extent and to improve the emulsifying activity and emulsion stability.	[38]
Soy protein isolate–gum acacia	Dry heating, 60 °C, RH 79%, 6 days	The soy protein isolates–gum acacia (SPI-GA) conjugates films containing essential oils showed the highest radical scavenging activity and antibacterial activity.	[39]
Soy protein hydrolysate-dextran	Wet heating, 85 °C, 1 h	The soy protein hydrolysate-dextran conjugates produced through the wet method under optimal conditions showed the lowest creaming index and the best freeze–thaw stability.	[36]
Soy protein isolate–Okara dietary fibre (ODF)	Dry heating, 60 °C, RH 78%, 6–72 h	The resulting ODF-SPI conjugates were thermally stable and exhibited excellent Pickering emulsion stabilization potentials.	[40]
Soy glycinin–soy polysaccharide	Dry heating, 60 °C, RH 78%, 24 h and 72 h	The glycation with soy soluble polysaccharide (SSPS) greatly improved the emulsification performance of soy glycinin, the gel network formation and stability (against heating or freeze–thawing) of the resultant high internal phase emulsions.	[41]
Soy protein–pectin	Wet heating, pH 4.5, 95 °C, 30 min	With the addition of glycyrrhizic acid nanofibrils, self-standing soy protein-pectin nanoparticles (SPNPs) stabilized emulsion gels with small droplet size, homogeneous appearance, and microstructure were obtained.	[42]
Soy protein isolate–pectin	Dry heating, 60 °C, RH 79%, 1–7 days	The solubility and emulsifying properties were improved after the Maillard reaction and the strong steric-hindrance effect of pectin facilitated the stability of the emulsion.	[43]

3.5. Fermentation Modification

As an ancient processing approach, lactic fermentation has long been applied on soy milk or soy protein modification for improved sensory properties [44]. However, few studies focused on the effect of fermentation on the emulsifying properties of soy protein. The fermentation process would not only undergo enzymatic hydrolysis on soy proteins, but also affect the conformation of protein molecules through acidification, resulting in a comprehensive impact on the functional properties of the protein. Meinlschmidt et al. [45] studied the effect of liquid state *lactobacillus helveticus* fermentation on the solubility, emulsifying capacity, and foaming activity of soy protein isolate. The fermentation significantly decreased the SPI solubility at pH 7.0, while increased the solubility at pH 4.0. Non-fermented SPI exhibited the highest emulsifying capacity of 660 mL/g, while fermentation resulted in a significantly decreased emulsifying capacity (475–483 mL/g), which was ascribed to the decreased solubility at pH 7.0. The foaming activity of SPI was nearly doubled after fermentation, while the foam density decreased after fermentation.

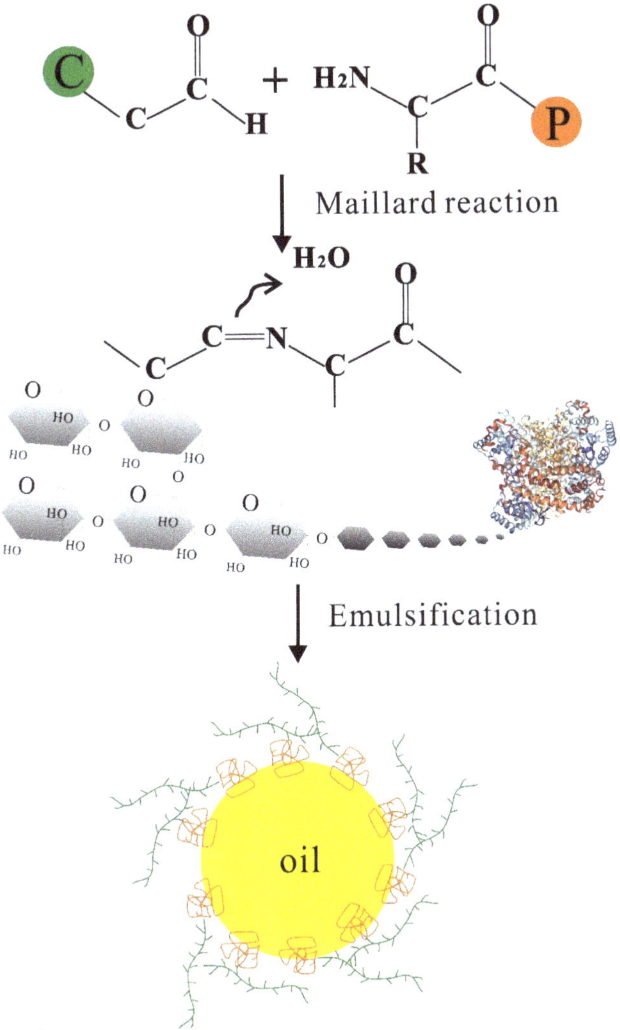

Figure 2. The schematic illustration of the formation of the soy protein–polysaccharide complex for emulsion stabilization.

It should be noted that the modification method is not limited to those discussed above. Indeed, each modification method has some advantages and disadvantages that affect the utilization in the food industry. For example, the physical modification has the advantage of high productivity, while usually showed low effectiveness on the emulsifying properties improvement. Enzymatic hydrolysis was widely recognized as a cost-effective way to modify the functionalities of proteins due to its controllability and minimal formation of by-products. However, the deeply hydrolyzed protein is generally accompanied by the production of bitter peptides, which has an adverse effect on the product flavor. Traditional chemical modification usually has safety and environmental concerns, especially when applied in the food matrix. The non-enzyme glycation is environmentally friendly, safe, and easily acceptable by consumers. The most recent studies on the modification of soy proteins for improved emulsifying properties focused on the combination of novel technologies with traditional modification methods. Wen et al. [46] applied a novel slit divergent ultrasound to facilitate the formation of soy protein isolate–lentinan conjugates via Maillard reaction. The results showed that ultrasonic treatment (40 min) markedly increased the degree of grafting (26. 48%) compared with the traditional heating method (2 h, 13.89%). The hydrophobicity, emulsifying activity, and emulsion stability were doubled after ultrasonic treatment for 40 min and the SPI-lentinan conjugates stabilized emulsions were stable against the various environmental stress (pH, temperature, and ionic strength).

4. Soy Polysaccharides

Soybean soluble polysaccharide (SSPS), a by-product of isolating soybean proteins, has been reported to be used as an emulsifier for the emulsification of beverages owing to the acidic nature of polysaccharide. Rhamnogalacturonan backbone is present in the SSPS structure, which is branched by β-1,4-galactan and homogalacturonan, α-1,3-, or α-1,5-arabinan chains [47] (Figure 3). It has been reported as a source of dietary fibers in fortified foods as well as a functional ingredient for food and pharmaceutical applications. The conformation of SSPS is not easily affected by pH and ionic strength which results in environmental stability of the SSPS stabilized emulsions. The structure of glycoprotein present in SSPS is almost similar to that of the Wattle Blossom Model suggested for gum arabic. The attachment of carbohydrate functionality of the polysaccharide on the oil–water interface is mainly due to the protein fraction of SSPS. Therefore, the hydrophilic portion of SSPS forms a 30 nm thick hydrated layer that retard the chance of coalescence and stabilizes the oil droplets by steric repulsion [48] (Figure 3).

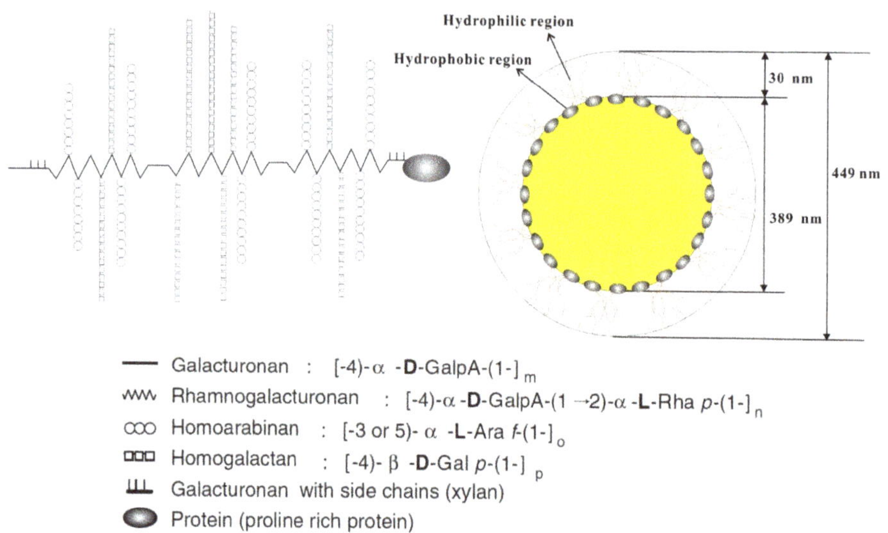

Figure 3. The chemical structure of soybean soluble polysaccharide (SPSS) (**left**) and a schematic diagram of SSPS-stabilized emulsion droplets (**right**).

Generally, the emulsifying ability of SSPS is affected by the protein fraction, molecular weight, and extraction conditions. Nakamura, Takahashi [47] studied the emulsifying potential of three different types of SSPS and observed that all the soy polysaccharide stabilized emulsions showed stability against creaming for 30 days at pH 3.5–5.0 when the polysaccharide concentration was above 4%. The soy polysaccharide that was extracted at pH 3.0 and 120 °C for 2 h showed the best emulsifying ability, in which the protein fraction had an inevitable effect. Nakamura et al. [49] separated the SSPS into two fractions, i.e., high molecular and low molecular, and found that the 2.2% protein facilitated the emulsifying properties of the high molecular fraction but did not exhibit a similar effect on the low molecular fraction of SSPS. The emulsions stabilized by the high molecular fraction showed no change when heated at 90 °C and pH 3.0–7.0 or in the presence of <10 mM $CaCl_2$, while the low molecular fraction stabilized emulsions undergo aggregation when heated at pH 7.0 [50]. The SPSS was further enzyme-digested by pectinases (polygalacturonase (PGase), hemicellulases (galactosidase (GPase), and rhamnogalacturonase (RGase)), and arabinosidase (Afase). The Rgase digested SSPS showed improved emulsifying properties while the others compromised the emulsifying potential [51]. The additions of SSPS to the protein-stabilized emulsions have also been reported to progress the stability against thermal treatment, low pH, and under simulated gastric conditions. Yin et al. [52] fabricated the stable nano-sized emulsions with the help of soy protein and SPSS complexes, which formed the interfacial films under the influence of the temperature by the process of electrostatic complexation of the denatured protein and soy polysaccharide. The interfacial fixed polysaccharide chains are also able to stabilize the oil droplets in an aqueous medium even in the unfavorable condition of soy protein in which they undergo aggregation.

Recently SPSS has shown potential applications in food emulsion products, such as beverages and mayonnaise. Nakamura et al. [53] used SSPS as a stabilizer in the dispersions of acid milk and suggested the comparable stabilizing potential of SSPS with pectin. It was also found that SSPS did not show interaction with casein at pH > 4.6, but exhibited better stabilizing ability at pH < 4.2 than high methoxyl pectin [54]. Chivero et al. [55] examined the ability of SSPS to produce O/W mayonnaise-like emulsions and observed that SSPS could stabilize emulsions with a maximum oil content of about 60 wt%, and the emulsions remained stable after 30 days. The improved stability was observed when SSPS was combined with a thickening agent (xanthan gum) to induce a stronger network. Xu et al. [56] fabricated casein and SSPS compact complex aggregates of 133 nm (Figure 4a), leading to the stabilized emulsions having the stability of more than 500 days with a curcumin loading efficiency of 99.9% and droplet diameter of about 324 nm (Figure 4b). It was also found that the absorption of curcumin was more effective compare with the absorption of the curcumin/Tween 20 suspension group, resulting in 11-fold higher oral bioavailability of curcumin in the emulsion group (Figure 4c). Zhan et al. [57] studied the SSPS effect on the functional characteristics of pea protein isolate (PPI), and found that SSPS adhered to PPI by means of hydrophobic interaction and hydrogen bonding which resulted in decreased hydrophobicity of the surface of reconstituted PPI particles and enhanced the stability of emulsions. It was also suggested that the incorporation of SSPS rearranged and interconnected the modified particles, resulting in the improved interfacial and rheological properties of the emulsions.

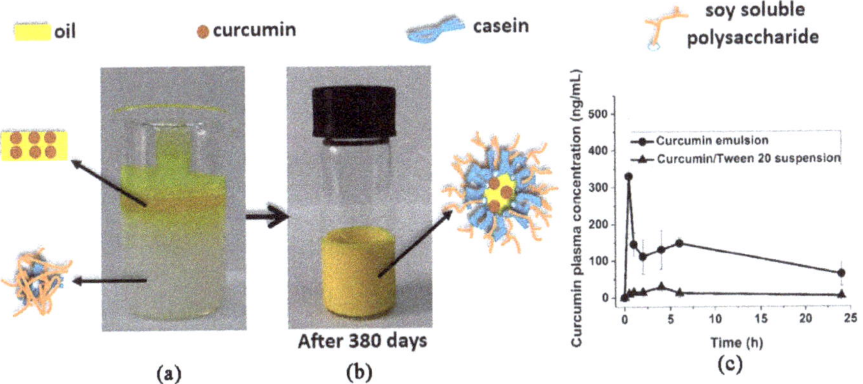

Figure 4. Diagrammatic representation of formation of SSPS-casein complex (**a**), the stabilized emulsion with a long-term stability (**b**), and the significantly improved loading and bioavailability of SSPS-casein stabilized curcumin emulsion compared to that stabilized by Tween 20 (**c**) [56].

Apart from SSPS, the insoluble soy polysaccharide (ISP) containing cellulose, hemicellulose, lignin, and a protein fraction, has attracted attention due to its potential as a Pickering emulsion stabilizer. Porfiri et al. [58] performed the acidic extraction at pH 3.5, 120 °C and extracted insoluble soybean polysaccharide (ISPS) from insoluble okara. The pretreatments (high-pressure homogenization or sonication) are assumed to expose the internal site of the structure of protein and polysaccharide to enhance the superficial hydrophobicity. This in turn facilitates the formation of the outer layer and/or absorption of the macromolecules at the oil–water interface, hence providing increased rigidity of the interfacial film. Particularly, the molecules that resulted from high-pressure homogenization exhibited promising emulsifying potential and showed stability against the pH variation of the emulsions. Mwangi et al. [59] found that ISP dispersions under a high power ultrasonication treatment result in the breakdown of polysaccharide fibers and allow the preparation of the nanoparticles with a size range of 127–221 nm. The fabricated nanoparticles exhibited remarkable potential towards emulsification and allow the formation of Pickering emulsions. It was further reported Yang et al. [60] also fabricated ultrasound-induced insoluble soy polysaccharide nanoparticles with a size of about 160 nm, which can be served as an outstanding Pickering stabilizer for the emulsions having a high internal phase due to the formed gel network (Figure 5). The high internal phase Pickering emulsions showed high stability against environmental stress. The gel structure can be maintained over the pH range 2.0–12.0 and ionic strength range 0–500 mM. All the high internal phase Pickering emulsion gels exhibited excellent stability against prolonged storage and heating, as well as unique reversibility of freeze–thawing-destabilization/re-emulsification.

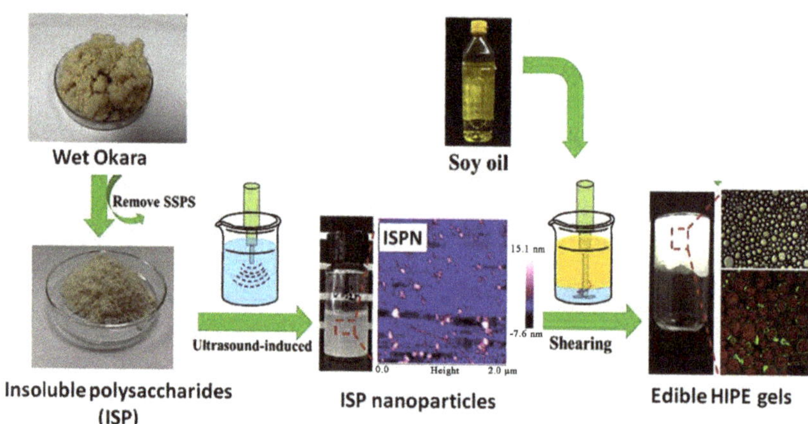

Figure 5. Synthesis of ultrasound-induced insoluble soy polysaccharide (ISP) based nanoparticles for developing and edible O/W high internal phase Pickering emulsion gels [60].

5. Soy Lecithin

Soy lecithin has been an important emulsifier for the production of food emulsion products. Commercial soy lecithin contains 34% triglycerides, 65–75% phospholipids, and small amounts of pigments, carbohydrates, sterol glycosides, and sterols. The commonly found phospholipids include phosphatidylcholine (29–46%), phosphatidylethanolamine (21–34%), and phosphatidylinositol (13–21%). Due to their amphiphilic nature, they can easily be adsorbed onto the surface or interface with the hydrophobic tail of fatty acid facing oil phase while the polar head group facing aqueous phase that results in decreased surface or interfacial tension. The phosphatidylinositol stabilizes the emulsion by serving as a barrier at the surface of oil or water droplets. The phosphatidylcholine and phosphatidylethanolamine contain the positively charged choline and ethanolamine groups, and the negatively charged phosphate and carbonyl groups. The soy phospholipids may form liposomes, micelles, lamellar structures, or bilayer sheets in an aqueous medium depending on the hydration, temperature, and concentration [10] (Figure 6). The described self-assembly systems are considered as potential delivery vehicles for bioactive and nutrients. Since commercial soy lecithin is a mixture of various phospholipids and other numerous constituents, its surface activity is a combined effect of all the surface-active substances. Though lecithin is not usually considered a suitable material to stabilize either oil-in-water or water-in-water emulsions, it can only be utilized for the preparation of emulsions at appropriate salt concentration, pH, temperature, and oil/water ratio.

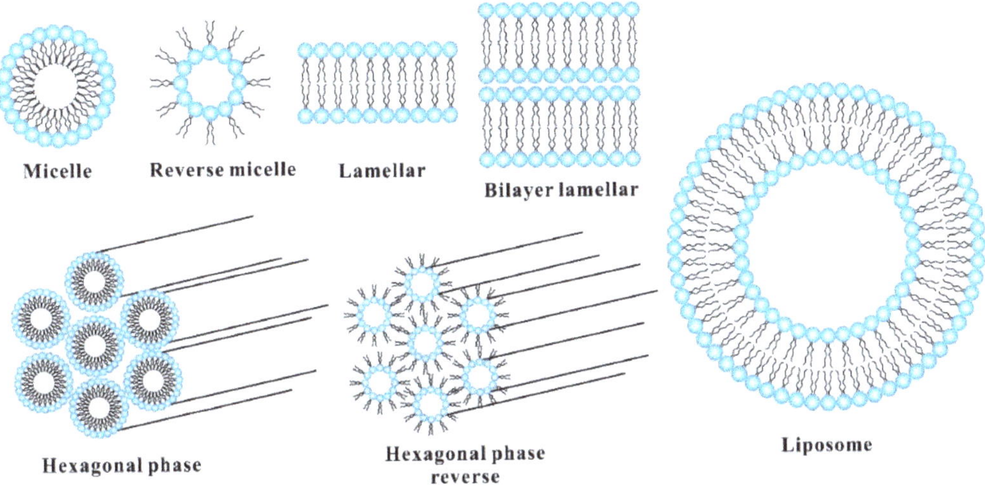

Figure 6. Diagrammatic presentation of various structures resulting from the self-assembly of phospholipids derived from soy lecithin.

Soy lecithin has been an effective emulsifier to fabricate delivery systems for enzymes, nutraceuticals, vitamins, flavors, pesticides, and antimicrobials. Flores-Andrade et al. [61] compared the emulsifying properties of soy lecithin and gum arabic for the fabrication of paprika oleoresin nanoemulsions and observed the remarkable efficacy of soy lecithin to form nano range droplets (d < 150 nm) rather than gum arabic (d < 539.6 nm). Yang et al. [62] prepared the lecithin stabilized emulsions with a droplet size range of 62.5–105 nm for co-delivery of essential oils and curcumin. It was found that the solubility of curcumin was significantly increased by 1700-fold, and its in vitro bioaccessibility was 4.79–10.6 fold higher than that of free curcumin. Koo et al. [63] prepared the emulsions stabilized by 1.5% soy lecithin, 0.5% sodium caseinate, and a combination of both (0.5% sodium caseinate and 0.5% lecithin). The sodium caseinate stabilized emulsions containing and the emulsion stabilized by the mixture of both (sodium caseinate and soy lecithin) undergo destabilization at pH 5 or below because of the aggregation of sodium caseinate near its isoelectric point. While the soy lecithin stabilized emulsion showed stability over a pH range of 3.5 to 7 because of increased repulsion among the droplets.

Soy lecithin has also been added to the protein stabilized emulsions to improve the emulsion stability through surfactant-protein interactions. García-Moreno et al. [64] investigated the influence of a combination of casein and phospholipid (0.3% and 0.5%, respectively) on the oxidative and physical characteristics of 10% fish oil-based emulsions at pH 7. Three different phospholipids were used to conduct the analysis which include lecithin, phosphatidylethanolamine, and phosphatidylcholine. The lecithin stabilized emulsion exhibited the best physical stability as they possess larger negative zeta potential with smaller droplet size. Additionally, they possess a smaller degree of oxidation which may be attributed to the combined effect of casein and lecithin, which results in favorable thickness and structure of the interfacial layer capable to prevent the oxidation of emulsion lipid. Wang et al. [65] added soy lecithin (0.5–1.0%) in the whey protein stabilized emulsions, and found improved emulsion stability because of the surfactant–protein interactions at the interface, resulting in a higher encapsulation efficiency of the spray-dried microcapsules with good re-dispersibility in water. Shen et al. [66] investigated the interactions of soybean lecithin with egg yolk granules, and they observed that incorporation of lecithin destroyed the aggregated structure of egg yolk granules, leading to better stability of emulsion because of lower particle size and higher surface charge.

Recently, a few studies have focused on constructing some novel soy lecithin-based emulsions with interesting functionalities. Sandoval-Cuellar et al. [67] fabricated the high oleic palm oil nanoemulsions, which were stabilized by the whey protein and soy lecithin and observed the less release of free fatty acid in in vitro gastrointestinal digestion as compared to non-encapsulated control. This is because of the lipase inhibition potential of soy lecithin. Zhuang et al. [68] inoculated *Bifidobacterium lactis* and *Lactobacillus acidophilus* into the lecithin-based oleogel emulsions prepared by using 20 wt% oleogelators (stearic acid: soy lecithin = 5:5), 70% canola oil, and 10% water. The semi-solid oleogel emulsions based on soy lecithin improved the viability of the encapsulated probiotics and prevented the progress of lipid oxidation. Jiang et al. [69] designed edible Pickering emulsions of high internal phase bearing a double-emulsion morphology (Figure 7). They dissolved the lecithin in squalane oil, and the dispersion of zein nanoparticles was prepared in water (w1). When the system allowed to emulsify for the first time, a w1/o emulsion was formed. The resulting primary w1/o emulsion was added with the second dispersion of zein nanoparticles (w2), which on emulsification leads to a w1/o/w2 double emulsion, the total ratio of oil and water was maintained at 3:1. It was found that soy lecithin enhances the surface elasticity of the interfacial films and resulted in highly stable emulsions.

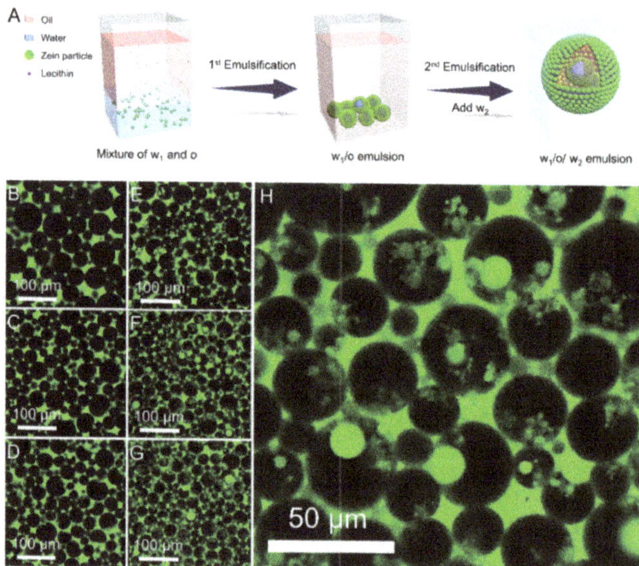

Figure 7. (**A**) Illustration of a two-step process for the preparation of lecithin based w/o/w high internal phase Pickering emulsions; (**B–G**) CLSM images of the emulsions stabilized with zein nanoparticles and lecithin, and the lecithin concentrations are 0%, 0.1%, 0.25%, 0.5%, 1%, and 2%, respectively; (**H**) CLSM image of the emulsion with 1% lecithin (**F**) at high magnification [69].

6. Food Applications

6.1. Bioactive Encapsulation and Delivery

Recently the interest in the fabrication of novel bioactive delivery systems increased continuously [70]. In food applications, various emulsion-based delivery systems are used but they still required long-time physical stability in various environmental conditions because the breakdown of emulsion significantly alter the texture, color, flavor, and shelf life of the products. Xu, Wang [56] found that the curcumin oral bioavailability of the casein–soy soluble polysaccharide complex stabilized emulsions was 11-fold higher compared with curcumin/Tween 20 suspension. Wang et al. [71] found that the addition of soy polysaccharides (soy hull polysaccharide and soy soluble polysaccharide) is capable to decrease the influence of simulated gastric fluid (i.e., pepsin, ionic strength, and pH) on the stability of emulsions.

In recent years, a few novel emulsions such as Pickering emulsions [60], emulsion gels [72], oleogel emulsions [68], high internal phase emulsions, multiple emulsions [69,73], nanoemulsions, and microemulsions [74] have been fabricated for encapsulation and delivery of bioactive compounds (Figure 8). Physically stabilized Pickering emulsions with solid particles that were moderately wetted by oil and water showed improved stability against steric mechanism-based flocculation and coalescence [59]. Soy-based emulsifiers are considered promising Pickering stabilizers due to easy availability, the ability to form nano-aggregates, and health effects. Liu and Tang [75] reported the heat-treated soy glycinin stabilized gel-like Pickering emulsions capable of sustained release β-carotene which was confirmed by the in vitro experiment for intestinal digestion which indicated that the formation of a gel-like network significantly slowed down the release of β-carotene. Muñoz-González et al. [76] produced four emulsion gels containing soy protein, olive oil, and alginate-based cold gelling agent to encapsulate polyphenol. The emulsion gels with added polyphenols exhibited the presence of gallic acid, flavanol monomers, and their derivatives, which play a vital role to make it a suitable system for the delivery of various bioactive compounds. Zhuang, Gaudino [68] fabricated novel soy-lecithin based W/O oleogel emulsions for improved lipid stability and probiotic viability. The oleogel emulsion was composed of 20 wt% oleogelators (soy lecithin: stearic acid, 1:1), 70% canola oil, and 10% water. Flores-Andrade, Allende-Baltazar [61] compared the O/W paprika oleoresin nanoemulsions which were stabilized by whey protein concentrates, soy lecithin, and gum arabic under high-pressure homogenization, and found that soy lecithin was the most effective emulsifier for nanoemulsion preparation.

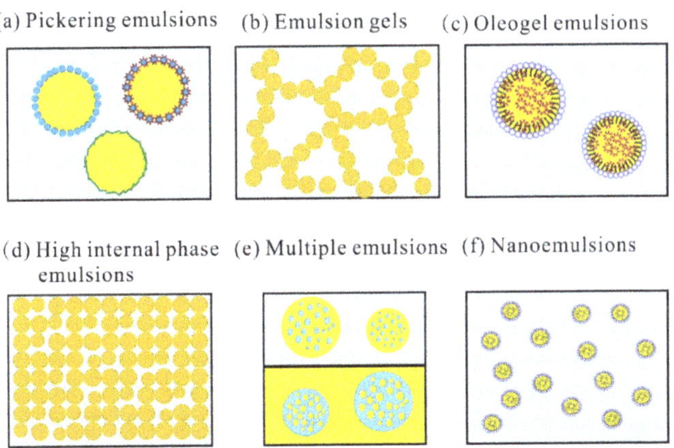

Figure 8. Structures of Pickering emulsions (**a**), emulsion gels (**b**), oleogel emulsions (**c**), high internal phase emulsions (**d**), multiple emulsions (**e**), and nanoemulsions (**f**).

6.2. Fat Replacer

To mimic the properties of animal fat, soy-based emulsified systems have been researched as a promising alternative to replace animal fat [77,78]. Emulsion gel structures based on soy protein have the potentials to replace fat in sausages [79,80]. Pintado et al. [81] fabricated the soy protein stabilized emulsion gels that contained two different solid extracts of polyphenols obtained from grape seed and olive to be used as an animal fat replacer in the development of frankfurters.

The addition of SPI in ice creams may decrease the influence of heat on the recrystallization and melting rate of ice. Chen et al. [82] reported that larger-sized SPI nanoparticles with higher surface hydrophobicity and enhance the potential of packing at the oil–water interface are more suitable to form Pickering emulsions with enhanced freeze–thaw stability. The addition of soy protein in ice cream reduces the heat shock effect on the melting rate and recrystallization of ice. Functional optimization of the structure of soy protein is currently attracted great interest to achieve a high degree of fat globule partial coalescence in the preparation of ice cream. Soy protein under selective hydrolysis provides sufficient fat partial coalescence and good melt-down rates. Chen et al. [83] compared the effect of commercial soy protein isolate (CSPI), native soy protein isolate (NSPI), soy protein hydrolyzed by pepsin (SPHPe), and skim milk powder (SMP), and soy protein hydrolyzed by papain (SPHPa) on ice cream mix stability and melt-down properties. Among all these, the highest melt-down rates were observed in the ice cream made with SPHPe, which ranged from 1.23% min^{-1} to 2.05% min^{-1}. While the CSPI, SMP, and SPHPa based ice creams exhibited the lowest melt rate with no significant difference from each other. The highest fraction of protein at the fat globules during aging and freezing was SPHPa mix (22.6%) which was almost similar to SMP (21.8%).

Currently, the soybean has been considered as popular among all the plant-based sources used for yogurt production, because of its quantity, quality, and functional characteristics of the protein. Such plant-based systems upon acidification cause destabilization of soy proteins, which may result in the formation of a non-continuous, weak gel [84]. SPNPs stabilized emulsions have been studied as functional non-dairy yogurts. Sengupta et al. [85] found that the addition of SPNPs distinctly improved the quality of soy yogurts with enhanced radical scavenging activity and ferric reducing property. The SPNPs (72.42–586.72 nm) incorporated soy yogurts exhibited significantly enhanced oxidative stability against peroxidation of lipids, indicating the applicability in the development of functional yogurts.

Consumers are purchasing plant-based "milks" and "creams" frequently because of their environmental and perceived health benefits. Hence, the food companies have been developing a wide range of plant-based milk and creamer products including soy products. Chung et al. [86] applied soy lecithin (1–5%) to stabilize 10% O/W emulsions, which remained physically stable upon addition of an acidic hot solution of coffee and retard the phase separation or increment in particle size. The soy lecithin stabilized oil droplets exhibited the enhanced surface negativity, hence they exert the strong electrostatic repulsion among the droplet and retard their aggregation. Koo, Chung [63] found the ability of soy lecithin to replace the caseinate in coffee creamers. The model O/W emulsions stabilized with a mixture of emulsifiers based on sodium caseinate (0.5%) and soy lecithin (0.5%) showed physical stability over a pH range of 5.5 to 7.

7. Challenges and Future Trends

It should be noted that the commercial soy-based products are changing along with the development of pro-cessing technologies. The functional properties of soy-based emulsifiers are sensitive to the processing history, such as enzymatic, thermal, and jet cooking processes [87–89], which makes the soy ingredients used in literatures varied. Hence, understanding the physiochemical states of various soy-based emulsifiers is necessary for better application of that in food products. Lee, Ryu [88] compared the solubility of commercial soy protein from different manufacturers and they found the soy protein isolates can be classified into three groups. One group had high solubility near the pl. Another group had low solubility near the pl, but high solubility at pH 11. The third group had low solubility even at pH 11. Recently, Zheng, Wang [89] selected 20 soy protein isolate samples from three manufacturers and they found that the best sample was Chiba tofu due to high hardness and springiness, and excellent quality. Hence, the researchers should identify their soy ingredients and select the most appropriate starting materials for their future academic or product development work.

Recently, various studies focused on the effect of novel technologies on the conformation, molecular interaction, and functional properties of soy proteins, soy polysaccharides, and soy lecithin, which provide fundamental understanding of the mechanism of their functional properties in various food processing conditions. However, there are still some challenges on the modification process on the soy-based materials. In future studies, the combination of the physical, chemical, enzyme, or biological methods to improve the emulsifying properties and other functional properties will be a trend for the development of novel food emulsifiers. However, the main challenge for the development of such novel emulsifiers is to select a stable fraction with high functionality under certain environmental conditions for the desirable formulated products. Therefore, more studies are needed to understand the relationship between the functional properties and molecular structure of these components before use in practice. The combination of soy based low molecular weight emulsifiers and high molecular emulsifiers (proteins and polysaccharides) will be a challenge in extending their applications to new fields.

Funding: This work was supported by the Natural Science Foundation of Hubei Province of China (Grant No. 2020CFB204), the Key Research and Development Program of Hubei Province of China (Grant No. 2020BAB078), and the Foundation of Xinjiang Uygur Autonomous Region of China (Grant No. 2020D01B08).

Conflicts of Interest: The author declares no conflict of interest.

References

1. McClements, D.J.; Gumus, C.E. Natural emulsifiers—Biosurfactants, phospholipids, biopolymers, and colloidal particles: Molecular and physicochemical basis of functional performance. *Adv. Colloid Interface Sci.* **2016**, *234*, 3–26. [CrossRef] [PubMed]
2. Mcclements, D.J. Critical review of techniques and methodologies for characterization of emulsion stability. *Crit. Rev. Food Sci. Nutr.* **2007**, *47*, 611–649. [CrossRef] [PubMed]
3. Pearce, K.N.; Kinsella, J.E. Emulsifying properties of proteins: Evaluation of a turbidimetric technique. *J. Agric. Food Chem.* **1978**, *26*, 716–723. [CrossRef]
4. Ashaolu, T.J. Applications of soy protein hydrolysates in the emerging functional foods: A review. *Int. J. Food Sci. Technol.* **2020**, *55*, 421–428. [CrossRef]
5. Kuroiwa, T.; Kobayashi, I.; Chuah, A.M.; Nakajima, M.; Ichikawa, S. Formulation and stabilization of nano-/microdispersion systems using naturally occurring edible polyelectrolytes by electrostatic deposition and complexation. *Adv. Colloid Interface Sci.* **2015**, *226*, 86–100. [CrossRef]
6. Rivas, H.; Sherman, P. Soy and meat proteins as emulsion stabilizers. 4. The stability and interfacial rheology of O/W emulsions stabilised by soy and meat protein fractions. *Colloids Surf.* **1984**, *11*, 155–171. [CrossRef]
7. Liu, M.; Lee, D.-S.; Damodaran, S. Emulsifying properties of acidic subunits of soy 11S globulin. *J. Agric. Food Chem.* **1999**, *47*, 4970–4975. [CrossRef]
8. Utsumi, S.; Matsumura, Y.; Mori, T. Structure-function relationships of soy proteins. In *Food Proteins and Their Applications*; CRC Press: Boca Raton, FL, USA, 2017; pp. 257–292.
9. Molina, E.; Papadopoulou, A.; Ledward, D. Emulsifying properties of high pressure treated soy protein isolate and 7S and 11S globulins. *Food Hydrocoll.* **2001**, *15*, 263–269. [CrossRef]
10. Xu, Q.; Nakajima, M.; Liu, Z.; Shiina, T. Soybean-based surfactants and their applications. In *Soybean-Applications and Technology*; IntechOpen: London, UK, 2011; pp. 341–364.
11. Tang, C.-H. Emulsifying properties of soy proteins: A critical review with emphasis on the role of conformational flexibility. *Crit. Rev. Food Sci. Nutr.* **2017**, *57*, 2636–2679. [CrossRef]
12. Shen, P.; Zhou, F.; Zhang, Y.; Yuan, D.; Zhao, Q.; Zhao, M. Formation and characterization of soy protein nanoparticles by controlled partial enzymatic hydrolysis. *Food Hydrocoll.* **2020**, *105*, 105844. [CrossRef]
13. Nishinari, K.; Fang, Y.; Guo, S.; Phillips, G. Soy proteins: A review on composition, aggregation and emulsification. *Food Hydrocoll.* **2014**, *39*, 301–318. [CrossRef]

14. Zang, X.; Liu, P.; Chen, Y.; Wang, J.; Yu, G.; Xu, H. Improved freeze-thaw stability of o/w emulsions prepared with soybean protein isolate modified by papain and transglutaminase. *LWT Food Sci. Technol.* **2019**, *104*, 195–201. [CrossRef]
15. Corredig, M. Heat-induced changes in oil-in-water emulsions stabilized with soy protein isolate. *Food Hydrocoll.* **2009**, *23*, 2141–2148.
16. Iwabuchi, S.; Watanabe, H.; Yamauchi, F. Observations on the dissociation of. beta.-conglycinin into subunits by heat treatment. *J. Agric. Food Chem.* **1991**, *39*, 34–40. [CrossRef]
17. Iwabuchi, S.; Watanabe, H.; Yamauchi, F. Thermal denaturation of. beta.-conglycinin. Kinetic resolution of reaction mechanism. *J. Agric. Food Chem.* **1991**, *39*, 27–33. [CrossRef]
18. Singla, M.; Sit, N. Application of ultrasound in combination with other technologies in food processing: A review. *Ultrason Sonochem.* **2021**, *73*, 105506. [CrossRef] [PubMed]
19. Huang, L.; Jia, S.; Zhang, W.; Ma, L.; Ding, X. Aggregation and emulsifying properties of soybean protein isolate pretreated by combination of dual-frequency ultrasound and ionic liquids. *J. Mol. Liq.* **2020**, *301*, 112394. [CrossRef]
20. Jambrak, A.R.; Lelas, V.; Mason, T.J.; Krešić, G.; Badanjak, M. Physical properties of ultrasound treated soy proteins. *J. Food Eng.* **2009**, *93*, 386–393. [CrossRef]
21. Ma, X.; Yan, T.; Hou, F.; Chen, W.; Miao, S.; Liu, D. Formation of soy protein isolate (SPI)-citrus pectin (CP) electrostatic complexes under a high-intensity ultrasonic field: Linking the enhanced emulsifying properties to physicochemical and structural properties. *Ultrason Sonochem.* **2019**, *59*, 104748. [CrossRef]
22. Albano, K.M.; Cavallieri, Â.L.F.; Nicoletti, V.R. Electrostatic Interaction between Soy Proteins and Pectin in O/W Emulsions Stabilization by Ultrasound Application. *Food Biophys.* **2020**, *15*, 297–312. [CrossRef]
23. Ren, X.; Li, C.; Yang, F.; Huang, Y.; Huang, C.; Zhang, K.; Yan, L. Comparison of hydrodynamic and ultrasonic cavitation effects on soy protein isolate functionality. *J. Food Eng.* **2020**, *265*, 109697. [CrossRef]
24. Puppo, M.; Speroni, F.; Chapleau, N.; de Lamballerie, M.; Añón, M.; Anton, M. Effect of high-pressure treatment on emulsifying properties of soybean proteins. *Food Hydrocoll.* **2005**, *19*, 289–296. [CrossRef]
25. Torrezan, R.; Tham, W.P.; Bell, A.E.; Frazier, R.A.; Cristianini, M. Effects of high pressure on functional properties of soy protein. *Food Chem.* **2007**, *104*, 140–147. [CrossRef]
26. Wang, X.-S.; Tang, C.-H.; Li, B.-S.; Yang, X.-Q.; Li, L.; Ma, C.-Y. Effects of high-pressure treatment on some physicochemical and functional properties of soy protein isolates. *Food Hydrocoll.* **2008**, *22*, 560–567. [CrossRef]
27. Li, Y.; Chen, Z.; Mo, H. Effects of pulsed electric fields on physicochemical properties of soybean protein isolates. *LWT Food Sci. Technol.* **2007**, *40*, 1167–1175. [CrossRef]
28. Mozafarpour, R.; Koocheki, A.; Milani, E.; Varidi, M. Extruded soy protein as a novel emulsifier: Structure, interfacial activity and emulsifying property. *Food Hydrocoll.* **2019**, *93*, 361–373. [CrossRef]
29. Wang, Y.; Zhang, A.; Wang, X.; Xu, N.; Jiang, L. The radiation assisted-Maillard reaction comprehensively improves the freeze-thaw stability of soy protein-stabilized oil-in-water emulsions. *Food Hydrocoll.* **2020**, *103*, 105684. [CrossRef]
30. Kinsella, J.; Whitehead, D. Emulsifying and foaming properties of chemically modified proteins. In *Advances in Food Emulsions and Foams*; Elsevier Applied Science Publishers Ltd.: London, UK, 1988; pp. 163–188.
31. Kutzli, I.; Weiss, J.; Gibis, M. Glycation of Plant Proteins Via Maillard Reaction: Reaction Chemistry, Technofunctional Properties, and Potential Food Application. *Foods* **2021**, *10*, 376. [CrossRef]
32. Liu, J.; Ru, Q.; Ding, Y. Glycation a promising method for food protein modification: Physicochemical properties and structure, a review. *Food Res. Int.* **2012**, *49*, 170–183. [CrossRef]
33. Naik, R.R.; Wang, Y.; Selomulya, C. Improvements of plant protein functionalities by Maillard conjugation and Maillard reaction products. *Crit. Rev. Food Sci. Nutr.* **2021**, 1–26. [CrossRef]
34. Sivapratha, S.; Sarkar, P. Multiple layers and conjugate materials for food emulsion stabilization. *Crit. Rev. Food Sci. Nutr.* **2018**, *58*, 877–892. [CrossRef]
35. Yang, H.; Su, Z.; Meng, X.; Zhang, X.; Kennedy, J.F.; Liu, B. Fabrication and characterization of Pickering emulsion stabilized by soy protein isolate-chitosan nanoparticles. *Carbohydr. Polym.* **2020**, *247*, 116712. [CrossRef] [PubMed]
36. Zhang, A.; Yu, J.; Wang, G.; Wang, X.; Zhang, L. Improving the emulsion freeze-thaw stability of soy protein hydrolysate-dextran conjugates. *LWT Food Sci. Technol.* **2019**, *116*, 108506. [CrossRef]
37. Li, R.; Wang, X.; Liu, J.; Cui, Q.; Wang, X.; Chen, S.; Jiang, L. Relationship between molecular flexibility and emulsifying properties of soy protein isolate-glucose conjugates. *J. Agric. Food Chem.* **2019**, *67*, 4089–4097. [CrossRef] [PubMed]
38. Xu, W.; Zhao, X.H. Structure and property changes of the soy protein isolate glycated with maltose in an ionic liquid through the Maillard reaction. *Food Funct.* **2019**, *10*, 1948–1957. [CrossRef]
39. Xue, F.; Gu, Y.; Wang, Y.; Li, C.; Adhikari, B. Encapsulation of essential oil in emulsion based edible films prepared by soy protein isolate-gum acacia conjugates. *Food Hydrocoll.* **2019**, *96*, 178–189. [CrossRef]
40. Ashaolu, T.J.; Zhao, G. Fabricating a pickering stabilizer from Okara dietary fibre particulates by conjugating with soy protein isolate via maillard reaction. *Foods* **2020**, *9*, 143. [CrossRef] [PubMed]
41. Hao, Z.Z.; Peng, X.Q.; Tang, C.H. Edible pickering high internal phase emulsions stabilized by soy glycinin: Improvement of emulsification performance and pickering stabilization by glycation with soy polysaccharide. *Food Hydrocoll.* **2020**, *103*, 105672. [CrossRef]

42. Li, Q.; He, Q.; Xu, M.; Li, J.; Liu, X.; Wan, Z.; Yang, X. Food-grade emulsions and emulsion gels prepared by soy protein-pectin complex nanoparticles and glycyrrhizic acid nanofibrils. *J. Agric. Food Chem.* **2020**, *68*, 1051–1063. [CrossRef]
43. Ma, X.; Chen, W.; Yan, T.; Wang, D.; Hou, F.; Miao, S.; Liu, D. Comparison of citrus pectin and apple pectin in conjugation with soy protein isolate (SPI) under controlled dry-heating conditions. *Food Chem.* **2020**, *309*, 125501. [CrossRef]
44. Meinlschmidt, P.; Schweiggert-Weisz, U.; Eisner, P. Soy protein hydrolysates fermentation: Effect of debittering and degradation of major soy allergens. *LWT Food Sci. Technol.* **2016**, *71*, 202–212. [CrossRef]
45. Meinlschmidt, P.; Ueberham, E.; Lehmann, J.; Schweiggert-Weisz, U.; Eisner, P. Immunoreactivity, sensory and physicochemical properties of fermented soy protein isolate. *Food Chem.* **2016**, *205*, 229–238. [CrossRef] [PubMed]
46. Wen, C.; Zhang, J.; Qin, W.; Gu, J.; Zhang, H.; Duan, Y.; Ma, H. Structure and functional properties of soy protein isolate-lentinan conjugates obtained in Maillard reaction by slit divergent ultrasonic assisted wet heating and the stability of oil-in-water emulsions. *Food Chem.* **2020**, *331*, 127374. [CrossRef] [PubMed]
47. Nakamura, A.; Takahashi, T.; Yoshida, R.; Maeda, H.; Corredig, M. Emulsifying properties of soybean soluble polysaccharide. *Food Hydrocoll.* **2004**, *18*, 795–803. [CrossRef]
48. Kontogiorgos, V. Polysaccharides at fluid interfaces of food systems. *Adv. Colloid Interface Sci.* **2019**, *270*, 28–37. [CrossRef]
49. Nakamura, A.; Yoshida, R.; Maeda, H.; Furuta, H.; Corredig, M. Study of the role of the carbohydrate and protein moieties of soy soluble polysaccharides in their emulsifying properties. *J. Agric. Food Chem.* **2004**, *52*, 5506–5512. [CrossRef] [PubMed]
50. Nakamura, A.; Maeda, H.; Corredig, M. Influence of heating on oil-in-water emulsions prepared with soybean soluble polysaccharide. *J. Agric. Food Chem.* **2007**, *55*, 502–509. [CrossRef] [PubMed]
51. Nakamura, A.; Maeda, H.; Corredig, M. Emulsifying properties of enzyme-digested soybean soluble polysaccharide. *Food Hydrocoll.* **2006**, *20*, 1029–1038. [CrossRef]
52. Yin, B.; Deng, W.; Xu, K.; Huang, L.; Yao, P. Stable nano-sized emulsions produced from soy protein and soy polysaccharide complexes. *J. Colloid Interface Sci.* **2012**, *380*, 51–59. [CrossRef]
53. Nakamura, A.; Furuta, H.; Kato, M.; Maeda, H.; Nagamatsu, Y. Effect of soybean soluble polysaccharides on the stability of milk protein under acidic conditions. *Food Hydrocoll.* **2003**, *17*, 333–343. [CrossRef]
54. Nakamura, A.; Yoshida, R.; Maeda, H.; Corredig, M. The stabilizing behaviour of soybean soluble polysaccharide and pectin in acidified milk beverages. *Int. Dairy J.* **2006**, *16*, 361–369. [CrossRef]
55. Chivero, P.; Gohtani, S.; Yoshii, H.; Nakamura, A. Assessment of soy soluble polysaccharide, gum arabic and OSA-Starch as emulsifiers for mayonnaise-like emulsions. *LWT Food Sci. Technol.* **2016**, *69*, 59–66. [CrossRef]
56. Xu, G.; Wang, C.; Yao, P. Stable emulsion produced from casein and soy polysaccharide compacted complex for protection and oral delivery of curcumin. *Food Hydrocoll.* **2017**, *71*, 108–117. [CrossRef]
57. Zhan, F.; Shi, M.; Wang, Y.; Li, B.; Chen, Y. Effect of freeze-drying on interaction and functional properties of pea protein isolate/soy soluble polysaccharides complexes. *J. Mol. Liq.* **2019**, *285*, 658–667. [CrossRef]
58. Porfiri, M.C.; Vaccaro, J.; Stortz, C.A.; Navarro, D.A.; Wagner, J.R.; Cabezas, D.M. Insoluble soybean polysaccharides: Obtaining and evaluation of their O/W emulsifying properties. *Food Hydrocoll.* **2017**, *73*, 262–273. [CrossRef]
59. Mwangi, W.W.; Lim, H.P.; Low, L.E.; Tey, B.T.; Chan, E.S. Food-grade Pickering emulsions for encapsulation and delivery of bioactives. *Trends Food Sci. Technol.* **2020**, *100*, 320–332. [CrossRef]
60. Yang, T.; Li, X.T.; Tang, C.H. Novel edible pickering high-internal-phase-emulsion gels efficiently stabilized by unique polysaccharide-protein hybrid nanoparticles from Okara. *Food Hydrocoll.* **2020**, *98*, 105285. [CrossRef]
61. Flores-Andrade, E.; Allende-Baltazar, Z.; Sandoval-González, P.E.; Jiménez-Fernández, M.; Beristain, C.I.; Pascual-Pineda, L.A. Carotenoid nanoemulsions stabilized by natural emulsifiers: Whey protein, gum Arabic, and soy lecithin. *J. Food Eng.* **2021**, *290*, 110208. [CrossRef]
62. Yang, Q.-Q.; Sui, Z.; Lu, W.; Corke, H. Soybean lecithin-stabilized oil-in-water (O/W) emulsions increase the stability and in vitro bioaccessibility of bioactive nutrients. *Food Chem.* **2020**, *338*, 128071. [CrossRef] [PubMed]
63. Koo, C.K.W.; Chung, C.; Fu, J.T.R.; Sher, A.; Rousset, P.; McClements, D.J. Impact of sodium caseinate, soy lecithin and carrageenan on functionality of oil-in-water emulsions. *Food Res. Int.* **2019**, *123*, 779–789. [CrossRef] [PubMed]
64. García-Moreno, P.J.; Horn, A.F.; Jacobsen, C. Influence of casein–phospholipid combinations as emulsifier on the physical and oxidative stability of fish oil-in-water emulsions. *J. Agric. Food Chem.* **2014**, *62*, 1142–1152. [CrossRef]
65. Wang, S.; Shi, Y.; Tu, Z.; Zhang, L.; Wang, H.; Tian, M.; Zhang, N. Influence of soy lecithin concentration on the physical properties of whey protein isolate-stabilized emulsion and microcapsule formation. *J. Food Eng.* **2017**, *207*, 73–80. [CrossRef]
66. Shen, Y.; Chang, C.; Shi, M.; Su, Y.; Gu, L.; Li, J.; Yang, Y. Interactions between lecithin and yolk granule and their influence on the emulsifying properties. *Food Hydrocoll.* **2020**, *101*, 105510. [CrossRef]
67. Sandoval-Cuellar, C.E.; de Jesus Perea-Flores, M.; Quintanilla-Carvajal, M.X. In-vitro digestion of whey protein- and soy lecithin-stabilized High Oleic Palm Oil emulsions. *J. Food Eng.* **2020**, *278*, 109918. [CrossRef]
68. Zhuang, X.; Gaudino, N.; Clark, S.; Acevedo, N.C. Novel lecithin-based oleogels and oleogel emulsions delay lipid oxidation and extend probiotic bacteria survival. *LWT Food Sci. Technol.* **2021**, *136*, 110353. [CrossRef]
69. Jiang, H.; Zhang, T.; Smits, J.; Huang, X.; Maas, M.; Yin, S.; Ngai, T. Edible high internal phase Pickering emulsion with double-emulsion morphology. *Food Hydrocoll.* **2021**, *111*, 106405. [CrossRef]
70. Tang, C.-H. Nanostructured soy proteins: Fabrication and applications as delivery systems for bioactives (a review). *Food Hydrocoll.* **2019**, *91*, 92–116. [CrossRef]

71. Wang, S.; Shao, G.; Yang, J.; Zhao, H.; Qu, D.; Zhang, D.; Zhu, D.; He, Y.; Liu, H. Contribution of soybean polysaccharides in digestion of oil-in-water emulsion-based delivery system in an in vitro gastric environment. *Food Sci. Technol. Int.* **2020**, *26*, 444–452. [CrossRef]
72. Paglarini, C.D.S.; Martini, S.; Pollonio, M.A.R. Using emulsion gels made with sonicated soy protein isolate dispersions to replace fat in frankfurters. *LWT Food Sci. Technol.* **2019**, *99*, 453–459. [CrossRef]
73. Balcaen, M.; Steyls, J.; Schoeppe, A.; Nelis, V.; Van der Meeren, P. Phosphatidylcholine-depleted lecithin: A clean-label low-HLB emulsifier to replace PGPR in w/o and w/o/w emulsions. *J. Colloid Interface Sci.* **2020**, *581*, 836–846. [CrossRef]
74. Jalali-Jivan, M.; Abbasi, S. Novel approach for lutein extraction: Food grade microemulsion containing soy lecithin & sunflower oil. *Innov. Food Sci. Emerg. Technol.* **2020**, *66*, 102505.
75. Liu, F.; Tang, C.-H. Soy glycinin as food-grade Pickering stabilizers: Part. III. Fabrication of gel-like emulsions and their potential as sustained-release delivery systems for β-carotene. *Food Hydrocoll.* **2016**, *56*, 434–444. [CrossRef]
76. Muñoz-González, I.; Ruiz-Capillas, C.; Salvador, M.; Herrero, A.M. Emulsion gels as delivery systems for phenolic compounds: Nutritional, technological and structural properties. *Food Chem.* **2021**, *339*, 128049. [CrossRef] [PubMed]
77. Dreher, J.; Blach, C.; Terjung, N.; Gibis, M.; Weiss, J. Influence of protein content on plant-based emulsified and crosslinked fat crystal networks to mimic animal fat tissue. *Food Hydrocoll.* **2020**, *106*, 105864. [CrossRef]
78. Dreher, J.; Blach, C.; Terjung, N.; Gibis, M.; Weiss, J. Formation and characterization of plant-based emulsified and crosslinked fat crystal networks to mimic animal fat tissue. *J. Food Sci.* **2020**, *85*, 421–431. [CrossRef] [PubMed]
79. Utama, D.T.; Jeong, H.S.; Kim, J.; Barido, F.H.; Lee, S.K. Fatty acid composition and quality properties of chicken sausage formulated with pre-emulsified perilla-canola oil as an animal fat replacer. *Poult. Sci.* **2019**, *98*, 3059–3066. [CrossRef] [PubMed]
80. de Souza Paglarini, C.; de Figueiredo Furtado, G.; Honório, A.R.; Mokarzel, L.; da Silva Vidal, V.A.; Ribeiro, A.P.B.; Cunha, R.L.; Pollonio, M.A.R. Functional emulsion gels as pork back fat replacers in Bologna sausage. *Food Struct.* **2019**, *20*, 100105. [CrossRef]
81. Pintado, T.; Muñoz-González, I.; Salvador, M.; Ruiz-Capillas, C.; Herrero, A.M. Phenolic compounds in emulsion gel-based delivery systems applied as animal fat replacers in frankfurters: Physico-chemical, structural and microbiological approach. *Food Chem.* **2021**, *340*, 128095. [CrossRef]
82. Chen, Y.B.; Zhu, X.F.; Liu, T.X.; Lin, W.F.; Tang, C.H.; Liu, R. Improving freeze-thaw stability of soy nanoparticle-stabilized emulsions through increasing particle size and surface hydrophobicity. *Food Hydrocoll.* **2019**, *87*, 404–412. [CrossRef]
83. Chen, W.; Liang, G.; Li, X.; He, Z.; Zeng, M.; Gao, D.; Qin, F.; Goff, H.D.; Chen, J. Effects of soy proteins and hydrolysates on fat globule coalescence and meltdown properties of ice cream. *Food Hydrocoll.* **2019**, *94*, 279–286. [CrossRef]
84. Grasso, N.; Alonso-Miravalles, L.; O'Mahony, J.A. Composition, physicochemical and sensorial properties of commercial plant-based yogurts. *Foods* **2020**, *9*, 252. [CrossRef] [PubMed]
85. Sengupta, S.; Bhattacharyya, D.K.; Goswami, R.; Bhowal, J. Emulsions stabilized by soy protein nanoparticles as potential functional non-dairy yogurts. *J. Sci. Food Agric.* **2019**, *99*, 5808–5818. [CrossRef]
86. Chung, C.; Sher, A.; Rousset, P.; Decker, E.A.; McClements, D.J. Formulation of food emulsions using natural emulsifiers: Utilization of quillaja saponin and soy lecithin to fabricate liquid coffee whiteners. *J. Food Eng.* **2017**, *209*, 1–11. [CrossRef]
87. Egbert, W.R. *Isolated Soy Protein: Technology, Properties, and Applications. Soybeans as Functional Foods and Ingredients*; Liu, K., Ed.; AOCS Press: Champaign, IL, USA, 2004; pp. 134–162.
88. Lee, K.; Ryu, H.; Rhee, K. Protein solubility characteristics of commercial soy protein products. *J. Am. Oil Chem. Soc.* **2003**, *80*, 85–90. [CrossRef]
89. Zheng, L.; Wang, Z.; Kong, Y.; Ma, Z.; Wu, C.; Regenstein, J.M.; Teng, F.; Li, Y. Different commercial soy protein isolates and the characteristics of Chiba tofu. *Food Hydrocoll.* **2021**, *110*, 106115. [CrossRef]

MDPI
St. Alban-Anlage 66
4052 Basel
Switzerland
Tel. +41 61 683 77 34
Fax +41 61 302 89 18
www.mdpi.com

Actuators Editorial Office
E-mail: actuators@mdpi.com
www.mdpi.com/journal/actuators

www.ingramcontent.com/pod-product-compliance
Lightning Source LLC
LaVergne TN
LVHW071028170726
843515LV00006B/1227

9783036544090